D1594423

HIPPOCRATES

XI

LCL 538

HIPPOCRATES

VOLUME XI

EDITED AND TRANSLATED BY
PAUL POTTER

HARVARD UNIVERSITY PRESS
CAMBRIDGE, MASSACHUSETTS
LONDON, ENGLAND
2018

First published 2018

LOEB CLASSICAL LIBRARY® is a registered trademark
of the President and Fellows of Harvard College

Library of Congress Control Number 2018938156
CIP data available from the Library of Congress

ISBN 978-0-674-99657-1

*Composed in ZephGreek and ZephText by
Technologies 'N Typography, Merrimac, Massachusetts.
Printed on acid-free paper and bound by
Maple Press, York, Pennsylvania*

CONTENTS

LIST OF HIPPOCRATIC WORKS SHOWING THEIR DIVISION INTO VOLUMES IN THIS EDITION

Volume I (LCL 147)

Ancient Medicine. Airs Waters Places. Epidemics I. Epidemics III. Oath. Precepts. Nutriment.

Volume II (LCL 148)

Prognostic. Regimen in Acute Diseases. Sacred Disease. Art. Breaths. Law. Decorum. Dentition.

Volume III (LCL 149)

Wounds in the Head. In the Surgery. Fractures. Joints. Instruments of Reduction.

Volume IV (LCL 150)

Nature of Man. Regiment in Health. Humors. Aphorisms. Regimen I–III. Dreams.

Volume V (LCL 472)

Affections. Diseases I. Diseases II.

GENERAL INTRODUCTION[1]

The texts in this volume represent the most extensive account in the Hippocratic Collection of pathological conditions of the female reproductive organs, although many other treatises published in Loeb *Hippocrates* IX and X exhibit similar, and in some cases even verbatim identical, texts. Indeed, these shared texts, together with a number of explicit cross-references, have led scholars to hypothesize various identities of authorship and/or source material among the gynecological treatises: however, in most cases the state of the evidence makes it difficult to draw definite conclusions.[2]

Whatever the true relationship between *Diseases of Women I–II* themselves may have been at their original composition, they have existed in their present form and with their present title (literally, "Matters of Women") as far back as they are known.[3] The first-century AD medi-

[1] The two treatises are analyzed in more detail in their particular introductions.

[2] See, for example, Ermerins, vol. 2, lxxviii–xciii; Bourbon, pp. viii–lx.

[3] The title Περὶ Γυναικείων means, literally, "Matters of Women," without any closer indication what those matters might include (see τὰ γυναικεῖα/gynaikeia, Loeb *Hippocrates* vol. 10, xii).

cal glossator Erotian lists Γυναικείων α' β' in the "therapy by regimen" section of his census of Hippocratic writings and includes over fifty words from each book in his *Glossary*, referring explicitly to α' Γυναικείων five times.[4] Pliny the Elder adopts in his *Natural History*[5] many prescriptions drawn from both books, and Soranus of Ephesus in his *Gynecology* attributes views expressed in *Diseases of Women* to Hippocrates and his followers (οἱ περὶ τὸν Ἱπποκράτην).[6] Galen of Pergamum refers in his *Glossary* seven times to πρῶτον τῶν Γυναικείων and twelve times to δεύτερον τῶν Γυναικείων, while providing glosses for over fifty lemmata from each of the treatises; he also makes reference to two passages in *Diseases of*

The title *Diseases of Women* seems to have been become current in the middle ages (cf. Ullmann, pp. 246–48: *Kitāb Auǧā' an-nisā'*), presumably either on the basis of references found in the Hippocratic treatises *Generation* 4 (vol. 10, 14: ἐν τῆσι γυναικείησι νούσοισι), *Nature of the Child* 4 (vol. 10, 42: ἐν τοῖσι γυναικείοισι νοσήμασιν), *Diseases IV* 26 (vol. 10, 184: ἐν τοῖσι γυναικείοισι νοσήμασι), and *Barrenness* 1 (vol. 10, 332: ἐν τοῖσι γυναικείοισι νοσήμασιν), or under the influence of the first words of the text of *Diseases of Women I*.

The account of testimonies to the treatises that follows is based on the exhaustive collection of relevant sources given in Anastassiou/Irmer vol. 1, 338–59; vol. 2.1, 341–57; vol. 2.2, 256f.; vol. 3, 286–94.

[4] Cf. Nachmanson, pp. 415–37.

[5] E.g., *Natural History* XX 27 ≈ *Diseases of Women II* 80; *NH* XX 139 ≈ *DW I* 91(6); *NH* XX 220 ≈ *DW II* 63; *NH* XX 252 ≈ *DW II* 87(11); *NH* XXXII 131 ≈ *DW I* 91(6).

[6] E.g., *Gynecology* 3,29 ≈ *Diseases of Women II* 14 and 22; *Gyn* 4,13 ≈ *DW I* 77; *Gyn* 4,36 ≈ *DW II* 35.

Women I in his commentary on the Hippocratic *Aphorisms*. In the fifth century Hesychius of Alexandria glosses about twenty-five terms originating directly or indirectly from each of the treatises, while the sixth-century Byzantine medical compiler Aëtius of Amida takes over one prescription from *Diseases of Women I*.[7]

Diseases of Women I–II also appear to have been a part of the medical curriculum in Alexandria in the sixth century: Stephen of Athens (the Philosopher) quotes from them four times in his commentary on the *Aphorisms*, and he refers to them as the Γυναικεῖα both there and in his commentary on *Prognostic*; John of Alexandria refers to the Γυναικεῖα in his commentary on the *Nature of the Child*; and the contemporary Aristotle commentator John Philoponus locates a term he is discussing in the Γυναικεῖα of Hippocrates.

Two early Latin manuscripts, St. Petersburg F. v. VI. 3 (VIII/IX c.) and Parisinus Latinus 11219 (IX c.), contain gynecological compendia that incorporate textual material translated from the Greek text of *Diseases of Women I–II*. Because some of these translations display similarities in language and technique with extant Latin translations of Oribasius' medical texts often assumed to have been made in the neighborhood of Ravenna in the sixth century, scholars have proposed the same place and date of origin for the *Diseases of Women* translations.[8] Closest to the Greek text is the Latin *De conceptu* (*De mulierum affectibus* according to Vázquez Buján), whose thirty-three

---

[7] Aëtius 4,18 ≈ *Diseases of Women I* 92(2).
[8] See Mazzini/Flammini, pp. 43–46.

chapters consist of complete or partial translations of *Diseases of Women I* chapters 1 and 7–38;[9] more distant and fragmentary, and probably by a different translator, are chapters 28–39, 51, 75, and 78–79 of the Latin work *De diversis causis mulierum* containing texts translated from chapters 12–15, 17–26, 30, 53, 57, and 78–79 of *Diseases of Women II*.[10]

It seems improbable that an Arabic translation of *Diseases of Women I–II* was ever made, but excerpts from Arabic translations of two Greek commentaries on chapters 1–11 of *Diseases of Women I* are transmitted by three Arabic medical writers—ʿAlī ibn Riḍwān, Maimonides, and ibn abī Usaibiʿa—one attributed erroneously to Galen and the other to an unidentifiable Asclepius.[11]

Finally, the twelfth-century Latin manuscript Brussels 1342–50 contains an apocryphal biography of Hippocrates entitled *Yppocratis genus, vita, dogma*, which includes, in two slightly differing lists of Hippocratic works, *"et duos de genecia, id est mulieribus,"* where *genecia* is a transliteration of Γυναικεῖα.[12]

[9] Cf. Mazzini/Flammini. The Latin text is edited in both Mazzini/Flammini and Vázquez Buján.

[10] Mazzini/Flammini, p. 10f. The Latin translation is edited provisionally in W. Brütsch, *De diversis causis mulierum nach einer Petersburger Handschrift aus dem IX. Jahrhundert zum ersten Mal gedruckt,* Diss. med. (Freiburg in B., 1922).

[11] See Ullmann, pp. 245–62.

[12] See Pinault, pp. 24–28 and 133 (ll. 93–94).

HISTORY AND CONSTITUTION
OF THE GREEK TEXT[13]

Θ	Vindobonensis Medicus Graecus 4 (X/XI c.)
M	Marcianus Venetus Graecus 269 (X/XI c.)
V	Vaticanus Graecus 276 (XII c.)
I	Parisinus Graecus 2140 (XIII c.)
H	Parisinus Graecus 2142
Ha	older part (XII c.)
Hb[14]	newer part (XIV c.)
R	Vaticanus Graecus 277 (XIV c.)
Recentiores	approximately twenty manuscripts (XV/XVI c.)

The *stemma codicum* appearing as Figure 1 provides an overview of the interdependencies among the manuscripts containing *Diseases of Women I–II*. Two notable irregularities in this general scheme exist in *Diseases of Women I*:

1. the manuscript Θ reads a section of text in chapter 3 ἢν δὲ ῥόος μὴ γίνηται . . . ὑγιὴς δὲ γίνεται ἐν κόσμῳ

---

[13] For detailed accounts of the Hippocratic manuscripts and their relationships, see Irigoin, pp. 191–210 ("Hippocrate et la Collection hippocratique") and pp. 211–36 ("Hippocrate, Galien et quelques autres médecins grecs"); B. Mondrain, "Lire et copier Hippocrate . . . au XIVe siècle," in Boudon et al., pp. 359–410; B. Mondrain, "La place de la *Collection Hippocratique* à Byzance d'après les manuscrits," in Jouanna/Zink, pp. 385–99.

[14] The Greek text of *Diseases of Women I–II* is transmitted in this part of the manuscript on folios 351 to 406.

Figure 1. *Stemma Codicum*

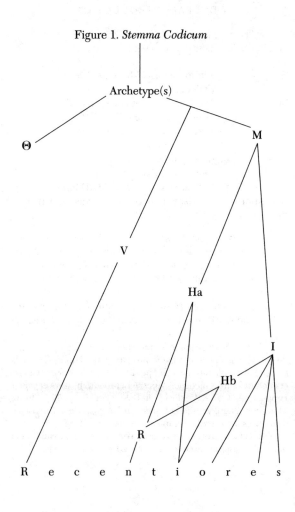

ἰηθεῖσα not only in the same place as M and V but also a second time inserted between chapters 9 and 10 (= Θ′): since Θ′ contains readings that differ significantly from the Θ text, it must be seen as the representative of an independent strand of the transmission, equal in authority with each of the other three. At the same location between chapters 9 and 10, the manuscript V adds a brief summary of Θ's added passage: ἢν δέ οἱ ῥόος μὴ γένηται, συμβήσεταί οἱ ὥστε δοκέειν ἐν γαστρὶ ἔχειν· καὶ ταῦτα πείσεται ἅπερ τόκῳ τὰ καταμήνια ὅδον οὐκ ἠδύνατο εὑρεῖν· ἢ παχέα καὶ γλίσχρα καὶ κολλώδεα ἴῃ.

2. The manuscript M and its descendants omit the text of *Diseases of Women I*, chapters 92–109, leaving Θ and V here as the sole independent witnesses.

Three Greek papyri contain fragments of the text of *Diseases of Women I–II*: P. Rylands 3.531 (III/II c.);[15] P. Köln 7.311 (II/III c.);[16] P. Antinoopolis 3.184 (VI c.).[17]

The Latin translations described above (= Lat.) have only a very limited value for establishing the Greek text of *Diseases of Women I–II*, both because the original sixth-century translations were not very literal and also because the translations appear to have been reworked before they were copied into the extant manuscripts.[18]

The complete Greek text of *Diseases of Women I–II* was last edited by Littré in 1853 and Ermerins in 1862, neither of whom had direct access to the three indepen-

---

[15] See Marganne, pp. 283–86.
[16] See Marganne/Mertens, p. 16f.; Hanson, p. 149f.
[17] See Marganne, p. 124f.
[18] See Vázquez Buján, p. 9f.

dent witnesses to the text, Θ, M, and V.[19] The present edition is based on a collation of the three manuscripts from microfilm.

This volume has benefited in many passages from corrections and suggestions made by my spouse, Beate Gundert.

NOTE ON TECHNICAL TERMS[20]

The following terms require some clarification as they are employed in this volume with special scientific or medical significance.

ἰκμάς: this noun is often used to denote one of the many specific nutritive essences present in human food and drink and also in the earth for plants. Such an essence is imagined in general to be a moisture (although it can on occasion be "dry," cf. *Affections* 51) and to emanate or be exhaled from its source, being then attracted and taken

---

[19] Littré did, however, possess a complete and accurate collation of the manuscript Θ commissioned before 1849. See Littré vol. 6, 139, n. 1: "M. Villemain . . . eut la bonté de faire collationner pour moi les deux livres des Maladies des femmes . . . qui n' aient pas été collationnées par Mack dans son édition d' Hippocrate. . . . Ce travail a été fait avec une extrême exactitude et un soin parfait par M. le docteur Poeschl de Vienne." As representatives of M, Littré had the Parisini Graeci 2140 (= I) and 2142 (= Hb), and for the V tradition he had Parisinus Graecus 2146 (= C).

[20] See also Loeb *Hippocrates* vol. 8, 10f. and vol. 10, x–xiii for the explanation of other technical terms occurring in the present volume.

up by the gastrointestinal tract (κοιλίη) or the body as a whole, or by a plant's roots; cf. Lonie, pp. 269f.

καταρρηγνύναι: in gynecological texts this verb refers most frequently to a downward discharge of the menses, but when applied to the uterus or breasts it means "to involute": cf. the *Sydenham Soc. Lex.* definition: "Involution: the retrograde change which occurs . . . in some organ when its permanent or temporary purpose has been fulfilled, as in the uterus after the cessation of menstruation or after delivery."

REFERENCES

EDITIONS, TRANSLATIONS, AND COMMENTARIES

Hippocratic Collection

Early Works

Calvus   *Hippocratis Coi . . . octoginta volumina . . . per M. Fabium Calvum, Rhavennatem . . . latinitate donata . . .* Rome, 1525.

Aldina   *Omnia opera Hippocratis . . . in aedibus Aldi & Andreae Asulani soceri.* Venice, 1526.

Froben   *Hippocratis Coi . . . libri omnes . . . [per Ianum Cornarium].* Basel, 1538.

Cornarius   *Hippocratis Coi . . . opera . . . per Ianum Cornarium . . . Latina lingua conscripta.* Venice, 1546.

Cordaeus   *Liber prior de morbis mulierum . . . Mauricio Cordaeo . . . interprete & explicatore.* Paris, 1585.

Foes   *Magni Hippocratis . . . opera quae extant . . . Latina interpretatione & annotationibus illustrata, Anutio Foesio . . . authore.* Geneva, 1657–1662.

REFERENCES

Linden   *Magni Hippocratis Coi opera omnia graece & latine edita . . . industria & diligentia Joan. A. Vander Linden.* Leiden, 1665.

Post-Eighteenth Century

Bourbon   Bourbon, F. *Hippocrate. Nature de la femme.* Budé XII (1). Paris, 2008.

Countouris   Countouris, N. *Hippokratische Gynäkologie.* Diss. med. Hamburg, 1985.

Ermerins   Ermerins, F. Z. *Hippocratis . . . reliquiae.* Utrecht, 1859–1864.

Fuchs   Fuchs, R. *Hippokrates, sämmtliche Werke, ins Deutsche übersetzt. . . .* Munich, 1895–1900.

Gardeil   Gardeil, J. B. *Traduction des Œuvres médicales d' Hippocrate, sur le texte grec, d' après l' édition de Foës.* Toulouse, 1801.

Grensemann   Grensemann, H. *Hippokratische Gynäkologie.* Wiesbaden, 1982.

Littré   Littré, E. *Oeuvres complètes d' Hippocrate.* Paris, 1839–1861.

Lonie   Lonie, I. M. *The Hippocratic Treatises "On Generation," "On the Nature of the Child," "Diseases IV": A Commentary.* Berlin, 1981.

Mazzini/   Mazzini, I., and G. Flammini. *De conceptu.*
Flammini   *Estratti di un' antica traduzione latina del* Περὶ γυναικείων *pseudoippocratico l. I.* Bologna, 1983.

Vázquez   Vázquez Buján, M. E. *El De Mulierum Af-*
Buján   *fectibus del Corpus Hippocraticum: Estu-*
            dio y edición crítica de la antigua traduc-
            ción latina. Santiago de Compostela, 1986.

Other Authors

Aëtius   Olivieri, A. *Aetii Amideni Libri medicina-*
            les I–VIII. Corpus Medicorum Graecorum
            VIII 1–2. Leipzig, 1935–1950.

Berendes   Berendes, J. *Dioskurides. Arzneimittellehre*
            . . . übersetzt und mit Erklärungen versehen.
            Stuttgart, 1902.

Dioscorides Wellmann, M. *Pedanii Dioscuridis . . . De*
            materia medica. Berlin, 1906–1914.

Erotian   Nachmanson, E. *Erotiani Vocum Hippo-*
            craticarum collectio. Gothenburg, 1918.

Galen   Kühn, C. G. *Claudii Galeni opera omnia*
            . . . Leipzig, 1825–1833.

Galen *Gloss.* Perilli, L. *Galeni Vocum Hippocratis Glos-*
            sarium. Corpus Medicorum Graecorum V
            13,1. Berlin, 2017.

Hesychius Latte, K. et al. *Hesychii Alexandrini Lexi-*
            con, Copenhagen, 1953–1966, and Berlin,
            2005–2009.

John of   Westerink, L. G. et al. *John of Alexandria.*
Alexandria *Commentary on Hippocrates' Nature of the*
            Child. Corpus Medicorum Graecorum XI
            1,4. Berlin, 1997.

REFERENCES

Nachmanson   Nachmanson, E. *Erotianstudien*. Uppsala, 1917.

Oribasius   Raeder, I. *Oribasii Collectionum Medicarum Reliquiae*. Corpus Medicorum Graecorum VI 1,1–2,2. Leipzig, 1928–1933.

Paulus   Adams, F. *The Seven Books of Paulus Aegineta*. London, 1844–1847.

Pinault   Pinault, J. R. *Hippocratic Lives and Legends*. Leiden, 1992.

Pliny   Rackham, H. et al. *Pliny. Natural History with an English translation*. London, 1938–1963.

Soranus   Ilberg, I. *Soranus*. Corpus Medicorum Graecorum IV. Leipzig, 1927.

Temkin, O. *Soranus' Gynecology*. Baltimore, 1956.

Stephanus   Duffy, J. M. *Stephanus the Philosopher. A Commentary on the Prognosticon of Hippocrates*. Corpus Medicorum Graecorum XI 1,2. Berlin, 1983.

Westerink, L. G. *Stephanus of Athens. Commentary on Hippocrates' Aphorisms*. Corpus Medicorum Graecorum XI 1,3,3. Berlin, 1995.

GENERAL WORKS

Anastassiou/   Anastassiou, A., and D. Irmer. *Testimonien*
Irmer   *zum Corpus Hippocraticum*. Göttingen, 1997–2012.

Boudon et al. Boudon-Millot, V., A. Garzya, J. Jouanna, and A. Roselli, eds. *Ecdotica e ricezione dei testi medici greci.* Naples, 2006.

Frisk Frisk, H. *Griechisches etymologisches Wörterbuch.* Heidelberg, 1973–1979.

v. Grot v. Grot, R. "Über die in der hippokratischen Schriftensammlung enthaltenen pharmakologischen Kenntnisse." *Kobert's Historische Studien aus dem Pharmakologischen Institute der Kaiserlichen Universität Dorpat* 1 (1889): 58–133.

Hanson Hanson, A. E. "Papyri with Medical Content." *Society for Ancient Medicine. Newsletter* 19 (1991): 149f.

Index Hipp. Kühn, J.-H., U. Fleischer et al. *Index Hippocraticus.* Göttingen, 1986–1999.

Irigoin Irigoin, J. *Tradition et critique des textes grecs.* Paris, 1997.

Jouanna/Zink Jouanna, J., and M. Zink, eds. *Hippocrate et les hippocratismes: médecine, religion, société. XIVe Colloque International Hippocratique.* Paris, 2014.

Marganne Marganne, M.-H. *Inventaire analytique des papyrus grecs de médecine.* Geneva, 1981.

Marganne/ Mertens Marganne, M.-H., and P. Mertens. "Medici et Medica, 2e édition." In 'Specimina' per il Corpus *dei Papiri Greci di Medicina*, edited by I. Andorlini, 3–71. Florence, 1997.

Sydenham Soc. Lex. Power, H., and L. W. Sedgwick. *The New Sydenham Society's Lexicon of Medicine and the Allied Sciences.* London, 1881–1899.

REFERENCES

Ullmann   Ullmann, M. "Zwei spätantike Kommentare zu der hippokratischen Schrift 'De morbis muliebribus.'" *Medizinhistorisches Journal* 12 (1977): 245–62.

GENERAL BIBLIOGRAPHY

Indispensable for any study of the transmission of the Hippocratic Collection are the four volumes of Anastassiou/Irmer and the literature they cite.

EDITIONS, TRANSLATIONS, AND COMMENTARIES

Craik, E. M. *Hippocrates.* Places in Man, Oxford, 1998.

———. *Two Hippocratic Treatises:* On Sight *and* On Anatomy. Leiden, 2006.

———. *The Hippocratic Treatise* On Glands. Leiden, 2009.

Ecca, G. *Die hippokratische Schrift Praecepta.* Wiesbaden, 2016.

García Gual, C. et al. *Tratados hipocráticos . . . introducciones, traducciones y notas.* Madrid, 1983–2003.

Giorgianni, F. *Hippokrates. Über die Natur des Kindes* (De genitura *und* De natura pueri). Wiesbaden, 2006.

Hanson, M. *Hippocrates. On Head Wounds.* Corpus Medicorum Graecorum I 4,1. Berlin, 1999.

Joly, R. *Hippocrate. De la génération, De la nature de l'enfant, Des maladies IV . . .* Budé XI. Paris, 1970.

Jouanna, J. *Hippocrate. Airs, Eaux, Lieux.* Budé II (2). Paris, 1996.

————. *Hippocrate. De la maladie sacrée.* Budé II (3). Paris, 2003.

————. *Hippocrate. Pronostic.* Budé III (1). Paris, 2013.

Jouanna, J., and M. D. Grmek. *Hippocrate. Epidémies V et VII.* Budé IV (3). Paris, 2000.

Jouanna, J., and C. Magdelaine. *Hippocrate. L' Art de la médecine.* Paris, 1999.

Lienau, C. *Hippokrates. Über Nachempfängnis, Geburtshilfe und Schwangerschaftsleiden.* Corpus Medicorum Graecorum I 2,2. Berlin, 1973.

Lloyd, G. E. R., ed. *Hippocratic Writings.* Harmondsworth, 1978.

Overwien, O. *Hippokrates. Über die Säfte.* Corpus Medicorum Graecorum I 3,1. Berlin, 2014.

Roselli, A. *Ippocrate. La malattia sacra.* Venice, 1996.

Sakalis, D. T. Ἱπποκράτους Ἐπιστολαί. Ἔκδοση κριτική και ερμηνευτική. Ioannina, 1989.

Schiefsky, M. *Hippocrates. On Ancient Medicine.* Leiden, 2005.

Smith, W. D. *Hippocrates. Pseudepigraphic Writings: Letters – Embassy – Speech from the Altar – Decree.* Leiden, 1990.

GENERAL WORKS

Bliquez, L. *The Tools of Asclepius.* Leiden, 2014.

Craik, E. M. *The 'Hippocratic' Corpus: Content and Context.* London, 2015.

Dean-Jones, L. A. *Women's Bodies in Classical Greek Science.* Oxford, 1994.

Dean-Jones, L. A., and R. M. Rosen, eds. *Ancient Concepts of the Hippocratic.* Leiden, 2015.

Demand, N. *Birth, Death and Motherhood in Classical Greece.* Baltimore, 1994.

Diepgen, P. *Die Frauenheilkunde der alten Welt.* Munich, 1937.

Eijk, Ph. J. van der, ed. *Hippocrates in Context: Papers Read at the XIth International Hippocrates Colloquium.* Leiden, 2005.

Eijk, Ph. J. van der. *Medicine and Philosophy in Classical Antiquity.* Cambridge, 2005.

Eijk, Ph. J. van der, H. F. Horstmanshoff, and P. H. Schrijvers, eds. *Ancient Medicine in Its Socio-cultural Context.* Amsterdam, 1995.

Fasbender, H. *Entwickelungslehre, Geburtshülfe und Gynäkologie in den hippokratischen Schriften.* Stuttgart, 1897.

Flashar, H. *Hippokrates. Meister der Heilkunst.* Munich, 2016.

Föllinger, S. *Differenz und Gleichheit. Das Geschlechterverhältnis in der Sicht griechischer Philosophen des 4. bis 1. Jahrhunderts v. Chr.* Stuttgart, 1996.

Garofalo, I., A. Lami, D. Manetti, and A. Roselli, eds. *Aspetti della terapia nel Corpus Hippocraticum. Atti del IXe Colloque International Hippocratique.* Florence, 1999.

Harris, W. V., ed. *Mental Disorders in the Classical World.* Leiden, 2013.

Heidel, W. A. *Hippocratic Medicine: Its Spirit and Method.* New York, 1941.

Horstmanshoff, H. F., ed. *Hippocrates and Medical Education: Selected Papers Read at the XII International Hippocrates Colloquium.* Leiden, 2010.

Jouanna, J. *Hippocrates.* Translated by M. B. DeBevoise. Baltimore, 1999.

———. *Greek Medicine from Hippocrates to Galen.* Leiden, 2012.

King, H. *Hippocrates' Woman: Reading the Female Body in Ancient Greece.* London, 1998.

Leven, K.-H., ed. *Antike Medizin. Ein Lexikon.* Munich, 2005.

Nutton, V. *Ancient Medicine.* London, 2004.

Oser-Grote, C. M. *Aristoteles und das Corpus Hippocraticum. Die Anatomie und Physiologie des Menschen.* Stuttgart, 2004.

Pormann, P. E., ed. *Epidemics in Context. Greek Commentaries on Hippocrates in the Arabic Tradition.* Berlin, 2012.

Thivel, A., and A. Zucker, eds. *Le normal et le pathologique dans la Collection hippocratique. Actes du Xème colloque international hippocratique.* Nice, 2001.

Totelin, L. M. V. *Hippocratic Recipes: Oral and Written Transmission of Pharmacological Knowledge in Fifth- and Fourth-Century Greece.* Leiden, 2009.

Wittern, R., and P. Pellegrin, eds. *Hippokratische Medizin und antike Philosophie. Verhandlungen des VIII. Internationalen Hippokrates-Kolloquiums.* Hildesheim, 1996.

DISEASES OF WOMEN I

INTRODUCTION

The chapters of *Diseases of Women I* can be arranged under seven main headings: the first five of these deal with disorders that occur in each of the successive stages of female reproductive life—menstruation, conception, pregnancy, lochial cleaning, the puerperium; the sixth is centered on conditions of the uterus; and the seventh is a collection of treatments for gynecological and other pathologies.

A. Menstrual disorders

B. Barrenness

These chapters vary widely in the internal arrangement, sequence, and relative fullness of their component parts, and no standard template is employed: name, etiology, symptoms, course, prognosis, and treatment may each re-

ceive fuller or more summary attention, and no set order of presentation is followed.[1]

Diseases of Women I appears in all the collected editions and translations of the Hippocratic Collection, and the treatise was the subject of one special study in the late sixteenth century: M. Cordaeus, *Liber prior de morbis mulierum . . . interprete et explicatore* (Paris, 1585). More recently, H. Grensemann includes an edition, translation, and commentary on many chapters of the treatise in his *Hippokratische Gynäkologie*, as does N. Countouris.[2]

---

[1] On the role nosology plays in the organization of a comparable Hippocratic treatise, cf. P. Potter, "Nosology and Organization in *Barrenness*," in Jouanna/Zink, pp. 59–68.

[2] Grensemann, pp. 88–141; Countouris, pp. 47–80.

ΠΕΡΙ[1] ΓΥΝΑΙΚΕΙΩΝ Α

VIII 10
Littré

1. Τὰ δὲ ἀμφὶ γυναικείων νούσων· φημὶ γυναῖκα ἄτοκον ἐοῦσαν ἢ τετοκυῖαν χαλεπώτερον καὶ θᾶσσον ἀπὸ
καταμηνίων νοσέειν· ὅταν γὰρ τέκῃ, εὐρωότερά οἱ τὰ
φλέβιά ἐστιν ἐς τὰ καταμήνια, εὔροα δέ σφε ποιέει ἡ
λοχείη κάθαρσις. καὶ ἢν καταρραγὴ ⟨ἢ⟩[2] τοῦ σώματος, τὰ πλησιάζοντα μάλιστα τῆς τε κοιλίης καὶ τῶν
μαζῶν καταρρήγνυται· καταρρήγνυται δὲ καὶ τὸ ἄλλο
σῶμα· ὑφ᾽ ὅτευ δὲ γίνεται, εἴρηταί μοι ἐν τῇ Φύσει
τοῦ Παιδίου τοῦ ἐν Τόκῳ. καταρραγέντος δὲ τοῦ σώματος, ἀνάγκη τὰς φλέβας μᾶλλον στομοῦσθαι καὶ
εὐρωτέρας γίνεσθαι ἐς τὰ καταμήνια, καὶ τὰς μήτρας
μᾶλλον στομοῦσθαι, οἷα τοῦ παιδίου χωρήσαντος διὰ
σφέων καὶ βίην καὶ τόνον[3] παρασχόντος· καὶ τούτων
ὧδε ἐχόντων, τὰ καταμήνια ἀκαματώτερον ἀποκαθαίρεται ἡ γυνή, ἐπὴν λοχείων ἔμπειρος γένηται. εἰ δὲ
καί τι πάθημα τῇ γυναικὶ γένοιτο τῇ ἤδη τετοκυίῃ,
ὥστε τὰ καταμήνια μὴ δύνασθαι καθαρθῆναι, ῥηϊτέρως τὸν πόνον οἴσει ἢ εἰ ἄτοκος ἦν· ἠθάδες γὰρ αἱ

---

[1] ΠΕΡΙ om. MV.
[2] Add. Potter.

DISEASES OF WOMEN I

1. Concerning diseases of women: I assert that a woman who has not borne children becomes ill from her menses more seriously and sooner than one who has borne children. For when a woman has given birth, her small vessels allow a freer flow for her menses, since the lochial cleaning makes them fluent. Also, if there has been an involution of her body, it is the parts nearest to the cavity and breasts that involute, although the rest of the body involutes too—why this occurs I have explained in *Nature of the Child in Childbirth*.[1] As the body involutes, the vessels are forced to dilate more and to allow a freer flow for the menses; the uterus too must dilate more, as the child moving through it causes straining and stretching. These things being so, a woman will clean out her menses with less difficulty once she has experienced the lochia. Also, if some condition that prevents the menses from being cleaned befalls a woman who has previously borne children, she will bear the trouble more easily than if she had not borne children, since her uterus and the rest of her

[1] Cf. *Nature of the Child* 19, Loeb *Hippocrates* vol. 10, 91.

---

3 τόνον Potter: πόνον codd.; cf. *Diseases I* 20: (sc. φλέβιον) σπᾶται δ᾽ ὑπὸ τόνου καὶ βίης.

μῆτραι καὶ τὸ σῶμά ἐστι πληροῦσθαι, ἅτε ἐν γαστρὶ
ἐχούσῃ, καὶ ἅμα εὐρυχωρίη ἐστὶν ἐν τῷ σώματι
πλείων τῷ αἵματι, ἐπὴν τέκῃ, οἷα τοῦ σώματος καταρραγέντος, καὶ ἐν εὐρυχωρίῃ ἐὸν τὸ αἷμα ἀπονώτερον
γίνεται, ἢν μὴ ὑπερπιμπλῶνται αἱ φλέβες καὶ ὑπερ-
τονέωσιν.

12   Ἀτόκῳ | δὲ ἐούσῃ, τοῦ τε σώματος οὐ ξυνήθεος
ἐόντος, ἐπὴν πληρωθῇ, ἰσχυροῦ[4] καὶ στερεωτέρου καὶ
πυκνοτέρου ἐόντος ἢ εἰ λοχείων ἔμπειρος γένοιτο, καὶ
τῶν μητρέων ἀστομωτέρων ἐουσέων, τὰ ἐπιμήνια ἐπι-
πονωτέρως χωρέει, καὶ τὰ παθήματα προσπίπτει
πλείονα, ὥστε τὰ καταμήνια ἀποφράσσεσθαι, ἐπὴν
ἄτοκος ᾖ. ἔχει δὲ ὧδε ὥς μοι καὶ πρῶτον[5] εἴρηται·
φημὶ τὴν γυναῖκα ἀραιοσαρκοτέρην καὶ ἁπαλωτέρην
εἶναι ἢ τὸν ἄνδρα· καὶ τούτου ὧδε ἔχοντος, ἀπὸ τῆς
κοιλίης ἕλκει τὴν ἰκμάδα καὶ τάχιον καὶ μᾶλλον τὸ
σῶμα τῆς γυναικὸς ἢ τοῦ ἀνδρός.

Καὶ γὰρ εἴ τις ὑπὲρ ὕδατος ἢ καὶ χωρίου ὑδρηλοῦ
δύο ἡμέρας καὶ δύο εὐφρόνας θείη εἴρια καθαρὰ καὶ
εἶμα καθαρὸν καὶ βεβυσμένον εὖ, σταθμῷ[6] ἴσον τοῖσιν εἰρίοισιν, ἀνελὼν εὑρήσει στήσας πολλὸν βαρύ-
τερα τὰ εἴρια ἢ τὸ εἶμα· ὅτι δὲ τοῦτο γίνεται, αἰεὶ
ἀποχωρέει ἐς τὸ ἀνεκὰς ἀπὸ τοῦ ὕδατος ἐν ἀγγείῳ
εὐρυστόμῳ ἐόντος, καὶ τὰ μὲν εἴρια, ἅτε ἀραιά τε καὶ
μαλθακὰ ἐόντα, ἀναδέξεται τοῦ ἐπιχωρέοντος πλέον,
τὸ δὲ εἶμα, ἅτε πλῆρες ἐὸν καὶ βεβυσμένον, ἀποπλη-
ρώσεται τὸ πολλὸν οὐκ ἐπιδεχόμενον τοῦ ἐπιχωρέοντος.

body are accustomed to being full, just as they are in a woman who is pregnant, and at the same time there is more open space for blood in her body after she has given birth, since her body has involuted, so that the blood, being in an open space, will be less troublesome, as long as the vessels do not overfill and overstretch.

Since in a woman who has not given birth the body is not accustomed to being filled up (sc. with blood), but is robust, solider and denser than if she had experienced the lochia, and her uterus has not been dilated, her menstrual flow will be accompanied by more pain, and more troubles will be present: i.e., her menses will be obstructed when she has not given birth. This is so for the reason I first indicated when I contended that a woman is more porous and softer than a man; this being so, a woman's body draws what is being exhaled from her cavity more quickly and in a greater amount than does a man's.

Thus, if someone sets both some clean flocks of wool and a clean densely woven carpet of exactly the same weight as the flocks over water or a moist location for two days and two nights, on removing them he will discover, on weighing them, that the flocks have become much heavier than the carpet. This happens because (sc. moisture) always moves up away from water present in a wide-necked vessel, and flocks, being porous and soft, take up a greater quantity of what is moving away, while a carpet, being compact and densely woven, becomes saturated without accepting much of what is moving toward it.

---

⁴ ἰσχ. ΘV: ἰσχυροτέρου M. ⁵ πρῶτον Θ: πρὶν MV.
⁶ εὖ, σταθμῷ Index Hipp. s.v. εὔσταθμος: εὐστάθμῳ ΘV: ἐνστάθμως M.

Οὕτω δὲ καὶ ἡ γυνή, ἅτε ἀραιοτέρη ἐοῦσα, εἵλκυσε
πλέον ἀπὸ τῆς κοιλίης τῷ σώματι τῆς ἱκμάδος καὶ
θᾶσσον ἢ ὁ ἀνήρ, καὶ ἅτε ἀπαλοσάρκῳ ἐούσῃ τῇ
γυναικί, ἐπὴν πλησθῇ τοῦ αἵματος τὸ σῶμα, ἢν μὴ
ἀποχωρήσῃ ἀπὸ αὐτοῦ, πληρευμένων τῶν σαρκῶν
καὶ θερμαινομένων, πόνος γίνεται· θερμότερον γὰρ τὸ
αἷμα ἔχει ἡ γυνή, καὶ διὰ τοῦτο θερμοτέρη ἐστὶ τοῦ
ἀνδρός· ἢν δὲ τὸ πλεῖον ἐπιγενόμενον ἀποχωρέῃ, οὐ
14 γίνεται ὁ πόνος[7] πρὸς τοῦ αἵματος. ὁ δὲ | ἀνὴρ στερε-
οσαρκότερος ἐὼν τῆς γυναικὸς οὔτε ὑπερπίμπλαται
τοῦ αἵματος[8] τόσσον, ὥστ' ἢν μὴ ἀποχωρέῃ τι τοῦ
αἵματος καθ' ἕκαστον μῆνα, πόνον γίνεσθαι, ἕλκει τε
ὅσον ἐς τροφὴν τοῦ σώματος,[9] τό τε σῶμά οἱ οὐχ
ἀπαλὸν ἐὸν οὐχ ὑπερτονέει, οὐδὲ θερμαίνεται ὑπὸ
πληθώρης ὡς τῇ γυναικί· μέγα δὲ ξυμβάλλεται ἐς
τοῦτο τῷ ἀνδρί, ὅτι ταλαιπωρέει μᾶλλον τῆς γυναι-
κός· ἡ γὰρ ταλαιπωρίη ἀπάγει τι τῆς ἱκμάδος.

2. Ὅταν οὖν γυναικὶ ἀτόκῳ ἐούσῃ κρυφθῇ τὰ ἐπιμήνια καὶ μὴ δύνηται ἔξω ὁδὸν εὑρεῖν, νοῦσος γίνεται, τοῦτο δὲ συμβαίνει, ἢν τῶν μητρέων τὸ στόμα
μεμύκῃ ἢ ἰδνωθῇ, ἢ ξυστραφῇ[10] τι τοῦ αἰδοίου· ἢν
γάρ τι τούτων ᾖ, οὐ δυνήσεται ἔξω εὑρεῖν ὁδὸν[11] τὰ
ἐπιμήνια, πρὶν αἱ μῆτραι ἐς τὴν φύσιν τὴν ὑγιεινὴν
μεταστέωσι. γίνεται δὲ τὸ νόσημα τοῦτο μάλιστα
ᾗσιν στενοστόμους τὰς μήτρας ἐχούσῃσιν[12] ἢ τὸν

<hr />

[7] Add. καὶ ἡ θέρμη MV.   [8] Add. ἐς τὸ ἄνω τῆς κοιλίης
MV.   [9] σώμ. ΘV: αἵματος M.

In the same way, a woman, being more porous, will draw into her body more of what is being exhaled from her cavity, and more quickly, than a man does. Also, because a woman's flesh is softer, when her body fills up with blood, unless the blood is then discharged from her body, the filling and warming of her tissues that ensue will provoke pain: for a woman has hotter blood, and for this reason she herself is hotter than a man; if, however, most of the blood that was added is subsequently discharged, no pain will arise from it. A man, having solider flesh than a woman, will never overfill with so much blood that, unless some of it is discharged each month, he feels pain, and besides he takes in only as much (sc. blood) as is necessary for the nourishment of his body, and his body—lacking softness as it does—is never overstretched or heated by fullness as a woman's is. A great amount of this is also due in a man to his exerting himself physically more than a woman, which consumes a part of the exhalation (sc. rising from his food).

2. Now when, in a woman who has not given birth, the menses fail to appear and cannot find their way out, a disease arises, and this happens if the mouth of the uterus is closed or folded over, or some part of the vagina has become constricted; for if any of these things happens, the menses will be unable to find their way out until the uterus returns to a natural healthy state. This disease generally occurs in women whose uterus has a narrow mouth or its

10 ἢ ξυστραφῇ om. Θ.

11 ἔξω εὑρ. ὁδὸν Θ: ἔξοδον εὑρεῖν MV.

12 αἷστισι . . . ἐχούσῃσιν Potter: αἵτινες . . . ἔχουσιν ΘV: ταύταις αἵτινες . . . ἔχουσι M.

αὐχένα πρόσω τοῦ αἰδοίου κείμενον· ἢν γὰρ τούτων
θάτερον ᾖ, καὶ μὴ μίσγηται ἡ γυνὴ τῷ ἀνδρί, καὶ
κενωθῇ ἡ κοιλίη μᾶλλον τοῦ καιροῦ ὑπό τευ παθήματος, στρέφονται αἱ μῆτραι· οὔτε γὰρ ἰκμαλέαι εἰσὶ
κατὰ σφέας, οἷα μὴ λαγνευομένης, εὐρυχωρίη τε σφίσιν ἐστίν, ἅτε τῆς κοιλίης κενοτέρης ἐούσης, ὥστε
στρέφεσθαι ἅτε ξηροτέρας καὶ κουφοτέρας ἐούσας[13]
τοῦ καιροῦ. καὶ ἔστιν ὅτε στρεφομένων σφέων τυγχά-
νει τὸ στόμα πρόσω[14] παραστραφέν, οἷα[15] τοῦ αὐχένος
πρόσω τοῦ αἰδοίου κειμένου· ἢν γὰρ ἰκμαλέαι ἔωσιν
16 αἱ μῆτραι ἀπὸ λαγνείης καὶ | ἡ κοιλίη μὴ κενῶται, οὐ
ῥηϊδίως στρέφονται. τοῦτο οὖν γίνεται αἴτιον ὥστε
αὐτὰς συμμύειν, οἷα μὴ λαγνευομένης τῆς γυναικός.

Ἐν δὲ τοῖσι τρίτοισιν ἄριστα μὲν πείσεται, ἢν οἱ
κατελθόντα ἐξαγάγῃ τὰ προϋπάρχοντα· εἰ δὲ μή, πείσεται τάδε ἡ γυνή· πνὶξ τέ οἱ ἄλλοτε καὶ ἄλλοτε συμ-
πεσεῖται, καὶ πῦρ λήψεται ἄλλοτε καὶ ἄλλοτε καὶ
φρίκη καὶ ὀσφύος ἄλγημα. ταῦτα πείσεται ἐν τοῖσι
τρίτοισιν ἐπιμηνίοισιν, ἢν μή οἱ ἐξίῃ· ἐν δὲ τοῖσι τε-
τάρτοισιν, ἢν μή οἱ ἐξίῃ τοῖσί τε προτέροισιν ἔξοδον
ποιήσῃ, τά τέ μιν[16] τρίτα πονήματα πάντα μάλιστα[17]
πονήσει, καὶ μάλιστα ἐν τῷ χρόνῳ τῶν καταμηνίων,
ἔπειτα ἧσσον, πολλάκις δὲ καὶ δοκέει ἄπονος εἶναι·
ἔσται δὲ ἐπὶ τοῖσι καὶ τάδε ἕτερα σημεῖα· οὐρήσει
πολὺ παχὺ[18] ἄλλοτε καὶ ἄλλοτε, καὶ ἡ γαστὴρ αὐτῆς[19]

---

[13] ξηροτέρας . . . κουφοτέρας ἐούσας Ermerins: -τέρης . . .
-τέρης ἐούσης ΘΜ: om. V.

neck lying further forward into the vagina. For if either of these be the case, and the woman does not have intercourse with her husband, and her cavity is more empty than it should be as the result of some disease, the uterus turns aside. For it has no moistness of its own, since the woman is not having intercourse, and there is an open space for it since the cavity is too empty, so that it turns aside because it is drier and lighter than it should be. And sometimes as it turns aside its mouth becomes displaced too far to one side because its neck is lying too far into the vagina. For if the uterus is moist as the result of intercourse and the cavity is not empty, it is not likely to turn aside. This is why the uterus closes as the result of a woman not having intercourse.

Over three cycles, it will be best for a woman if what moves down draws out what has arrived there before; if this does not happen, she will suffer the following: suffocation will befall her from time to time, as will fever along with shivering and pain of the lower back. She will suffer this during the third menstrual cycle unless the material is discharged; also during the fourth cycle, unless a discharge occurs then and makes an exit for the previous material. She will feel the pains of the third cycle even more, especially at the time of the menses, but after that less, and often she will even become free of pain. During these cycles the following other signs will also be present: the woman will pass copious thick urine from time to time,

---

[14] πρόσω om. MV.

[15] οἷα Θ: οἷα τε M: ἅτε V.

[16] μιν ΘM: μὴν V.

[17] μάλιστα Θ: μᾶλλον MV.

[18] πολὺ παχὺ M: παχὺ Θ: πολὺ V.

[19] αὐτῆς om. Θ.

σκληρὴ ἔσται καὶ μείζων ἢ τὸ πρόσθεν, καὶ βρύξει
τοὺς ὀδόντας, καὶ ἀσιτήσει, καὶ ἀγρυπνήσει. τοιαῦτα
δὲ πείσεται ἐπὶ τοῖσι τετάρτοισι καταμηνίοισι· μελεδαινομένη δὲ καὶ ἐν τούτοισιν ὑγιαίνει. καὶ ἐν τοῖσι
πέμπτοισιν, ἢν μή οἱ πολλὰ τὰ καταμήνια κατίῃ, καὶ
πόνος ἰσχυρὸς προσπίπτει.

Ἐν δὲ τοῖσιν ἕκτοισιν ἤδη ἀνίητος ἔσται. καὶ τὰ
μὲν πρότερον σημεῖα μᾶλλον πονήσει, ἐπέσται δ᾽ ἐπ᾽
αὐτοῖσι καὶ τάδε· ἀλύξει τε καὶ ῥίψει ἑωυτὴν ἄλλοτε
καὶ ἄλλοτε, καὶ λιποθυμήσει, καὶ ἐμέεται φλέγμα, καὶ
δίψα ἰσχυρή μιν[20] λήψεται, ἅτε καιομένης τῆς κοιλίης
ὑπὸ τῶν μητρέων πληρέων ἐουσέων αἵματος, καὶ ψαυομένη ἀλγήσει, καὶ μάλιστα τὸ ἦτρον, καὶ πῦρ ἕξει
ἄλλοτε καὶ ἄλλοτε ὀξέως, καὶ βορβορύξουσιν αἱ |
18 μῆτραι ἄλλοτε καὶ ἄλλοτε, ἅτε τοῦ αἵματος ἐγκλονευ-
μένου καὶ[21] διαχωρέοντος ἐν αὐτῇσι, καὶ ἡ κοιλίη οὐ
διαχωρήσει κατὰ τρόπον, οὔτε ἡ κύστις διηθήσει τὸ
οὖρον, ἐπὴν οἱ αἱ μῆτραι προσπέσωσι πρὸς τὸν στόμαχον νευρώδεα ἐόντα, ἢ ἐς τὴν κοιλίην ἐμβάλωσιν·
ἀλγέει τε τὴν ῥάχιν καὶ τὸ νῶτον πᾶν, καὶ χαλινοῦται,
καὶ γλῶσσα ἀσαφής· καὶ λιποθυμίη, καὶ ἔστιν ᾗσιν
ἀφωνίη· καὶ δάκνεται τὸν στόμαχον, καὶ χολὴ ξανθὴ
ἔξεισι, καὶ πνεῦμα προσπαῖον,[22] καὶ ἀλύει, καὶ ῥίπτει
ἑωυτὴν, καὶ ἐμπίμπραται. ἐπὴν δὲ μεταστέωσι[23] καὶ
εἰρύσῃ ἡ κύστις τὸ λεπτὸν τοῦ αἵματος τοῦ ἀπὸ τῶν
μητρέων, τὸ οὖρον τότε διουρέεται ἐρυθρόν, καὶ πονέει
μὲν καὶ τὸ ἄλλο σῶμα, μάλιστα δὲ[24] τὸν τράχηλον

her belly will be constipated and larger than it was before, she will grind her teeth, turn away from food, and suffer insomnia. These are the things she experiences while the fourth menses are flowing, but if she is treated even during these, she will recover. In the fifth menses, unless these are very copious, the patient will also have violent pains.

But by the sixth menses she will already be untreatable, and will suffer the signs mentioned above but more intensely, and subsequently the following will happen: she will be restless and cast herself about from time to time, she will lose consciousness, she will vomit phlegm, a pressing thirst will befall her as a result of her cavity burning due to her uterus being filled with blood, she will feel pain when she is touched, especially on her lower abdomen, occasional acute fever will come on, her uterus will sometimes rumble, since the blood gurgles as it passes through it, her cavity will fail to pass stools through it in the proper manner and her bladder will not pass urine, since her uterus has fallen against the sinewy orifice (sc. of the bladder) or impinged upon the cavity. The patient feels pain in her back and along her whole spine, she is tongue-tied, and her speech is unclear; there is loss of consciousness and in some patients speechlessness; she feels a stinging pain in the orifice, yellow bile is passed, her breathing is broken, she is restless and casts herself about, and she is inflamed. When the uterus changes its position, and the bladder draws the thin component of the blood from it, the urine passed is red and the woman suffers pain in her whole body, especially at the throat, spine,

---

^{20} μιν om. Θ. ^{21} Add. οὐ V. ^{22} προσπαῖον ΘΜ:
πρὸς πλεῖον V. ^{23} μεταστήσῃ Θ. ^{24} δὲ MV: μὲν Θ.

17

καὶ τὴν ῥάχιν καὶ τὴν ὀσφῦν, τούς τε βουβῶνας, καὶ
ἐς τοῦτο ἐλθούσῃ αὐτῇ ἥ τε γαστὴρ αἴρεται, καὶ τὰ
σκέλεα ὑπὲρ τὸ χρεὼν διοιδίσκεται καὶ αἱ κνῆμαι καὶ
οἱ πόδες, καὶ θάνατος ἔπεισι. καὶ περὶ μὲν ταύτης
οὕτω τελευτᾷ ἐς ἓξ μῆνας τὰ ἐπιμήνια ἀδηλεύμενα.

Γίνεται δὲ καὶ τάδε· ἔστιν ᾗσι τῶν γυναικῶν, ἐπὴν
δίμηνα ᾖ τὰ καταμήνια ἐν τῇσι μήτρῃσι πολλὰ
ἐόντα, ἔρχεται ἐς πλεύμονα, ἐπὴν ἀποληφθῇ, καὶ πάσχει πάντα ἅπερ ἐν φθινάδι εἴρηται, καὶ οὐχ οἵη τέ
ἐστι περιεῖναι. γίνεται δὲ καὶ τάδε· ἔστιν ᾗσι διάπυα
γίνεται τὰ ἐπιμήνια χρονίσαντα, ἐπὴν γένηται δίμηνα ἢ τρίμηνα· τοῦτο δὲ μάλιστα γίνεται, ἢν συγ-
καῇ ὑπὸ τοῦ πυρός. σημεῖα δέ ἐστιν, ἢν διάπυα ᾖ·
ὀδύναι ἐμπίπτουσιν ἐς τὸ ἦτρον ἰσχυραὶ καὶ σφύ-
ξιες,[25] καὶ ψαυομένη οὐκ ἀνέχεται, καὶ ἢν μέλλῃ βελ-
τιόνως ἔχειν, ῥήγνυται αὐτῇ τὰ ἐπιμήνια κατὰ τὸ
αἰδοῖον, καὶ χωρέει πῦον καὶ αἷμα· ὀζόμενον δὲ χωρέει ἐφ᾽ ἡμέρας ἑπτὰ[26] ἢ ἐννέα· ἐν δὲ τῷ πρὶν χρόνῳ
20 πονέεται, | ὡς εἴρηται πρόσθεν· ἐπὴν δὲ ἀποκαθαρθῇ,
ἄριστον μὲν εἰ μὴ γένοιτο ἕλκεα· ἢν δ᾽ ὑπολείπηται
ἕλκεα, πλέονος δεήσει θεραπείης ὅπως τὰ ἕλκεα μὴ
μυδήσῃ καὶ κάκοδμα γένηται· ἄφορος δ᾽ ἔσται καὶ ἢν
ῥαΐσῃ, ἢν μεγάλα ᾖ τὰ ἕλκεα γενόμενα ἐν τῇσι
μήτρῃσιν. ἢν δὲ μή οἱ[27] κατὰ τὸ αἰδοῖον χωρήσῃ τὰ
ἐπιμήνια διάπυα γενόμενα, συμβήσεται ὑπὲρ τοῦ
βουβῶνος κατὰ τὴν λαπάρην ῥαγῆναι, ἄτερ φύματος,
ἅτε τοῦ πύου διαφαγόντος, κἀκείνῃ χωρήσει πυώδεα

18

lower back and groins, and when she has progressed to this stage her belly becomes raised, her legs swell up excessively with edema, including her shins and her feet, and death arrives. In such a woman's case, death occurs in this manner after the menses have failed to appear for six months.

It also happens to some women, when the menses in their uterus are excessive for two months, that on being blocked off these move to the lungs, and the women suffer all the signs described as belonging to consumption, and inevitably succumb. The following can occur too: in some women the menses become purulent after they have failed to pass for two or three months, in most cases as the result of their being burned together by fever. In the case of purulence, here are the signs: severe pains occupy the lower abdomen together with throbbing, and if the woman is touched there she cannot stand it. If she is going to improve, the menses will break out from her vagina with the discharge of pus and blood; this ill-smelling flux continues over seven or nine days. In the time before this the patient suffers the pain described above. When the cleaning is taking place, it is best if no ulcers form, but if ulcers are left behind, the patient will require further treatment, in order that the ulcers do not weep and become ill-smelling. Barrenness will result, even if the woman mends, if the ulcers in the uterus are large. If the menses that have become purulent are not discharged through the vagina, they will go on to break out above the groin through the flanks without the formation of an abscess, the pus simply eating its way through the flesh, and ill-smelling, purulent

---

²⁵ καὶ σφ. om. Θ.   ²⁶ Add. ἢ ὀκτὼ M.   ²⁷ οἱ om. Θ.

ὀδμαλέα· καὶ ἢν τοῦτο γένηται, οὐ περιγίνεται ἡ
γυνή· ἢν δὲ καὶ περιγένηται, αἰεὶ ἄφορος ἔσται· ταύτῃ
γὰρ οἱ τὸ λοιπὸν ἡ ὁδὸς γίνεται τοῖσιν ἐπιμηνίοισιν
ἔξω· τὸ γὰρ στόμα τῶν μητρέων πρὸς τοῦτο τὸ χωρίον προσπέπτωκε.

Γίνεται δὲ καὶ τόδε· ἔστιν ᾗσιν ἐπὴν δίμηνα ἢ
τρίμηνα ἢ χρονιώτερα ᾖ τὰ ἐπιμήνια καὶ προσπέσῃ
πρὸς τὴν λαπάρην, μὴ διαπύων τῶν καταμηνίων ἐόντων, ὡς φῦμα γίνεται ὑπὲρ τοῦ βουβῶνος ἀκέφαλον,
μέγα, ἐρυθρόν. καὶ τῶν ἰητρῶν πολλοὶ ἤδη οὐκ εἰδό-
τες τοῦτο οἷόν ἐστιν ἔταμον καὶ ἐς κίνδυνον ἤγαγον
οὕτω. τὸ δὲ ὡς φῦμα γινόμενον γίνεται τρόπῳ τοιῷδε·
ἐπαυρίσκεται τοῦ αἵματος ἡ σάρξ, ἅτε προσκειμένου
τοῦ στόματος τῶν μητρέων τῇ λαπάρῃ, καὶ πίμπλα-
ται ἀπ' αὐτοῦ, καὶ ἀφίσταται[28] ἅτε πληρευμένη τοῦ
αἵματος ἡ σάρξ· καὶ ἔστιν ὅτε, ἢν μεταστῇ τὸ στόμα
τῶν μητρέων καὶ γένηται κατὰ τὸ αἰδοῖον, καὶ χωρέῃ
διὰ τοῦ αἰδοίου τὰ καταμήνια, καθίσταται τὸ ἐξεστη-
κὸς κατὰ τὴν λαπάρην, διαδιδοῖ[29] γὰρ ἐς τὰς μήτρας,
αἱ δ' ἔξω ἐχάλασαν· ἢν δὲ μὴ στραφῇ κατὰ τὸ αἰδοῖον τὸ στόμα τῶν μητρέων, διαπυεῖ κατὰ τὴν λαπάρην, καὶ τότε ἤδη ὁδὸς γίνεται τοῖσι καταμηνίοισι,
καὶ οἱ κίνδυνοι οἱ αὐτοί εἰσιν αὐτῇ οἱ καὶ πρόσθεν
22 εἰρημένοι. τρέπεται δὲ καὶ ἐς | ἔμετον· ἔστιν ᾗσι καὶ
κατὰ τὴν ἕδρην, ὥσπερ μοι[30] εἴρηται ἐν τῇσι Παρθε-

---

[28] ἀφίστ. Θ: ἐξίστ. MV.   [29] ἐξεδιαδιδοῖ Θ.
[30] ὥσ. μοι M: ἔστιν ᾗσιν ὡς Θ: ὥσ. V.

menses will appear there. If this happens, the woman is not likely to survive, or if she does survive, she will be permanently barren: for in the future her menses will pass to the outside at that place, since the mouth of her uterus has become lodged exactly there.

The following, too, is possible: in some women when two or three months or even more time has passed without the menses appearing, and these fall against a flank (sc. internally) without suppurating, a kind of growth will form above the groin and this growth will be large and red, although lacking a head. Indeed, in the past many healers have in their ignorance of its nature lanced such a growth and thereby brought their patient into danger. This thing that appears as a growth is formed as follows: the flesh in that location takes in blood, since the mouth of the uterus is lying against the flank there, and becomes saturated with it, and being filled with this blood rises up. Sometimes, if the mouth of the uterus changes its position and becomes aligned with the vagina, and the menses pass through the vagina again, the protuberance in the flank will recede, since (sc. the moisture in the swelling) flows back through into the uterus, which allows it to flow out. But if the mouth of the uterus does not revert to the vagina, it will expel pus through the flank and then a passage for the menses is formed, and the dangers are the same as described above. The menses may also be derivated to vomiting. There are also women in whom they pass through the anus, as described in *Diseases of Girls*,[2] and

[2] No such description is present in *Girls*, Loeb *Hippocrates* vol. 9, 359–63. Cf. *Barrenness* 1, Loeb *Hippocrates* vol. 10, 339.

νίῃσι Νούσοισι, καὶ σημεῖα καὶ πόνους τοὺς αὐτοὺς
δείκνυσι τοῖσι κεῖθι εἰρημένοισιν· ἦσσον δὲ ταύτην
τὴν ὁδὸν ποιέεται τὰ ἐπιμήνια τῇσι γυναιξὶν ἢ τῇσι
παρθένοισιν.

3. Ἐπὴν δὲ τὰ ἐπιμήνια κρυφθῇ, ὀδύνη ἔχει τὴν
νείαιραν γαστέρα, καὶ δοκέει τι ἐγκεῖσθαι βάρος, καὶ
τὰς ἰξύας ἐκπάγλως πονέει καὶ τοὺς κενεῶνας. ἢν δὲ
τὰ ἐπιμήνια παντάπασι μὴ γίνηται ὑπὸ νούσου ἢ παχέα καὶ γλίσχρα καὶ κολλώδεα εἴη, πρῶτον χρὴ τὴν
κοιλίην καθῆραι ἄνω τε καὶ κάτω· ἔπειτα τὰς ὑστέρας
προσθέτῳ, ὑφ' οὗ αἷμα καθαίρεται, καὶ διαλείπειν, καὶ
πῖσαι ὑφ' οὗ αἷμα ἴῃ· πινέτω δὲ κρῆθμον τῷ οἴνῳ τῷ
ἀπὸ δαιδός. ἢν[31] δὲ ῥόος μὴ γίνηται, ἔσται ὥστε δοκέειν ἐγκύμονα εἶναι, καὶ μισγομένην ἀνδρὶ ἀλγεῖν
ὥστε δοκέειν ἐγκεῖσθαί τι, καὶ βρῖθος ἐν τῇ γαστρὶ
ἐγγίνεται, καὶ ἡ γαστὴρ πρόκειται, καὶ ἐπαίρεται
ἠδελφισμένως ὡς[32] ἐν γαστρὶ ἐχούσῃ, καὶ καρδιώσσει,[33] ἐπὴν ἡμέραι πεντήκοντα[34] μάλιστα ἔωσι, καὶ
πόνος ἴσχει ἄλλοτε καὶ ἄλλοτε τῆς γαστρὸς τὸ κάτω
τοῦ ὀμφαλοῦ, τόν τε τράχηλον καὶ τοὺς βουβῶνας
καὶ τὴν ὀσφῦν. καὶ ὅταν δύο μῆνες ἢ τρεῖς γένωνται,
ἔστιν ὅτε ἐρράγη οἱ κατὰ τὸ αἰδοῖον τὰ καταμήνια[35]
24 ἀθρόα, καὶ δοκέει ὥσπερ σαρκία εἶναι | τὰ ἀπιόντα ὡς
ἐκ διαφθορῆς καὶ μέλανα.

[31] The text ἢν δὲ ῥόος . . . (ch. 3) κόσμῳ ἰηθεῖσα is
transmitted a second time in Θ after ch. 9; see p. xiii f. above.

these women display the same signs and sufferings as described there: this route is taken by the menses less often in women than in girls.

3. When the menses cease to appear, pain occupies the lower belly, and a heaviness seems to be lying there; the woman suffers terribly in her loins and flanks. If the menses fail altogether as the result of a disease, or if they are thick, sticky and gluey, you must clean first the cavity both upward and downward, and then the uterus with a suppository that purges blood; after leaving an interval, give a potion that moves blood: have the patient drink samphire in wine mixed with pinewood. If no flux follows, the woman will appear to be pregnant, and when she has intercourse with her husband she will suffer a pain that seems to indicate that some object is lying there. A weight is present in her belly and it protrudes, rising up just as in a woman who is pregnant; she suffers heartburn, particularly after the fiftieth day, and pain from time to time occupies her belly down from the navel, as well as her neck, her groin and her lower back. After two or three months sometimes her menses break out in a mass through the vagina, and what comes out has a fleshy appearance, as if it were from an abortion, and is dark in color.

---

32 ἐπαίρ. ἠδελφ. ὡς M: εἰμείρεται ἠδελφ. Θ: μετεωρίζεται ὁμοίως ὥσπερ εἰ Θ': ἱμείρει ἠδελφ. ὡς V.

33 καρδιώσσει ΘM: -διάζει Θ': -διώξει V.

34 πεντήκ. ΘMV: τριάκ. Θ'.

35 αὐτῇ add. M.

Ἔστι δ᾽ ᾗσι καὶ ἕλκεα γίνεται ἐν τῇσι μήτρῃσι, καὶ δεήσεται τὴν μελέτην προσέχειν. πολλῇσι δὲ συμβαίνει ὥστε δοκέειν ἐξ μῆνας ἔχειν ἐν γαστρὶ ἢ ὀλίγῳ ἐλάσσονα χρόνον, καὶ ἡ γαστὴρ πρόκειται, καὶ τἄλλά οἱ γίνεται ὡς ἐν γαστρὶ ἐχούσῃ· ἔπειτα ἔστι μὲν ᾗσι διάπυα[36] ἐρράγη ὑπὲρ τοῦ βουβῶνος ἅμα πέμπτῳ ἢ ἕκτῳ μηνὶ καὶ ὁδὸν ταύτῃ ἐποιήσατο· ἔστι δὲ ᾗσι καὶ ἕλκεα ἐν τῇσι μήτρῃσι ἐγγίνεται κατὰ τὸ ὑπὲρ τοῦ[37] βουβῶνος, καὶ κινδυνεύσει ἀποθανεῖν· ἢν δὲ καὶ περιγένηται, ἄφορος ἔσται. ἔστι δὲ ᾗσι κατὰ τὸ αἰδοῖον ῥήγνυται, καὶ χωρέει[38] σεσηπότα τε καὶ πυώδεα, καὶ ἀπὸ τούτων ἕλκεα ἐγγίνεται ἐν τῇσι μήτρῃσι, καὶ κινδυνεύει, ἀλλὰ χρὴ ὅπως[39] μὴ τὰ ἕλκεα παλαιὰ γένηται, ἰητρεύειν προσέχοντα·[40] ἄφορος δὲ καὶ αὕτη καὶ[41] ἢν ἰηθῇ. ἢν δὲ μή οἱ ῥαγῇ τὰ καταμήνια διενεχθέντα ἐς ἐξ μῆνας, πείσεται πάντα ἅπερ ἀτόκῳ ἐούσῃ τὰ καταμήνια ὁδὸν οὐκ ἠδύνατο εὑρεῖν· καὶ ἢν μὲν θεραπευθῇ, ὑγιὴς ἔσται· εἰ δὲ μή, διενέγκασα καὶ ἐς ὀκτὼ μῆνας θνήσκει. πολλῇσι δὲ γίνεται, ἢν τὰ καταμήνια φλεγματώδεα χωρέῃ, ἐπὶ πολὺν χρόνον χωρέειν καὶ ἐλάσσονα εἶναι τῶν ὑγιηρῶν· ὑγιὴς δὲ γίνεται ἐν κόσμῳ ἰηθεῖσα.

4. Ἢν δὲ τὰ ἐπιμήνια γυναικὶ χωρέῃ μέν, ἐλάσσονα δὲ τοῦ δέοντος χωρέῃ, οἷα τοῦ στόματος τῶν

---

[36] διάπυα Θ′: διατείνοντα Θ: διατείναντα Μ: διαπύοντα V.
[37] κατὰ . . . τοῦ ΘV: καὶ περὶ Θ′: καὶ κάτω ὑπὲρ τοῦ Μ.
[38] Add. αὐτέῃ Μ: add. αὐταίῃσι V.

There are some women in whom ulcers arise in the uterus, too, and these require treatment. To many women it happens that they seem to be pregnant for six months or a little less, their belly protrudes, and they experience all the other signs present in pregnant women. Then in some of these purulent material breaks out above the groin about the fifth or sixth month, and makes a passageway for itself there, while in others ulcers form in the uterus in the area above the groin, and these women come into mortal danger; any that survive will be barren. In yet others, expulsion is through the vagina, and the flux is putrid and purulent; from this ulcers form in the uterus and the patients are in danger: it is imperative, lest the ulcers become established, to treat attentively, and this patient too will be barren, even if she is treated. If the menses do not break out, even when they have been deferred for six months, the woman will suffer all the same things as a woman who has not borne children in whom the menses have been unable to find a passage way out: if she is treated, she will recover her health, if not, she will go on until the eighth month, and then die. To many women it happens that, if their menstrual flux contains phlegm, it flows for a longer time and in less quantity than in a healthy woman: such a woman recovers her health on being treated in the proper manner.

4. If a woman's menses flow, but less than they should, due to the mouth of her uterus being slightly inclined away

---

39 καὶ κινδ. . . . ὅπως Θ: καὶ δεήσει μελεδώνης ὅκως Θ′: καὶ κινδυνεύει καὶ χρὴ ὅκως M: ἀλλὰ χρονικῶς V.

40 τὰ ἕλκεα π. . . . προσέχοντα ΘMV: οἱ τὰ ἕλκεα σαπρὰ γενόμενα θάνατον ἐπάξει Θ′.   41 καὶ Θ: om. Θ′MV.

26 μητρέων παρακεκλιμένου | ὀλίγον τοῦ αἰδοίου ἢ ἁρμῷ
μεμυκότος ἐς τοῦτο, ὥστε χωρέειν μέν, ἀποφράσσεσθαι δὲ ἀπ᾽ αὐτῶν † . . . καὶ αἱ δίοδοι αἱ περαιοῦσαι
. . . † ἐπὴν γὰρ⁴² κατέλθη ἐς τὰς μήτρας, τὰ μέτρια
τελείως τοῦ αἵματος ἐπικειμένου τῷ στόματι αἰεί,
προέρχεται⁴³ κατ᾽ ὀλίγον· ἔπειτα δὲ ἐπὴν αἱ ἡμέραι
παρέλθωσιν ἧσι καθαίρεσθαι μεμάθηκε, καὶ ἐρχθῇ
τὸ αἷμα ἐν τῆσι μήτρησι τὸ ὑπολειφθέν, καὶ ἕτερα
ἐπικατιόντα ἐπιμήνια μὴ ἐξωθῇ τὸ ἐρχθὲν αἷμα, ἀλλὰ
βαρύνῃ κατ᾽ ὀλίγον, ἔσται τῇ γυναικὶ τοὺς πρώτους
μῆνας ἐπὶ δύο ἢ τρεῖς μὴ ἐσαΐειν κάρτα. ἐπὴν δὲ οἱ
μῆνες πλέονες γένωνται, ἔτι μᾶλλον πονήσει, καὶ οὐκ
ἴσχει ἐν γαστρὶ μέχρι ἂν οὕτως ἔχῃ, καὶ πῦρ λήψεταί
μιν μάλιστα τὰς ἡμέρας ἐν ἧσι καθαίρεσθαι μεμαθήκει, ἠπεδανόν· εἰκὸς δέ ἐστι καὶ ἐν τῷ μεσηγὺ
χρόνῳ πυρεταίνειν καὶ φρίσσειν καὶ καρδιώσσειν καὶ
ἀλγέειν⁴⁴ ἐπὶ τὸ πλῆθος ἀνὰ πάσας ἡμέρας, καὶ⁴⁵ ἄλλοτε καὶ ἄλλοτε τὸ σῶμα, καὶ μάλιστα τὴν ὀσφῦν καὶ
τὴν ῥάχιν καὶ τοὺς βουβῶνας, τά τε ἄρθρα τῶν χειρῶν καὶ τῶν σκελέων ἀλγέειν.⁴⁶ ταῦτα δὲ οὐχ ὁμοῦ
ἀλγέει, ἀλλ᾽ ἄλλοτε ἄλλο, ὅπη ἂν βρίσῃ τὸ αἷμα τὸ
ἀποκεκριμένον καὶ μὴ δυνάμενον εἶναι ἐν τῆσι μή-
τρησι· καὶ ὅπη ἂν στηρίξῃ τοῦ σώματος,⁴⁷ οἴδημα
ἔστιν ὅτε γίνεται καὶ σπασμὸς ἰσχυρὸς τῶν ἄρθρων
τοῦ σώματος, καὶ τῶν ἄλλων σημείων τῶν προειρη-
μένων φαίνεταί οἱ ἄλλοτε ἄλλο.

---

⁴² γὰρ om. MV.   ⁴³ δὴ add. M: ἤδη add. V.

from her vagina, or closed although still having a gap, so that although they flow they are obstructed from these things † . . . and the passages which penetrate through . . . † for when they have come down into the uterus, since exactly the middling amount of blood is always pressing on the mouth, they flow a little at a time. Then, when the days arrive on which she has been accustomed to be cleaned, and blood is left behind enclosed in her uterus, and additional menses coming down on top of this do not expel the blood that is closed in, but this gradually weighs the woman down, it is still possible that she will not perceive this very much for the first two or three months. But as months pass by, her suffering will continually increase, and she will not become pregnant as long as this persists; fever will seize her, especially on the days when she was accustomed to be cleaned, although of a mild sort. It is likely in the intervening times, too, for her to have fever, chills, and heartburn, and to suffer pain generally in her body at various times of the day, especially in her loins, back and groins; in the joints of her arms and legs she will also suffer pain. These things the patient does not suffer all at one time, but now and then, in whichever part the blood that is being excreted weighs her down since it cannot enter her uterus. Wherever blood becomes fixed in the body, edema occasionally forms, together with violent spasms of the body's joints; also, some of the other signs described above appear in the woman at one time or another.

---

44 ἀνάγειν V.   45 ἀλγέει add. M: ἀλγέειν add. V.
46 ἀλγέειν Θ: ἀλγέει M: om. V.
47 σώμ. M: αἵματος ΘV.

Αὕτη ἢν μὲν θεραπευθῇ κατὰ τρόπον, ὑγιὴς γίνε-
ται· εἰ δὲ μή, ἡ νοῦσος ἑπτάμηνος καὶ χρονιωτέρη
γενομένη θανατώσειεν ἄν, ἢ χωλεύσειεν, ἢ ἀκρατέα
28 τινὰ τῶν μελέων | ποιήσειεν, ἢν ὑπὸ ῥίγεος καὶ ἀσι-
τίης τὸ αἷμα, ἔνθα ἂν ἐπέλθῃ, πῆξιν ἔχῃ περὶ τὰ
νεῦρα. τοῦτο δὲ γίνεται τὸ νόσημα μᾶλλον τῇσιν
ἀνάνδροισιν· εἰ δὲ[48] ἐμπειροτόκων ταῦτα τὰ νοσήματα
προσπέσοι τὰ εἰρημένα ἢ ἄσσα μέλλει εἰρήσεσθαι,
πολυχρονιώτερά τε ἔσται καὶ ἧσσον ἐπίπονα· τὰ δὲ
σημεῖα ταὐτὰ καὶ τελευταὶ αἱ αὐταὶ γίνονται τῇ τε
ἀτόκῳ καὶ τῇ λοχίων ἐμπείρῳ, ἢν μὴ θεραπεύωνται·
χρὴ δὲ αὐτίκα τὴν θεραπείην ποιέεσθαι· εἰ δὲ μή, ἐπιφαίνεται τὰ νουσήματα.

5. Ἢν δὲ τὰ ἐπιμήνια πλέονα τοῦ δέοντος χωρέῃ
καὶ παχύτερα, οἷα τοῦ σώματος φύσει τε εὐρέος ἐόντος καὶ τοῦ στομάχου τῶν μητρέων πλησίον τοῦ αἰδοίου κειμένου, καὶ ἐπὶ τούτοισιν ἀνδρί τε συνίῃ
πολλὰ καὶ εὐωχέηται ἐσάπαξ ποτέ, πολλὰ ἀλέα κατελθόντα καὶ χωρέοντα βύζην ἐπευρύνει μᾶλλον τὸ
στόμα τῶν μητρέων βιησάμενα. καὶ ἢν ἐπὶ τούτοισι
μὴ ἐπιπέσῃ κεναγγίη, ἀλλ᾽ αὖθις πολλὰ ἴῃ ἀλέα καὶ
τὸ στόμα εὐρὺ ποιέῃ, καὶ τὸ σῶμα εὐωχουμένης[49] καὶ
συνεούσης ἀνδρὶ εὔροον ᾖ ἐπὶ τὰς μήτρας, καὶ ἀλέα
ἐπιλείβηται,[50] ἄχροός τε ἔσται μέχρι ἂν οὕτως ἔχῃ,
καὶ ἢν οἱ ὕστερόν τι νόσημα ἢ πάθημα ἐπιπέσῃ ὥστε

48 μὴ add. M.   49 εὐωχ. Θ: ἅτε εὐωχ. ἱμειρομένης M:
τε εὐωχ. καὶ ἱμειρομένης V.

If this patient is treated in the proper way, she will recover, but if not, the disease will continue for seven months or longer, and she will die or become lame, or lose command over some of her limbs, if, as the result of chilling or her failure to eat, the blood, wherever it happens to go, congeals around the cords. This disease occurs more often in unmarried women: if on the other hand the diseases which have been, or are about to be described befall women who have borne children, they will be of longer duration but less troublesome. The same signs and outcomes pertain both to a woman who has not borne children and to one who has experienced the lochia, if they go untreated. Treatment must be applied immediately: if not, the diseases will become manifest.

5. If the menses flow more than they should and thicker, due to the woman's body being by nature wide open and the orifice of her uterus lying near her vagina, and besides she has frequent intercourse with her husband, and she then at some time overindulges herself (sc. at table), a great volume of menses will descend, and through their pressure forcefully widen the mouth of her uterus even more. And if no emptying of the vessels supervenes, but a further great flux occurs and widens the mouth yet more, and the patient's body—as the result of her overindulgence and having intercourse with her husband—has passages open to the uterus so that a massive flux occurs, she will remain pale as long as this state persists; and if later some disease or affection emaciates her

---

τρυχωθῆναι τὸ σῶμα, ὅμως αἵ τε μῆτραι κατὰ τὸ ἔθος
εὐρύστομοί εἰσι καὶ τὸ σῶμα εὔροον ἐπ᾽ αὐτάς ἐστι.

30   Καὶ μετὰ ταῦτα πῦρ ἔχει, καὶ | ἀσιτέει, καὶ ἀλυκτέι,
καὶ λεπτὴ καὶ ἀμενηνὸς ἐκ τῶν ἐπιμηνίων, καὶ τὴν
ὀσφῦν πονήσει, καὶ τοῦ χρόνου προϊόντος, ἢν μὴ θεραπευθῇ, πάντα μιν μᾶλλον πονήσει ἐν τῷ μεταξὺ
χρόνῳ, καί οἱ ἔσται κίνδυνος ἀφόρῳ γενέσθαι ἢ τρυ-
χωθείσῃ ὑπὸ χρόνου καὶ τῆς νούσου, ἤν τί οἱ συμ-
πέσῃ καὶ ἄλλο νόσημα, ἐπὶ τούτῳ θανεῖν.

6. Χωρέει δὲ τὰ καταμήνια παχύτατα καὶ πλεῖστα
τῶν ἡμερέων τῇσιν ἐν μέσῳ, ἀρχόμενα δὲ καὶ τελευ-
τῶντα ἐλάσσονα καὶ λεπτότερα. μέτρια δ᾽ ἐστὶ πάσῃ
γυναικὶ χωρέειν, ἢν ὑγιαίνῃ, τὰ ἐπιμήνια ἐλθόντα
ὅσον κοτύλαι δύο Ἀττικαὶ ἢ ὀλίγῳ πλέονα ἢ ἐλάσσονα, ταῦτα δὲ ἐφ᾽ ἡμέρας δύο ἢ τρεῖς· ὁ δὲ πλείων
χρόνος ἢ ἐλάσσων ἐπίνουσος καὶ ἄφορος.[51] τεκμαίρε-
σθαι δὲ χρὴ ἐς τὸ σῶμα τῆς γυναικὸς ὁρῶντα, καὶ
ἐρωτᾶν πρὸς τὰ πρότερα συμβαλλόμενα, εἴτε ἐπίνοσα
ἴοι ἢ μὴ ἐπίνοσα.[52] ἢν γὰρ ἐλάσσονας ἢ πλείονας
ἡμέρας τοῦ μάθεος φοιτᾷ, ἢ αὐτὰ ἐλάσσονα ἢ πλείονα ἴῃ, ἐπίνοσά ἐστιν, ἢν μὴ ἡ φύσις αὐτὴ νοσηλὴ
καὶ ἄφορος ᾖ.

Ἢν δὲ τοῦτο ᾖ καὶ μεθίστηται ἐπὶ τὸ ὑγιηρότερον,
ἄμεινον. χωρέει δὲ αἷμα οἷον ἀπὸ ἱερείου, καὶ ταχὺ
πήγνυται, ἢν ὑγιαίνῃ ἡ γυνή. ᾗσι δὲ ἐν φύσει ἐστὶ
πλέονα καθαίρεσθαι τεσσάρων ἡμερέων καὶ πολλὰ
κάρτα χωρέει τὰ ἐπιμήνια, αὗται λεπταὶ γίνονται, καὶ
τὰ ἔμβρυα αὐτῶν λεπτὰ καὶ ἀμαλδύνεται. ᾗσι δὲ

body, her uterus will continue in its usual widemouthed state, and her body will remain fluent with it.

After this, fever will set in, the woman will fail to eat, she will be distressed, she will become thin and feeble from her menses, and she will suffer pain in her lower back; with time, unless treatment is applied, her pains will increase in the time between the menses, she will run the risk of becoming barren and emaciated due to the disease and its chronicity, and if some other disease befalls her in addition, she will perish from it.

6. The menses flow thickest and in the greatest amount on the days in their middle, in less quantity and thinner when they are beginning and ending. In every healthy woman the amount of the menstrual flux to pass is equal to two Attic cotyles, or a little more or less—this over two or three days; if the time is more or less than this, it indicates disease and barrenness. You must gather evidence by inspecting the patient's body and asking her how her previous menses were in comparison, and whether or not they were unhealthy. For if they pass for more or less days than they are used to, or in a greater or less quantity, this indicates malignancy, unless the woman is by nature diseased or barren.

If the case is such, and then there is a change in the direction of better health, it will be better. If a woman is healthy, her menstrual blood will pass like that of a sacrificial animal, and it will quickly congeal. But women who are naturally cleaned for more than four days and who pass very copious menses will become thin and have fetuses

---

51 Add. ἐστι MV.   52 εἴτε ἐπίν. . . . ἐπίνοσα Θ: εἴτε ἐπί-
νοσος ἀεὶ ᾖ εἴτε μή M: εἴτ᾽ ἐπίνοσα εἴη εἴτε μή ἐπίνοσα V.

τριῶν ἡμερέων ἔλασσον ἡ κάθαρσις γίνεται ἢ ὀλίγα
χωρέει, αὗται παχεῖαί τε καὶ εὔχροοι ἀνδρικαί τε, οὐ
μνησίτοκοι δέ εἰσιν, αἵδε οὐδὲ κυΐσκονται. |

32   7. *Ην δὲ πνὶξ προστῇ ἐξαπίνης, γίνεται δὲ μάλιστα τῇσι μὴ συνεούσῃσιν ἀνδράσι καὶ τῇσι γεραιτέρῃσι μᾶλλον ἢ τῇσι νέῃσι· κουφότεραι γὰρ αἱ μή-
τραι σφέων εἰσί· γίνεται δὲ μάλιστα διὰ τόδε· ἐπὴν
κενεαγγήσῃ καὶ ταλαιπωρήσῃ πλέονα τῆς μαθήσιος,
θερμανθεῖσαι[53] αἱ μῆτραι ὑπὸ τῆς ταλαιπωρίης στρέφονται, ἅτε κεναὶ ἐοῦσαι καὶ κοῦφαι· εὐρυχωρίη γάρ
ἐστι σφὶν ὥστε στρέφεσθαι, ἅτε τῆς κοιλίης κενῆς
ἐούσης· στρεφόμεναι δὲ ἐπιβάλλουσι τῷ ἥπατι, καὶ
ὁμοῦ γίνονται, καὶ ἐς τὰ ὑποχόνδρια ἐμβάλλουσι· θέ-
ουσι γὰρ καὶ ἔρχονται ἄνω πρὸς τὴν ἰκμάδα, ἅτε ὑπὸ
τῆς ταλαιπωρίης ξηρανθεῖσαι μᾶλλον τοῦ καιροῦ· τὸ
δὲ ἧπαρ ἰκμαλέον ἐστίν· ἐπὴν δὲ ἐπιβάλωσι τῷ ἥπατι,
πνῖγα ποιέουσιν ἐξαπίνης ἐπιλαμβάνουσαι τὸν διάπνοον τὸν περὶ τὴν κοιλίην. καὶ ἅμα τε ἄρχονται
ἔστιν ὅτε προσβάλλειν πρὸς τὸ ἧπαρ, καὶ ἀπὸ τῆς
κεφαλῆς φλέγμα καταρρεῖ ἐς τὰ ὑποχόνδρια οἷα πνι-
γομένης, καὶ ἔστιν ὅτε ἅμα τῇ καταρρύσει τοῦ φλέγματος ἔρχονται ἐς χώρην ἀπὸ τοῦ ἥπατος, καὶ παύε-
ται ἡ πνίξ. κατέρχονται δὲ καθελκύσασαι ἰκμάδα καὶ
βαρυνθεῖσαι· τρυσμὸς δ᾽ ἀπ᾽ αὐτῶν γίνεται, ἐπὴν χω-
ρέωσιν ἐς ἕδρην τὴν σφῶν αὐτῶν· ἐπὴν δ᾽ ἔλθωσιν,
ἔστιν ὅτε ἡ γαστὴρ μετ᾽ ἐκεῖνα ὑγροτέρη γίνεται ἢ ἐν
τῷ πρὶν χρόνῳ. χαλᾷ γὰρ ἤδη ἡ κεφαλὴ τοῦ φλέγματος ἐς τὴν κοιλίην. ἐπὴν δὲ πρὸς τῷ ἥπατι ἔωσιν

that are thin and weak. Those in whom the cleaning occurs for less than three days or in a small amount will be stout, of a good color, and manly, but not prolific or likely to become pregnant.

7. If suffocation suddenly occurs, this happens mainly in women that are not having intercourse with men, and more in older women than in younger ones: for their uterus is lighter. Generally it comes about as follows: When a woman has empty vessels and exerts herself more than she is used to, as her uterus is warmed by the exertion, it turns to the side because it is empty and light: for there is open space into which it can turn, seeing that the cavity is empty. In turning to the side (sc. the uterus) falls against the liver and comes into contact with it, and then it falls against the hypochondria. For it moves rapidly upward toward the moisture, since it has been excessively dried by the exertion, and the liver is full of moisture. On falling against the liver, the uterus immediately produces suffocation by occupying the air space around the cavity. Sometimes, simultaneously with the uterus' first contact with the liver, phlegm also flows down from the head into the hypochondrium as a result of the woman's suffocation, so that sometimes with this simultaneous defluxion of phlegm, the uterus moves away from the liver back to its own location and the suffocation ceases. The uterus moves back down in this way on filling with moisture and becoming heavy, and a gurgling sound arises from it when it returns to its natural location. After it arrives there, sometimes the belly is moister than it was before; for the head has already released phlegm into the cavity. When the

---

53 θερμ. Θ: ἀνανθεῖσαι MV.

αἱ μῆτραι καὶ τοῖσιν ὑποχονδρίοισι, καὶ πνίγωσι, τὰ
λευκὰ τῶν ὀφθαλμῶν ἀναβάλλει, καὶ ψυχρὴ γίνεται·
εἰσὶ δὲ αἳ καὶ πελιδναὶ γίνονται ἤδη· καὶ τοὺς ὀδόντας βρύχει, καὶ σίελα ἐπὶ στόμα ῥέει, καὶ ἐοίκασι
τοῖσιν ὑπὸ τῆς Ἡρακλείης νούσου ἐχομένοισιν. ἢν δὲ
χρονίσωσιν αἱ μῆτραι πρὸς τῷ ἥπατι καὶ τοῖσιν ὑπο-
χονδρίοισιν, ἀποπνίγεται ἡ γυνή.

34   Ἔστι δ᾽ ὅτε, ἐπὴν | κενεαγγήσῃ ἡ γυνὴ καὶ ἐπιτα-
λαιπωρήσῃ, αἱ μῆτραι στρεφόμεναι πρὸς τῆς κύστιος
τὸν στόμαχον[54] προσπίπτουσι καὶ στραγγουρίην ποιέουσιν, ἄλλο δ᾽ οὐδὲν κακὸν ἴσχει, καὶ ἐν τάχει ὑγιαίνει θεραπευομένη, ἔστι δ᾽ ὅτε καὶ αὐτομάτη.[55] ἔστι
δ᾽ ᾗσιν ἐκ ταλαιπωρίης καὶ[56] ἀσιτίης πρὸς ὀσφὺν ἢ
πρὸς ἰσχία προσπεσοῦσαι πόνους παρέχουσιν.

8. Ἢν δὲ γυνὴ τὸ σῶμα φλαύρως ἔχῃ καὶ ἴῃ χολώδεα τὰ ἐπιμήνια, γνωστά[57] ἐστι τῷδε· μέλανά ἐστι
κάρτα, ἔστι δ᾽ ὅτε μέλανα[58] λαμπρά, καὶ κατ᾽ ὀλίγα
ἔρχεται, καὶ οὐ ταχὺ[59] πήγνυται, καὶ ἡ γονὴ[60] ἀμαλδύνεται ἀμφοῖν, τοῦ τε ἄρσενος καὶ τοῦ θήλεος, καὶ
οὐκ ἴσχει ἐν γαστρί, καὶ ἀρχομένης μὲν τῆς νούσου,
καθαίρεται τὰς ἡμέρας ἃς μεμαθήκει, οὐ πλέονας·
προϊόντων δὲ τῶν ἐπιμηνίων, πλέονάς τε[61] ἡμέρας
καθαίρεται καὶ ἐλάσσονα τὰ καταμήνια καθ᾽ ἑκάστην
ἡμέρην φαίνεται, πυρετοί τε ἐπιγίνονται πλάνητες
ὀξέες σὺν φρίκῃ, καὶ ἀσιτίη ἄλλοτε καὶ ἄλλοτε, καὶ

---

[54] τὸν στόμαχον ΘΜ: τὸ στόμα V.
[55] -μάτη V: -μάτῃσιν ΘΜ.

uterus is lying against the liver and the hypochondrium, and thereby provoking suffocation, the patient turns the whites of her eyes up, and becomes cold; some immediately turn livid. She may also grind her teeth, salivate in her mouth, and take on the appearance of people suffering from Hercules' disease (epilepsy). If a woman's uterus stays against her liver and hypochondria for a longer time, she chokes to death.

Sometimes, when a woman has empty vessels and she exerts herself, her uterus turns away and falls against the orifice of her bladder, bringing on strangury; but there is no further trouble, and on being treated she will quickly recover—sometimes even spontaneously. In some women, as the result of their exerting themselves and going without food, (sc. the uterus) falls against the loin or hips and causes pain.

8. If a woman is poorly in her body and bilious menses pass, this can be recognized by the following: the menses are very black—sometimes shiny black—they pass a little at a time, and they do not congeal at once. Both seeds—of the male and of the female—are weakened and the woman does not become pregnant. When the disease is beginning, she is cleaned on her accustomed days, and not more, but as the periods succeed one another she is cleaned for more days, with less menses appearing each successive day, acute wandering fevers set in with shivering, she loses her appetite from time to time and has heart-

56 καὶ ΘV: ἢ M.   57 γνω. ΘV: εὔγνω. M.
58 Add. ἢ M.   59 οὐ ταχὺ ΘV: ταχύτατα M.
60 ἡ γονὴ Θ: ὁ γόνος MV.
61 τε om. MV.

καρδιωγμός, καὶ πονήσει μάλιστα ἐπὴν πλησιάζῃ τὰ
καταμήνια αὐτῇ· ἐπὴν δὲ ἀποκαθαρθῇ, ῥαΐσει ἐπ᾽ ὀλίγον χρόνον πρὸς τὰ πρόσθεν, ἔπειτα δὲ αὖθις ἐς
τωὐτὸ καταστήσεται.

Μελεδαινομένη δὲ ἐν τάχει ὑγιαίνει. ἢν δὲ μὴ θεραπεύηται καὶ ὁ χρόνος προΐῃ, πάντα μὲν[62] μᾶλλον
πονήσει τὰ πρόσθεν εἰρημένα, καὶ ὀδύνη λήψεται,
τοτὲ μὲν τῆς γαστρὸς τὸ κάτω τοῦ ὀμφαλοῦ, τοτὲ δὲ
τοὺς βουβῶνας, τοτὲ δὲ τὴν ὀσφῦν τε καὶ κοχώνην,
τοτὲ δὲ τὸν τράχηλον, τοτὲ δὲ πνὶξ προσπεσεῖται
ἰσχυρή, καὶ πρὸ τῶν ὀφθαλμῶν ζόφος ἔσται οἷ καὶ
36   δῖνος, οἷα τῆς καθάρσιος ἄνω στελλομένης καὶ ἀνιού-
σης. ἢν γὰρ τὸ σῶμα φλαύρως ἔχῃ, γυναικὶ τὰ καταμήνια ἐλάσσονα γίνεται, καὶ ᾗσιν ἂν τὸ σῶμα ἔμ-
πλεον ᾖ, τὰ καταμήνια ταύτῃσι πλέα ἐστί· τῇ δὲ
χολώδει τὰ καταμήνια ἢν ἔχηται, ὀλιγοψυχίη ἐμπίπτει, καὶ ἀποσιτίη[63] ἄλλοτε καὶ ἄλλοτε, καὶ ἀλύκη,[64]
καὶ ἀγρυπνίη, καὶ ἐρυγγάνει θαμινά, καὶ οὐκ ἐθέλει
περιπατεῖν, καὶ ἀθυμέει, καὶ ἐμβλέπειν οὐ δοκέει, καὶ
δέδιε.

Καὶ ἢν μελεδαίνηται, ἐκ τούτων ὑγιὴς γίνεται· ἢν
δὲ ὁ χρόνος προΐῃ, ἐπὶ[65] μᾶλλον πονήσει· συμβαίη δ᾽
ἂν ἄριστα, εἰ ἔμετος χολώδης ἐπιγένοιτο· ἢν ἡ κοιλίη
ταραχθῇ μὴ ἰσχυρῶς[66] καὶ ὑπίῃ χολῶδες, καὶ ῥόος
ἐπιγένηται αὐτῇ μὴ ἰσχυρός· ἢν γὰρ τούτων τι ἰσχυρὸν ἐπιπέσῃ ἐπὶ σῶμα τετρυχωμένον, κινδυνεύσει· ἢν

---

[62] μὲν Θ: μιν MV.   [63] ὀλιγοψυχίη . . . ἀποσιτίη M:
ὀλιγοσιτίη ἐμπίπτῃ Θ: ὀλιγοσιτίη ἐμπίπτει V.

burn, suffering these things mainly when the menses are approaching. After she has been well cleaned, she is better for a little while, compared to before, but then the same condition returns again.

If the patient is treated, she will soon recover, but if she is not treated and time goes on, she will suffer everything that was described above and more severely: Pain will seize her at one time in the belly below the navel, at another in the groins, at another in the lower back and the perineum, and at another in the neck, and at another time powerful suffocation will befall her and there will be darkness before her eyes with vertigo because the (sc. menstrual) cleaning has turned upward and is moving in that direction. For if a woman's body is poorly, her menses will decrease, whereas if her body is hearty her menses will be full. If the menses of a woman who is bilious are held back, she will lose consciousness and have an aversion to food from time to time, as well as restlessness and insomnia; she also belches frequently, refuses to take a walk, is despondent, appears not to look at anything, and is frightened.

If this patient is cared for, she recovers from these things, but as time progresses she suffers more. It turns out best for her if she vomits bile, or if her belly is gently moved and excretes bilious stools, and she has a weak (menstrual) flux: for if any of these were to happen forcefully with the body in its disturbed state, it would bring

---

64 ἀλύκη ΘV: ἀλυσμός M.

65 ἐπὶ ΘV: ἔτι M.

66 ἢν ἡ κοιλίη . . . ἰσχυρῶς M: αὕτη μὴ ἰσχυρός Θ: ἡ κοιλίη ταραχθείη μὴ ἰσχυρῶς V.

δὲ ἠρεμέως ἀποκαθαίρεταί τι τοῦ χολώδεος ἢ πᾶν τὸ
λυπέον, ὑγιὴς γίνεται. ἢν δὲ μὴ μελεδαίνηται μήτε
μηδὲν τούτων γίνηται, ἀποθνήσκει ἡ γυνή.

Ὡς δὲ ἐπὶ τὸ πλέον συμβαίνει ῥόον ἐμπίπτειν
χολώδεα ἐκ τοιούτου νοσήματος. καὶ ἢν ῥόος ἐγγένηται, τὰ μὲν πρῶτα ὀλίγα οἱ τὰ φαινόμενα ἔσται, ἀνὰ
πάσας δὲ τὰς ἡμέρας ὡς ἐπίπαν πλέονα συμβαίνει·
ὅταν δὲ ὁ χρόνος προΐῃ, ἐπὶ πλέονα καὶ ἡ νοῦσος
ὀξέη γίνεται, καὶ αἱ μῆτραι δάκνονται ὑπὸ τῆς καθάρσιος τῆς χολώδεος χωρεούσης καὶ ἑλκοῦνται. ἔτι δὲ
καὶ ἐν τούτῳ ὑγιαίνει μελεδαινομένη, ἢν οἱ ἐρχθῇ ὁ
ῥόος· ἢν δὲ φλεγμαίνωσιν αἱ μῆτραι ὑπὸ τῶν ἑλκέων,
38 ἔτι ὀξυτέρη οἱ ἡ νοῦσος ἔσται | καὶ πολλά τε καὶ
ὁδμαλέα καὶ πυώδεα ἐλεύσεται ἀπ' αὐτῶν τῶν μητρέων, ἤδη ἀπιόντα καὶ ἑκάστοτε οἷον ἀπὸ κρεῶν
ἰχώρ, καὶ τὰ πρότερον εἰρημένα πάντα μιν μᾶλλον
πονήσει, καὶ τὰ ἕλκεα ἔτι μᾶλλον ἀγριώτερα ἔσται
μέχρι μιν ἀπενείκῃ· ἢν δὲ καὶ ἰηθῇ, ἄφορος ἔσται ἀπὸ
τῶν οὐλέων.

9. Ἢν δὲ γυνὴ τὸ σῶμα φλαύρως ἔχῃ καὶ φλεγματώδης ᾖ, τὰ καταμήνια χωρήσει οἱ φλεγματώδεα·
γνωστὸν δέ ἐστιν ἢν χωρέῃ φλεγματώδεα· ὑμενώδεα
γὰρ φαίνεται καὶ ὥσπερ ἀράχνια διατείνεται, καὶ
ὑπόλευκά ἐστι. τοῦτο δὲ γίνεται, ἢν οἱ τὸ σῶμα καὶ ἡ
κεφαλὴ φλέγματος πλέα ᾖ, καὶ τὸ φλέγμα[67] μὴ ὑπο-
καθαίρηται μήτε κατὰ τὰς ῥῖνας μήτε καθ' ἕδρην
μήτε κατ' οὐρήθρην, ἀλλ' ἐν τοῖσι καταμηνίοισι καὶ
ἐν τῷ ταράχῳ τοῦ αἵματος σὺν τῇ καθάρσει ἔξω ἴῃ.

danger. If some of the bilious material or all of what is pathological is gently cleaned out, the patient recovers, whereas if she is not cared for and none of this happens, she dies.

In most cases, a bilious flux follows from a disease of this kind, and, although when the flux first arrives there is little evidence of it, with the passing days it becomes more and more apparent. With the passage of time, the disease becomes still more evident, and the uterus is irritated by the evacuation of the bilious flux, and develops ulcers. Even in this case, on being cared for the patient recovers, if her flux is brought to an end. But if the uterus swells up with phlegm from the ulcers, the disease will have an exacerbation and copious ill-smelling pus will be expelled from the uterus itself, running out soon each time like the serum from meat; everything described above will trouble the patient more forcefully, and the ulcers will become even more malignant, eventually carrying the patient away. Even if treatment is successful, the patient will be barren on account of the uterine scars.

9. If a woman is poorly in her body and phlegmatic, her menstrual flux will also have phlegm. This can be recognized by the following: the menses contain membranous material spread out like cobwebs and they are whitish. This comes about if a woman's body and head are filled with phlegm, and the phlegm is not cleaned downward through the nostrils, the anus or the urethra, but passes out in the menses with the disturbance of the blood at the time of cleaning.

---

[67] τὸ φλέγμα Θ: τοῦτο MV.

Καὶ ἢν ταῦτα ὧδε ἔχῃ, ἐπὶ μὲν δύο ἢ τρεῖς μῆνας
οὐκ ἐσάει, ἐπὴν δὲ ὁ χρόνος πλέων γένηται καὶ μὴ
μελεδαίνηται, μᾶλλον πονήσει, καὶ πυρετὸς ἐπιλήψε-
ται πλάνος, καὶ ἀσιτήσει ἄλλοτε καὶ ἄλλοτε, καὶ καρδιώξει, καὶ πονήσει μάλιστα ἐπὴν οἱ πλησιάζῃ τὰ
ἐπιμήνια· ὅταν δὲ ἀποκαθαρθῇ, ῥηΐζει ἐπ᾽ ὀλίγον
χρόνον πρὸς τὰ πρόσθεν, ἔπειτα ἐς τωὐτὸ καθίσταται, καὶ ἢν μὴ μελεδαίνηται ἀλλὰ χρόνος προΐῃ,
γενήσεται πάνθ᾽ ὅσαπερ εἰ[68] χολώδεα ἐχώρεε τὰ
καταμήνια, μέχρι οἱ ὁ ῥόος εἴχετο.

40   Ἕπεται[69] δὲ καὶ ταύτῃ | ῥόον φλεγματώδεα γίνε-
σθαι, ἢ ἄλλα ἄσσα ἐγὼ ἐρέω ὀλίγον ὕστερον· καὶ ἢν
ῥόος ἐπιγένηται,[70] αἰεὶ πορεύεται ἀνὰ πάσας ἡμέρας,
ὁτὲ μὲν ἀθρόα, ὁτὲ δὲ ὀλίγα, καὶ ἔστιν ὅτε οἷον ἀπὸ
κρεῶν ὕδωρ χωρέει, ὁτὲ δὲ οἷον ἰχώρ, καὶ ἐν αὐτῷ
θρόμβοι πολλοὶ αἵματος ἐγγίνονται, καὶ ξύει τὴν γῆν
ὥσπερ ὄξος, καὶ δάκνει τῆς γυναικὸς ᾗ ἂν ἐπιψαύσῃ,
καὶ ἑλκοῖ τὰς μήτρας. καὶ ἐλθοῦσα ἐς τοῦτο τὰ μὲν
ἄλλα πάσχει ταὐτὰ ὥσπερ καὶ ἡ προτέρη· ἧσσον δὲ
τὴν κεφαλὴν ἐκείνης πονήσει, καὶ τὰ ἕλκεα οὔτε δυσειδέα[71] οὔτε μεγάλα οὔτε πυώδεα[72] οὔτε ὀδμαλέα
ὁμοίως κείνῃ γίνεται, ἀλλὰ ταύτῃ ἧσσον· ἐπιμελο-
μένη δὲ ὑγιαίνει καὶ προεληλυθυίης τῆς νούσου, καὶ
οὐ μάλα ἀποθνήσκει, φορὸς δὲ οὐ δύναται εἶναι ὧδε
ἔχουσα.[73]

---

[68] εἰ Littré: η Θ: τῇ MV.   [69] ἕπεται MV: ἔπειτα Θ.

[70] ῥόος ἐπιγ. ΘV: προσεπιγ. M.

[71] δυσειδέα M: δυσίδεα Θ: δυσίατα V.

If this is the case, for two or three months the patient does not become aware of it, but when the time goes beyond this and she is not cared for, her suffering increases: she has an irregular fever, loses her appetite for food from time to time, experiences heartburn, and suffers most when the menses are approaching. After she has been well cleaned, for a little while she is better than she was before, but then she falls back into the same state. If the patient is not looked after, with the further progression of time, all the same signs will appear that are present in a patient passing bilious menses, and continue until her flux is interrupted.

Such a woman will also have a flux of phlegm or the other things I describe below. If the flux begins, it continues day by day, sometimes appearing as a mass and at other times less copiously, having sometimes the quality of the fluid that runs out of meat and at other times resembling serum; it contains many blood clots, it raises bubbles on the earth the same way vinegar does, and it irritates the woman wherever it touches her; it also ulcerates her uterus. When she has reached this stage, she suffers the same things as the woman described above except for having less headache, and her ulcers are less irregular, extensive, purulent and ill-smelling than in the woman above. If a patient is treated, she will recover even if the disease is far advanced, and will not often die; but in this case she will be unable to bear any children.

---

[72] πυώδεα Θ: πυιώδεα M: πυρώδεα V.   [73] See note 31 above. V adds a brief summary of this same text here: ἢν δέ οἱ ῥόος μὴ γένηται, συμβήσεταί οἱ ὥστε δοκέειν ἐν γαστρὶ ἔχειν· καὶ ταῦτα πείσεται ἅπερ τόκῳ τὰ καταμήνια ὅδον οὐκ ἠδύνατο εὑρεῖν· ἢ παχέα καὶ γλίσχρα καὶ κολλώδεα ἴῃ.

10. Ὅσαι δὲ συνοικεῦσαι μὴ δύνανται ἐν γαστρὶ
ἔχειν, πυθέσθαι χρὴ εἰ σφίσιν ἐπιφαίνονται τὰ ἐπιμήνια ἢ οὔ, καὶ εἰ[74] αὐτίκα ἄπεισιν ἡ γονὴ ἢ τῇ ὑστεραίῃ ἢ τῇ τρίτῃ ἢ τῇ ἕκτῃ ἢ τῇ ἑβδόμῃ· ἢν μὲν οὖν
φῇ αὐτίκα ἀπιέναι ὅταν εὐνασθῇ, τὸ στόμα οὐκ ὀρθὸν
42 ἐστὶ | ἀλλ' ἰδνοῦται καὶ οὐ λάζεται τὴν γονήν· ἢν δὲ
φῇ δευτεραίη ἢ τρίτη, ἡ ὑστέρη ἐξυγρασμένη ἐστὶ
καὶ ἡ γονὴ ἐκπλύνεται· ἢν δὲ ἑκταίη καὶ ἑβδομαίη, ἡ
γονὴ κατασήπεται, κατασαπεῖσα δὲ ἀπέρχεται.

Τοῦ μὲν οὖν ἀρχὴν μὴ προσδέχεσθαι τὴν γονήν,
τῆς ὑστέρας τὸ στόμα[75] θεραπευτέον πρῶτον, ὅπως
ὀρθὸν ἔσται· τοῦ δὲ καταπλύνεσθαι δευτεραίη καὶ τριταίη, ἡ ὑστέρη καὶ ἡ κεφαλή· τοῦ δὲ κατασήπεσθαί
τε καὶ ἀπέρχεσθαι, καὶ ἡ ὑστέρη καὶ τὸ σῶμα ἅπαν
ἔνυγρον ὄν.[76] τούτων ἕκαστα γινώσκειν ὧδε χρή.

11. Ὁποίης δὲ χρῄζει καθάρσιος, γνώσῃ δὲ ὧδε·
ὅταν τὰ ἐπιμήνια γίνηται, ῥάκους πτύξας ὅσον σπι-
θαμήν, ἐπιτανύσαι ἐπὶ σποδιὴν λεπτήν· κἄπειτα ποιε-
ειν ὡς ἐπὶ τοῦτο ἐπιρρυῇ τὰ ἀπιόντα· εἶναι δὲ δύο τὰ
τρυχία χωρὶς ἑκάτερα, τὰ μεθ' ἡμέρην τε καὶ νύκτωρ·
τὰ μὲν ἡμερήσια πλύνειν χρὴ τῇ ὑστεραίῃ, τὰ δὲ
νύκτωρ, ὁπόταν αὐτοῖσιν ἡμέρη καὶ νὺξ γένηται, ἐπὶ
τῇ σποδιῇ κειμένοισιν· ἐν δὲ τῇ πλύσει σκέπτεσθαι
ὁποῖα ἄσσα γίνεται τὰ ῥάκεα πλυθέντα, ὅταν ἐν ἡλίῳ
τέρσηται· κράτιστον δ' ἐν σκοταίῳ χωρίῳ. ἢν μὲν οὖν

---

[74] εἰ ΘV: γίνεται πυώδεα ἢ ψυχρά· ἢν μὲν οὖν φῇ Μ.
[75] στόμα MV: σῶμα Θ.   [76] ἔνυγρον ὄν om. Θ: ὄν om. V.

10. Women who cohabit but are still unable to become pregnant you must question as to whether their menses appear or not, and whether the seed is immediately discharged (sc. after intercourse), or on the next day, or on the third, sixth or seventh day. Now if she says the seed is discharged immediately after intercourse, the mouth (sc. of her uterus) is not straight, but folded over and it fails to take up the seed. If she says this happens on the second or third day, then her uterus is full of moisture and the seed gets washed away. If it is on the sixth or seventh day, the seed is decomposing and being discharged in that state.

Now if the seed is not being taken up at the beginning, first the mouth of the uterus must be treated so that it will be straight. Then, if the seed is being washed away on the second or third day, the uterus and the head (sc. require treatment). Finally, when the seed is decomposing and being discharged, the uterus itself and the whole body (sc. require treatment), being full of moisture. Each of these conditions must be distinguished in this way.

11. Which cleaning is required you can determine as follows: While the menses are passing, fold a piece of cloth of one span, spread it out over some fine ashes, and then have the discharge run over this: there should be two separate rags, one to be used during the day, and another during the night. The rag used during the day should be washed on the following day, and the one used during the night should be washed after it has been lying on the ashes for a day and a night. In washing observe how the washed rags look when they are dried in the sun—actually best in a shaded place. Now if it is phlegm that is hindering (sc.

φλέγμα ᾖ τὸ κωλῦον, μυξώδεα τὰ ῥάκεα ἔσται· ἢν δὲ
ἅλμη τε καὶ χολή, πυρρά τε καὶ ὑποπέλιδνα.

Ταῦτ᾿ οὖν ἐσιδών, καὶ γνώμῃ σκεθρῇ βασανίσας,
ἐς ὅλον τὸ σῶμα ἀθρέειν, ἤν τε πολλῆς καθάρσιος
δοκέῃ δεῖσθαι, ἤν τε μή, ἀποσκεψάμενος ἐς τὴν
χροιὴν καὶ τὴν ἡλικίην καὶ ῥώμην καὶ ὥρην καὶ οἵῃ
διαίτῃ χρέονται· ἰητρείῃ δὲ καὶ τοῦ σώματος παντός,
44 καὶ τὰς ὑστέρας[77] καὶ τὸ στόμα | ἰῆσθαι· καὶ ἢν με-
μυκὸς ᾖ, ἀναστομῶσαι· ἢν δὲ λοξωθῶσιν, ἐξιθύνειν
χρή· ἢν δὲ ὑγραὶ ὦσιν, αὐαίνειν, καὶ τἆλλα δρᾶν
ἐναντίον.

Σχεδὸν δὲ πάσαις ἡ πολλὴ θεραπείη ἡ αὐτή ἐστι,
πλὴν τοῦ κατασπάσαι τὰ ἐπιμήνια· ὅσαις γὰρ γίνεται, οὐδὲν δεῖ ταύτῃσι κατασπᾶν, ἀλλ᾿ ὅ τι κακὸν ἐν
τοῖσιν ἐπιμηνίοισιν ἔνι, τοῦτο ἀφαιρέειν χρή, καὶ
ὅταν μὲν φλεγματώδεα καὶ ὑμενώδεα ἴῃ καὶ χολώδεα
καὶ ἰχωροειδέα καὶ λεπτὰ ἢ λευκὰ καὶ θρομβοειδέα,
καὶ ὅταν μέλανά τε καὶ ἀνθρακώδεα, ἢ ζοφοειδέα,
δριμέα, ἁλμυρά, θολερά, πυώδεα. αὗται πᾶσαι αἱ
προφάσιες ὑπεξαιρετέαι· κωλύουσι γὰρ λαμβάνειν ἐν
γαστρί.

Ὁπόσα μὲν οὖν φλεγματώδεά τε καὶ ὑμενοειδέα
τῶν ἐπιμηνίων ἐστί, καὶ αὗται σαρκώδεές εἰσι, ταύτῃσι τὸ στόμα ἔξυγρον, καὶ πτύαλον πολὺ καὶ γλί-
σχρον, καὶ ἢν ὀξέος γεύηται ἢ δριμέος, τὸ πτύαλον
ἐν τῷ στόματι πλαδωδέστερον ἔσται καὶ ἄναλτον, καὶ
πρὸς πάντα ὅτι δ᾿ ἂν φάγωσιν ἢ πίωσιν ὄχλος καὶ
προσίσταται αὐτῇσι, καὶ ἀείρεται[78] κοιλίη, καὶ ναυ-

conception), the rags will resemble mucus, whereas if saltiness and bile are to blame, they will look reddish and slightly livid.

After observing this evidence and examining it carefully, next inspect the patient's whole body in order to determine whether or not it seems to require a thorough cleaning, taking into account her skin color, age, and strength, as well as the season and the regimen being followed, and treat the whole body so as to heal the uterus and its mouth: if the mouth is occluded, dilate it, if the uterus is lying at an angle, straighten it, if it is moist, dry it, and similarly by applying opposite measures.

In almost all women, active treatment is the same, except for drawing down the menses; for in women in whom menstruation is occurring, there is no need to provoke it, but rather treatment must draw off whatever evil is present in the menses, such as when they contain phlegm and membranes, or bile and serum, or are thin, white and thrombotic, or black and coal-like, or dark, irritating, salty, turbid and purulent. All these causes must be removed, since they hinder conception.

In women whose menses contain phlegm and membranes, and who are themselves fleshy, their mouth will secrete copious sticky saliva, and if they have eaten something acidic or sharp, the saliva in their mouth will seem insipid and bland; in response to everything they eat or drink they feel an annoyance, their abdomen is raised, and they have nausea and great discomfort. A flux descends

---

[77] καὶ τὰς ὑστ. om. Θ.
[78] καὶ ἡ add. MV.

σίη, καὶ ἀλυσμὸς πολύς· ἀπό τε κεφαλῆς ῥεῦμα
καταρρέει, καὶ πάντ᾽ ἐπιπλάσσεται, καὶ πολλὴν ὑγρα-
σίην ἐπάγεται, καὶ τὰ ὑποφθάλμια πελιδνὰ καὶ πεφυ-
σημένα. ταύτας χρὴ ὅλας πυριᾶν, καὶ πυκινὰ ἐμέειν
καὶ ἀπὸ σιτίων καὶ νηστείας· μαλάσσειν δὲ τὴν κάτω
κοιλίην φαρμάκοισι πάμπαν κούφοισιν, ὅσα ἥκιστα
χολὴν ἄγει, καὶ μονοσιτέειν, καὶ γυμνάζεσθαι συχνά,
καὶ ὡς ξηροτάτῃ τροφῇ διαιτᾶσθαι, καὶ ποτῷ ἐλαχίστῳ ἀκρητεστέρῳ· κοιλίη δὲ εὔλυτος ἀμείνων διαφυ-
λάσσεσθαι. τὰς δ᾽ ὑστέρας χρή, ἢν μὴ πρὸς ταύτην
τὴν δίαιταν ἐνακούωσι, καθαίρειν φαρμάκοισιν ἀδή-
κτοισι προσθέτοισι· προπυριᾶν δὲ πρὸ τῶν καθαρσίων ἀεί, πρότερον μὲν τῇ πυρίῃ τῇ ἐκ τῶν μαράθων,
46 ἔπειτα δὲ τῇ ἐκ | τῶν θυμάτων. τὰς δὲ πυρίας ποι-
έεσθαι καὶ τὰς προσθέσιας τεκμαιρόμενον ὅπως
ἅπαντα πεποιήσεται καὶ κατὰ τρόπον ἔσται, ἐν ᾧ
χρόνῳ τὰ ἐπιμήνια εἴη.

Ἢν μὲν οὖν καθαρά τε καὶ[79] ἀκραιφνέα ᾖ καὶ
ἔναιμα γίνηται, οὕτως ἴτω παρὰ τὸν ἄνδρα ἐν ἀρχο-
μένοισι τοῖσιν ἐπιμηνίοισιν· ἄριστον δ᾽ ἐν ἀπολεί-
πουσι καὶ ἔτι ἰόντων μᾶλλον ἢ ἀφανέων. ὅταν δὲ
μέλλῃ ἰέναι παρὰ τὸν ἄνδρα, ὑποθυμιήσθω τι τῶν
θυμάτων τῶν στυπτικῶν· θυμιάσθω δὲ διὰ τοῦ κανείου καὶ τοῦ καλάμου, ἐπὶ σποδιὴν θερμὴν ἐπιπάσσουσα τὸ φάρμακον· ὅταν δὲ ἐπιπάσσῃ, καὶ τὸ κάνειον περιθεῖναι καὶ τὸν κάλαμον, καὶ καθιζομένην
πυριᾶσθαι. ὅταν δὲ οἵη τε ᾖ[80] θυμιᾶσθαι, τῷ μολιβδίῳ
χρήσθω, ὡς ἀνεῳγμένῳ τῷ στόματι θυμιᾶται· εἶθ᾽

from their head obstructing everything and attracting much moisture, and the areas beneath the eyes are livid and puffy. These women must be fomented over their whole body, vomit repeatedly after both eating and fasting, soften the lower cavity by using very light medications that draw the least bile, take one meal a day, exercise frequently, and employ a diet with very dry food and the minimum of drink taken quite undiluted. It is better to maintain the cavity in a fluent state. If the uterus does not respond to this regimen, you must clean the uterus with suppositories made of nonirritating medications. Always foment before the cleaning, first with a fomentation made from fennel and then with one made from incense. Make the fomentations and applications while seeing to it that everything necessary is being done so that the uterus will be as it should during the time when the menses are present.

Now if the menses are clean and pure, and then turn sanguine, have the woman go to her husband as they are starting, or even better when they are ending but still present and not already gone. When she is about to approach her husband, she should fumigate herself with one of the astringent agents; the fumigation should be through the lid and a reed, and made by sprinkling the medication over hot ashes: when she has sprinkled the medication and set the lid and the reed over it in place, she should seat herself over it and carry out the fomentation. When she has been prepared for the fomentation, have her employ a lead sound in order that the mouth (sc. of her uterus) will be open during the fomentation. Then, after she gets up, she

---

⁷⁹ καθ. τε καὶ om. Θ.   ⁸⁰ οἵη τε ᾖ Θ: δέῃ αὐτὴν MV.

ὅταν ἀνίστηται, πάλιν ἐν τῇ κλίνῃ προσθέσθω τὸ μο-
λύβδιον· εἶτα ἀφελομένη, αὐτίκα συνευναζέσθω τῷ
ἀνδρί, καὶ ἢν τὰ ἀπὸ τοῦ ἀνδρὸς μὴ δῆλά οἱ ᾖ, ἐκτεί-
νασα τὰ σκέλεα καὶ ἐπαλλάξασα ἠρεμείτω. νηστείη
δὲ ὄφελος ἐν ταύτῃ τῇ ἡμέρῃ, πλὴν κυκεῶνα, εἰ ἐθέ-
λοι, ἄναλτον ἐφ᾽ ὕδατι.

Τοῦτο δὲ ποιέειν τότε, ὅταν μέλλῃ θυμιᾶσθαι. ἢν
δὲ συγγενομένη τῷ ἀνδρὶ κατάσχῃ τῇ ὑστεραίῃ τὴν
γονὴν καὶ μηδ᾽ ἐς τὴν ἑτέρην ἀπίοι ἡμέρην, σιτίων
μὲν εἴργεσθαι καὶ λουτρῶν, πινέτω δὲ ἄλφιτον ἐν
ὕδατι ἄναλτον καὶ δὶς καὶ τρὶς τῆς ἡμέρης. αὕτη ἡ
δίαιτα[81] ἡμέρας ἕξ, ἀμείνων ἢ ἑπτά, ἢν μὴ ἀπίῃ τὰ
ἀπὸ τοῦ ἀνδρὸς ὅταν συγγένηται. ἀλουτείτω δὲ πάντα
τὸν χρόνον, καὶ ἀκινητέειν χρή· ἢν δὲ βούληται περιπατεῖν, ἐν ὁμαλῷ χωρίῳ καὶ λείῳ, πρὸς ἄναντες δὲ
μηδέν, μηδὲ κάταντες· καθέζεσθαι δὲ ἐπὶ μαλθακά, ἢν
48 συλλάβῃ· τὴν δὲ ἄλλην δίαιταν τὴν | αὐτὴν ποιέεσθαι
μέχρι ἡμερέων τριήκοντα. ἀλουσίη δέ· καὶ ὅτε δέοι,
ὀλίγῳ καὶ μὴ λίην θερμῷ· τὴν δὲ κεφαλὴν μὴ βρέχειν. σιτίοισι δὲ χρήσθω ἄρτοισι καὶ ἢν βούληται
μάζῃ· κρεῶν δὲ φάσσῃ καὶ τοῖσι παραπλησίοισι, θαλασσίων δὲ ὅσα κοιλίην ἵστησιν· εἴργεσθαι δὲ λαχάνων δριμέων· οἴνῳ δὲ μέλανι χρήσθω, τοῖσι κρέασιν
ὀπτοῖσι μᾶλλον ἢ ἐφθοῖσι, καὶ τῶν ἡμέρων καὶ τῶν
θηρίων.

12. Ταῦτα, ἢν μὲν συλλάβῃ, οὕτω χρὴ ποιέειν· ἢν
δὲ μὴ ξυλλάβῃ, ἀλλ᾽ οἴχηται δευτεραῖα ἢ τριταῖα ῥε-
όμενα παμπόλλη ὑγρασίη, δῆλον ὅτι ὑγρότεραί εἰσιν

should go back to bed and reinsert the sound; and then she should remove it again and immediately go to bed with her husband. If what comes from her husband does not reappear out of her, she should stretch out her legs, cross them, and lie at rest. Fasting is helpful on this day, except for a cyceon, if she wishes it, taken without salt after some water.

Do this when the woman is about to be fumigated. If after she has been with her husband she retains the seed on the following day, and it does not come out on the day after that, she should refrain from eating and bathing, but take barley meal without salt in water two or three times daily. Follow this regimen for six days, or better seven—as long as what came from her husband when she was together with him does not run out. She should avoid the bath for the whole time, and keep herself still. If she wishes to take a walk, it should be on even, level ground with no rises or dips. If she has conceived, she should rest on soft objects. The rest of her regimen should be kept the same for thirty days, but she should avoid the bath: when it must be, then it should be with very little water and that not hot; do not wet her head. As grains have her eat bread and, if she wishes, barley cake, of meats the ringdove and others like that, of sea foods whichever make the cavity firm; avoid irritating vegetables. She should employ dark wine, roasted meats in preference to boiled ones, and those both domesticated and wild.

12. This is what a woman must do if she conceives. If, however, she does not conceive, but a flux of copious moisture passes on the second or third day, it is clear that the

---

81 ἡ δίαιτα Θ: δὲ ἡ δίαιτα ἔστω MV.

αἱ ὑστέραι. θεραπεύειν οὖν χρὴ κατὰ τὸν ὑφηγημένον
λόγον,[82] μέχρι εὖ ξηραίνωνται· ὅταν δὲ δοκέωσι ξηραὶ
εἶναι, ἰητρείη οὖν ἀρίστη, φάρμακον μαλθακτήριον
κατόπιν τε καὶ ἐς τοὔμπροσθεν, μέχρι οὗ αὖαι κατὰ
φύσιν γένωνται· καὶ αὖθις ἴτω παρὰ τὸν ἄνδρα, ὅταν
τὰ ἐπιμήνια μηκέτι ἴῃ, ἀλλ᾽ ὀλίγα καὶ εὔχροια, καὶ
ὀργᾷ. καὶ ἐν τῇσιν ἄλλῃσιν ἡμέρῃσιν ἱμεροῦσθαι
χρὴ ἢν ἄριστα ἔχωσιν αἱ γυναῖκές τε καὶ[83] αἱ ὑστέραι.
ὅταν δὲ συγγένηται ἐν τῇσιν ἡμέρῃσι τῇσιν εἰρημέ-
νῃσιν, ἢν εὐτρεπισθῶσιν αἱ ὑστέραι, ἢν κατάσχῃ τὴν
γονήν, ἡμερέων δέκα ἢ δυώδεκα μὴ ἰέναι εἰς τὸν
ἄνδρα. ἢν δὲ μὴ ξυλλαμβάνῃ, ὑγιέες δὲ ὦσιν αἱ
μῆτραι—γίνεται γὰρ καὶ τοῦτο, ἐπειδὰν ἀκιδναὶ αἱ
50 μῆτραι, εὔτροφοι οὖσαι, ὑπὸ πάθεος γένωνται | ἢ[84]
ὑπὸ φαρμακείης τε καὶ πυρίης πολλῆς, καὶ οὐ δύνανται φέρειν τὴν γονήν, πρὶν ἠθάδες ἐῶσι καὶ ἰσχύωσι—
γνωστὸν δὲ καὶ τοῖς δ᾽ ἐστὶν ὧδε· ὅταν ἀπίῃ οἱ, ἄπεισι
δευτεραίη καὶ τριταίη καὶ ἔτι ἀνωτέρω, ἀπέρχεται δὲ
παχέα καὶ ξυνεστῶτα οἷον βλένναι, ἢν μὴ κακόν τι ᾖ
καὶ διὰ νοῦσον ἑτέρην ὑστερέων ἡ γονὴ ἀπαλλάσσηται.

Ὅταν οὖν τοιαῦτα φανῇ, θεραπείης μὲν ἀπόχρη
ὑστερέων· τοῦ δ᾽ ἄλλου σώματος ἐπιμελείην ἔχειν, ὡς
εὐεξίη τοιαύτη οἱ ᾖ, ὡς συνεσταλμένον τὸ σῶμα εἶναι
καὶ εὔογκον, λουτροῖσιν ὀλίγοισι, πόνοισι[85] πλείοσι·

82 λόγον Θ: τρόπον MV.   83 αἱ γυν. . . . . καὶ om. MV.
84 ἢ MV: μὴ Θ.   85 πόνοισι Θ: om. M: πόνοισι πρηέσι V.

uterus is too moist. Thus, treatment must be conducted according to the established principle, until it becomes good and dry. When it appears to be dry, the best treatment is with an emollient medication both posteriorly and toward the front, until the uterus becomes as dry as nature meant it to be. Then have the woman approach her husband again when her menses are no longer in full course, but are small in amount and of a good color, and she is eager. On the other days, too, the woman must feel desire, if she and her uterus are in the optimal state. When she has intercourse on the days indicated, if her uterus is well prepared and it retains the seed, she should not approach her husband again for ten or twelve days. If she does not conceive, although her uterus is healthy—for this too can happen, when in spite of being well nourished the uterus becomes feeble due to something that happens to it, or from purging and frequent fomentations, and it is unable to carry the seed, until it gets used to this and is strong— this can be recognized from the following: when the patient has a discharge, it runs out on the second and third day, and even later, and it comes out thick and compacted like mucus—as long as no trouble is present and the seed has not been discharged due to another disease of the uterus.

Now when such things show themselves, the treatment of the uterus will be sufficient. The rest of the body, though, needs attention that will make its condition healthy, so that it will be compact and of good bulk: prescribe few baths but many exercises; prohibit sharp and

δριμέων καὶ ἁλμυρῶν εἴργεσθαι· ἐμέτοισι χρῆσθαι
πρὸ τῶν ἡμερέων ὧν μεμάθηκε τὰ ἐπιμήνια γίνεσθαι,
καὶ αὖθις λιμοκτονέεσθαι, καὶ τἆλλα ἐπιτελεῖν ἄσσα
εἴρηται. αὕτη μὲν θεραπείη ἀμφὶ τῶνδε.

13. Ἤισι δέ, ὅταν συγγένηται, αὐτίκα ἄπεισι τὰ
ἀπὸ τοῦ ἀνδρός, ταύτῃσι τὸ στόμα τῶν ὑστερέων
πρόφασις. θεραπεύειν δὲ δεῖ· ἢν μὲν σφόδρα μεμυκὸς
ᾖ, ἀναστομῶσαι τοῖσι δαϊδίοισι καὶ τοῖσι μολυβδί-
οισι· πυριῆν δὲ μαλθακῇ πυρίῃ τῇ ἐκ τοῦ μαράθου,
καὶ καθαίρειν προσθέτοισιν ὅσα λεπτύνει τὰς ὑστέρας καὶ ἐς ἰθὺ καταστήσει· μετὰ δὲ τὰς καθαρσιάς τε
καὶ τὰς πυρίας κλύζειν τοῖσδεσιν ἄσσα ἐναντία τῇ
προφάσει.

Ὅσῃσι δὲ τὸ στόμα ἀπεστραμμένον ἐστὶ καὶ
προσπεπτωκὸς πρὸς τὸ ἰσχίον, γίνεται γὰρ καὶ
τοιαῦτα διακωλύοντα τὴν ὑστέρην τὴν γονὴν προσδέχεσθαι, ὅταν οὖν τι τοιοῦτον συμβῇ, πυριῆν χρὴ ταῖς
52 εὐώδεσι τῶν πυρίων· μετὰ δὲ τὴν | πυρίην παραφασ-
σαμένην[86] τῷ δακτύλῳ ἀποστῆσαι[87] τοῦ ἰσχίου· ὅταν
δὲ ἀποστήσῃ, ἐξορθοῦν τοῖσι δαϊδίοισι καὶ τῷ μολιβδίῳ κατὰ τὸν ἔμπροσθεν λόγον· ὅταν δὲ ἐξωρθω-
μέναι τε καὶ ἀνεστομωμέναι γένωνται, προσθέτοισι
μαλακοῖσι καθαίρειν, καὶ τἆλλα ποιέειν κατὰ τὸν
ὑφηγημένον λόγον.

Ὅταν ἀναστομῶνται μᾶλλον τοῦ δέοντος αἱ ὑστέραι, καθάρσιος δέονται· μετὰ δὲ τὰς καθάρσιας, κλυσμῶν καὶ θυμιημάτων.

Ἢν δ' ἐγγύτερον ἔωσι τοῦ δέοντος, ἐμέτων δέονται

salty foods, employ emetics before the days on which the menses usually appear, and then restrict the diet severely again and apply the measures that were indicated above. This is the treatment of these conditions.

13. In women who, after having intercourse, immediately discharge what has come from the man, the trouble is with the mouth of their uterus, and they require attention. If the mouth is tightly closed, open it with pine dilators and lead sounds. Apply a mild fomentation made from fennel, and clean the uterus with suppositories that will make it thinner and set it in a straight position. After the cleaning and the fomentations, give enemas counteracting the condition's cause.

In women whose cervix has turned to the side and is leaning against a hip—for this too may occur and prevent the uterus from receiving the seed: now when something like this has happened, you should foment with fragrant substances; after the fomentation, have the patient draw (sc. her cervix) aside with a finger and push it away from the hip. When she has done this, straighten (sc. the uterus) with pine twigs and a lead sound in the way described above. After the uterus has been straightened and dilated, clean it with mild suppositories, and attend to the rest according to established practice.

When the uterus is more dilated than it should be, it requires a cleaning, and after the cleaning the application of enemas and fumigations.

If the uterus is too close (sc. to the exterior), it requires

---

86 Linden after Foes in note 91: παρασπᾶσαι Θ: παρασπα-σαμένην MV.

87 ἀποστῆσαι Θ: ἀποσπᾶν MV.

αἱ ὑστέραι καὶ πυρίης δυσώδεος, μέχρι ὅτου ἐς χώρην ἔλθωσι· διαίτῃ δὲ τῇ ὑφηγημένῃ χρῆσθαι.

Ἢν δὲ τὸ στόμα πιμελῶδες ᾖ καὶ πάχετον, καὶ[88]
μὴ κυΐσκηται, νᾶπυ ἐφθὸν ἐσθίειν νῆστιν, καὶ ἄκρητον ἐπιπίνειν.

Πρόσθετον δὲ λίτρον ἐρυθρόν, κύμινον, ῥητίνην,
ἄριστον δ' ἐν ὀθονίῳ·[89] ἢ λίτρον σὺν σμύρνῃ καὶ ῥητίνῃ καὶ κυμίνῳ καὶ μύρῳ λευκῷ· ἢ ἐλάφου κέρας
καῦσαι καὶ μίξαι ὠμήλυσιν διπλασίην, ἐν οἴνῳ ἐπὶ
ἡμέρας τέσσερας πινέτω. ἢν δὲ μὴ ῥαΐζῃ, πράσα
ἑψεῖν καὶ ἀμφικαθίζεσθαι· ἢ φηγὸν τρίβουσα προστι-
θέσθω· σκόροδά τε τρώγειν νεαρά, καὶ μελίκρητον
πίνουσα ἐμείτω.

14. Ὅσῃσι δὲ ἐμμένει καὶ σήπεται καὶ ὄχλον ποιέει,
ἀνδραφάξιος ἀγρίης καρπὸν ἢ χυλὸν ἐν[90] μέλιτι ἢ
σὺν κυμίνῳ ἐλλικτὸν διδόναι.

15. Ὅταν δὲ δύσοδμα ἴῃ, οὐ κυΐσκεται·[91] ἀνδράχνην
καὶ χηνὸς ἔλαιον συμμίσγειν καὶ προστιθέναι. |

54　16. Ἧισιν ἑκταῖά τε καὶ ἑβδομαῖα τὰ ἀπὸ τοῦ
ἀνδρὸς κατασηπόμενα χωρέει, ταύτῃσιν εἰκὸς γίνεσθαι ὑπ' ἀμφοῖν ἐπιρροὴν χολῆς τε καὶ ἅλμης· θεραπεύειν δὲ χρὴ ὧδε, ἐλλεβόρῳ ἢ καὶ σκαμμωνίῃ ἢ
πεπλίῳ· καθαίρουσι γὰρ ἄνω τε καὶ κάτω φλέγμα τε
καὶ χολήν, καὶ φύσας ἄγει· πρὸ δὲ τῶν καθαρσίων
τῇσι πυρίῃσι τῇσιν ἐκ τῶν θυωμάτων· ὅταν δὲ πυριή-

---

[88] Add. διὰ τοῦτο MV.
[89] ἐν ὀθονίῳ Θ: οἴνῳ M: ἐν οἴνῳ V.

the application of emetics and an evil-smelling fomentation until it moves to its correct location: employ the established diet.

If the mouth (sc. of the uterus) is fatty and thick, and the woman does not become pregnant, have her eat boiled mustard in the fasting state, and then drink undiluted wine.

Suppository: red soda, cumin, and resin, best in a linen tent; or soda together with myrrh, resin, cumin and white unguent; or burn deer's horn and mix it with twice that amount of bruised meal of raw grain: have the patient drink this in wine for four days. If the patient does not improve, boil leeks and have her take a sitz bath; or grind acorn of the valonia oak and have her apply this in a suppository. Also have her eat fresh garlic, and then drink melicrat, and vomit.

14. To women in whom (sc. the seed) is retained, suppurates, and causes trouble, give an electuary of the seed or juice of wild orach in honey, or with cumin.

15. When the menses pass ill-smelling, a woman does not become pregnant: mix together purslane and goose oil, and apply in a suppository.

16. Women who on the sixth and seventh day discharge what has come from their husband in a decomposed state are likely to have a flux of both bile and salt. Treatment must be applied in the following way with hellebore or also scammony or wild purslane, since these medications clean both bile and phlegm both upward and downward; they also draw wind: preliminary to the cleaning (sc. employ)

---

90 ἐν ΘV: ἐμ- M.
91 οὐ κυΐσκεται ΘM: καὶ οὐ κυΐσκηται V.

σης, καθαίρειν προσθέτοισι τὸν αὐτὸν τρόπον ὡς ἐν
τῆσι πρόσθεν, καὶ μετὰ τὰς πυρίας[92] καὶ καθάρσιας
τοῖσι μαλθακτηρίοισι χρέο, καὶ τὸ ἐκ τῆς λινοζώστιος πρόσθετον, καὶ ἀρτεμισίης ποίης καὶ ἀνεμώνης
καὶ ἐλλεβόρου λευκοῦ ἢ μέλανος.

Τὰ μὲν τῶν φαρμακειῶν ταύτῃσιν ὧδε χρή· τὴν δὲ
δίαιταν ὑποσκεπτόμενον[93] τῆς ἀνθρώπου ἐς τὸ οὖλον
σῶμα ποιέειν, ἤντε αὐχμηροτέρη δοκέῃ εἶναι, ἤντε
σαρκωδεστέρη· ἢν μὲν γὰρ αὐχμηροτέρη, λουτροῖσι
πλείοσι καὶ ἐφθοῖσι πᾶσι καὶ τοῖσι θαλασσίοισι καὶ
τοῖσι κρέασιν, οἴνῳ δὲ ὑδαρεῖ, λαχάνοισιν ἐφθοῖσι
καὶ λιπαροῖσι[94] καὶ γλυκέσι· ταῦτα γὰρ ὡς ἐπὶ τὸ
πολὺ ποιέει ὑγρηδόνα εἶναι καὶ ἐν τῷ ἄλλῳ σώματι
καὶ ἐν τῇσιν ὑστέρῃσιν. ἢν δὲ ὑγροτέρη ᾖ, δέονται αἱ
τοιαῦται ὑστέραι τούτων οὐδέν, ἀλλὰ τἀντία· οὐ χρὴ
ψαύεσθαι οὐδὲ κλύζεσθαι νεοχμῶς οὐδὲ θυμιᾶσθαι·
πρὸς γὰρ τὸ κενούμενον[95] φιλεῖ ἰέναι. καὶ ἢν ἐξυγραν-
θῶσι μᾶλλον τῆς φύσιος, χρὴ δὴ ξηραίνειν καὶ θυ-
μιᾶν· ἢν δὲ χολώδεα ἴῃ ἐπὶ σφᾶς, ὅ τι χολὴν καθαίρει
διδόναι· ἢν δὲ ἀλμυρώδεα, γάλα ὄνειον καὶ οἶνον καὶ
τἄλλα ἀλεξητήρια. |

92 πυρίας Θ: προσθέσιας MV.
93 Θ: ἀποσκεπτ. MV.
94 Add. πᾶσι MV.
95 κενούμενον Potter: κεινεόμενον Θ: κινεύμενον MV.

fomentations with spices. After you have fomented, clean with suppositories in the same manner as in the cases above, and after the fomentations and cleanings use emollients and a suppository made from mercury herb and artemisia, anemone and white or black hellebore.

This is the pharmaceutic treatment to be used for such a woman; now follows the dietetic treatment, which should be applied with attention to her whole body as to whether it seems overly dry or fleshy. For if it is too dry, you should treat with rather frequent baths, with all sea foods and meats boiled, with dilute wine, and with vegetables boiled in oil with honey, since all these measures tend in general to produce moisture in both the uterus and the rest of the body. If the body is too moist, the uterus will require none of these agents, but the opposite kind; touching the patient or applying new enemas or fumigations are to be avoided, since moisture tends to go toward the place being emptied.[3] If the uterus is saturated with more moisture than is natural, it must be dried and fumigated;[4] if bilious material has gone to it, give cholagogues, if salty material, then ass's milk, wine, and other counteracting agents.

[3] Cf. J. T. Vallance's (*The Lost Theory of Asclepiades of Bithynia* [Oxford, 1990], pp. 58–79, esp. 58) remarks on Hippocratic assumptions regarding attraction as possibly prefiguring the πρὸς τὸ κενούμενον ἀκολουθία (following into what is emptied) principle of Strato of Lampsacus and Erasistratus. Here the idea seems to be that enemas or fumigations drying the uterus tend to draw new moisture to the place being emptied.

[4] The apparent contradiction between this statement and the preceding sentence prohibiting fumigation is resolved by Gardeil by specifying "fumigations sèches" here and "fumigations humides" above. But such inconsistencies may in fact be merely the results of the author's compilatory method.

56   17. Χρὴ δὲ σκεψάμενον τῶν νοσημάτων τὰς δυνά-
μιας καὶ ὑφηγεύμενον τὰς προφάσιας ὡς χρὴ ἐξ ὧν
αἱ νοῦσοι γίνονται, ὧδε ἐπὶ τὰ ἄλλα ἰέναι καὶ τὰ ἀμφὶ
τὰ χωρία ἰᾶσθαι. καὶ ὅσαι μὲν διὰ τὴν τῶν στομάτων
αἰτίην διακωλύονται μὴ[96] ἔχειν ἐν γαστρί, τούτων
εὐτρεπίζειν τὰ στόματα, ὡς εὐκρινέα ᾖ. ὁκόταν[97] δὲ ἡ
ὑγρηδὼν εἴργει, ταύτην ἐσορῆν ὡς μὴ διακωλύσῃ,
σκεπτόμενον ἐς ὅλα τὰ πρήγματα τῶν γυναικῶν, ἤντε
ἐξ ἅπαντος τοῦ σώματος δοκέῃ τι κινέεσθαι, ἤντε ἀπ᾽
αὐτῶν τῶν ὑστερέων, ἤντε ἀπ᾽ ἄμφω. τὰς δ᾽ ὑστέρας
ὧδε μελεδαίνειν, ὅκως μήτε ὑγραὶ ἔωσι μήτε λίην
αὖαι· ἀλλὰ τὰς μὲν ξηροτέρας εὐχύλως, ὅσῳ αὐχμη-
ρότεραι ἔωσι, τοιαύτῃ ἰκμάδι, ὡς πιαλέαι μᾶλλον
ἔσονται ἢ ἰσχναλέαι· τὰς δὲ ἐξυγρασμένας καὶ δια-
βρόχους ξηραίνειν ὑπολειπόμενον ἐνίκμους εἶναι καὶ
μὴ λίην ἀπεξηράνθαι· αἱ γὰρ ὑπερβολαὶ τούτων[98]
φευκτέαι. λαμβάνει δὲ ἐν γαστρὶ οὔτε ἡ ὑγρηδόνα
ἔχουσα, ἀτὰρ οὐδὲ ἡ αὐαινομένη, ἢν μή τι ἐν τῇ ἀρ-
χαίῃ φύσει τούτων ᾖ.

    Ἰέναι δὲ χρὴ παρὰ τὸν ἄνδρα, ὅταν τὰ ἀπὸ τῆς
θεραπείης καλῶς ἔχῃ, ληγόντων ἢ ἀρχομένων τῶν
ἐπιμηνίων· ἄριστον δὲ καὶ ἐπὴν παύσηται· μάλιστα
μὲν ἐν ταύτῃσι τῇσιν ἡμέρῃσι πειρηθῆναι, ἢν δύνηται κυΐσκεσθαι· αὗται γὰρ κυριώταται. ἢν δὲ μὴ
αὐτίκα συλλαμβάνῃ, τὰ ἄλλα δὲ καλῶς ἔχῃ, οὐδὲν τὸ
κωλῦον ἐν τῇσιν ἄλλῃσι τῶν ἡμερέων ξυνιέναι τῷ

17. After first examining the mechanisms of diseases and clarifying as necessary the causes from which they arise, you must proceed to other measures and treat the parts involved. Women who are prevented by a cause in the mouth (sc. of their uterus) from becoming pregnant require correction of the mouth so that it will be well opened. When moisture is the cause of such a blockage, examine this factor in order to prevent it from hindering (sc. conception), taking into account all aspects of the case, and determining whether there appears to be an afflux from the whole body, from the uterus by itself, or from both. Treat to make the uterus neither too moist nor too dry: when it is too dry, apply an amount of moisture matching its overdryness, in such a way as to make it succulent rather than withered; when it is saturated and sodden, apply drying measures but still leave it humid rather than overdrying it: for each of these extremes is to be avoided. Neither an overly-moist nor a parched woman can become pregnant, unless she has some of this tendency in her original nature.

When her treatment has brought her to a good state, she should go to her husband when her menses are just ending or beginning, or best after they have stopped: have her make an attempt to become pregnant especially on these days, as they are the most decisive. If she does not conceive at once, but everything seems to be as it should, there is nothing to prevent her from having intercourse

---

96 τὴν τῶν στομάτων . . . μὴ Θ: τῶν στομάτων προφανέων τὴν αἰτίην κωλυόνται μὴ Μ: τῶν στομ. προσφ. τὴν αἰτίην διακωλύειν ἢ V.   97 ὁκόταν MV: ὅσων Θ.

98 Add. πάντῃ MV.

ἀνδρί· προθυμίην γὰρ σφὶν ποιέει ἡ μελέτη, καὶ ἀναχαλᾶται τὰ φλέβια, καὶ ἢν τὰ ἀπὸ τοῦ ἀνδρὸς ἀπιόντα ὁμορροθῇ κατ᾽ ἴξιν τῷ ἀπὸ τῆς γυναικός, τα-
58 χύτερον κνήσει· | καὶ γὰρ τόδε αἴτιόν ἐστιν ᾗσιν, ἤντε
αὐτίκα ἤντ᾽ ἐν ὑστέρῳ χρόνῳ.⁹⁹

Ταῦτα μὲν ἀμφὶ τούτων λέλεκται.

18. Ἢν δὲ ὑγρότερον ᾖ τὸ στόμα τῶν ὑστερέων, οὐ
δύναται εἰρύσαι τὴν γόνην· προσθέτοισι δὲ δριμέσι
χρήσθαι· δηχθεὶς γὰρ καὶ φλεγμήνας ὁ στόμαχος
στερρὸς ἔστιν ὅτε γίνεται· κατὰ δέ τι ἠδέλφισται. καὶ
ἢν σκυρωθῶσιν· ἄρειον γὰρ τὰ δριμέα προστιθέναι,
ἀδαξῶντα γάρ, λεπτὰ καὶ πυρώδεα ἐόντα, τὸν σκύρον
διαχέει· ἢν δὲ λαπαχθῇ ὁ σκύρος, μαλθακοῖσιν ἰᾶσθαι καὶ ὅ τι μὴ δάξεται.

19. Ἢν δὲ διὰ παλαιοῦ μὴ¹⁰⁰ κυΐσκηται τῶν καταμηνίων ἐπιφαινομένων,¹⁰¹ ὅταν ᾖ τριταίη ἢ τεταρταίη,
στυπτηρίην λείην τρίψας, διεὶς μύρῳ, εἰρίῳ ἀνασπογ-
γίζων προστίθει, καὶ ἐχέτω ἡμέρας τρεῖς· τῇ δὲ
τρίτῃ,¹⁰² χολὴν βοὸς αὔην ἐν ἐλαίῳ ἀναζέσας, καὶ
ἄχνην ἀναδεύσας πρόσθες, καὶ ἐχέτω ἐπὶ ἡμέρας
τρεῖς· τῇ δ᾽ ὑστεραίῃ ἐξελέσθω, καὶ τῷ ἀνδρὶ συνίτω.

---

⁹⁹ Add. ἀπίῃ MV.
¹⁰⁰ μὴ M: οὐ μὴ V: om. Θ.
¹⁰¹ ἐπιφαιν. Θ: μὴ φαιν. MV.
¹⁰² τρίτῃ Θ: τετάρτῃ MV.

with her husband on the other days, for habitual practice will increase her desire, her small vessels will open up, and if what comes from her husband flows together in line with what comes from her, she will become pregnant more quickly, since in some women this brings about (sc. their pregna·cy), whether it happens at once or at a later time.

This is what I have to say about these matters.

18. If the mouth of a woman's uterus is too moist, it will not be able to attract the seed: employ sharp suppositories, for when the orifice becomes irritated and inflamed, it sometimes becomes hard. To a certain extent, this is related to induration, for which it is better to employ sharp suppositories, since irritating agents—being attenuant and fiery—dissolve indurations. If an induration is softened, heal the wound with gentle substances and anything that is nonirritant.[5]

19. If a woman does not become pregnant for a long time, although her menses are appearing,[6] on the third or fourth day grind alum fine, dissolve it in an unguent, soak it up with a piece of wool, and apply it as a suppository to be retained for three days. On the third day, thoroughly boil dried bull's gall in olive oil, soak some lint with this, apply, and have the patient retain it for three days. On the next day, have her remove the suppository and go to her husband.

[5] The logic of this chapter is difficult: first, sharp suppositories are applied to dry the cervix by causing induration, and, then, the same treatment is applied to dissolve indurations. Cf. *Nature of Women* 24.

[6] This is the text of Θ; M and V include a negative: "and her menses are not appearing."

20. Ἢν δὲ τὴν γονὴν μὴ δέχηται ἡ γυνὴ τῶν γυναικείων κατὰ φύσιν γινομένων, ἡ μῆνιγξ ἐπὶ πρόσθεν ἔσται· γίνεται δὲ καὶ ἐξ ἄλλων· γνώσει δὲ τόδε[103]
τῷ δακτύλῳ ⟨εἰ⟩[104] ἅψῃ τοῦ προβλήματος. πρόσθεμα
δὲ ποιῆσαι ῥητίνην καὶ ἄνθος χαλκοῦ, ἐν μέλιτι δι-
60 είς,[105] ὀθόνιον ἀρδαλώσας, | πρόσθες ῥάμμα ἐκδήσας
ἐκ τοῦ ἄκρου, ὡς ἐσωτάτω· ὅταν δὲ ἐξελκύσῃς, τὴν
μυρσίνην ἐν οἴνῳ ἀφέψων, τῷ οἴνῳ χλιαρῷ διανιζέ-
σθω· περιελεῖν δὲ τὸν χιτῶνα ἄμεινον.

21. Εἰσὶ[106] δὲ γυναῖκες αἵτινες λαμβάνουσι μὲν ῥηϊδίως ἐν γαστρί, ἐξενεγκεῖν δὲ οὐ δύνανται, ἀλλὰ
σφῶν τὰ παιδία φθείρονται ἅμα τῷ τρίτῳ μηνὶ ἢ τε-
τάρτῳ, οὐδεμιῆς βίης[107] ἐπιγενομένης, οὐδὲ βορῆς
ἀνεπιτηδείου.[108] καὶ ταύτῃσιν αἴτιόν ἐστι τῶν εἰρημέ-
νων·[109] καὶ μάλιστα ἐπὴν παραμεθίωσι τῆς αὔξης τῷ
ἐμβρύῳ αἱ μῆτραι. κοιλίη σφιν ταράσσεται, καὶ[110]
ἀσθενείη καὶ πυρετὸς σφοδρὸς καὶ ἀσιτίη ἐμπίπτει
τῷ χρόνῳ τούτῳ, ᾧ ἂν τὰ παιδία φθείρωσιν. ἔστι δὲ
καὶ τόδε αἴτιον, ἢν αἱ μῆτραι λεῖαι ἔωσι ἢ φύσει ἢ
ἑλκέων ἐν αὐτῇσιν ἐγγενομένων· ἢν γὰρ λεῖαι ἔωσιν,
ἔστιν ὅτε οἱ ὑμένες ἀπ᾽ αὐτῶν ἀφίστανται, ἐπὴν τὸ
παιδίον ἄρχηται κινέεσθαι, οἱ περιίσχοντες αὐτό, ἅτε
ἡσσόνως ἐχόμενοι τῶν μητρέων ἢ ὡς δεῖ, οἷα λείων

103 τόδε V: τῷδε Θ: om. M.
104 Add. Parisinus Graecus 2143 (s. XIV).
105 διείς Foes in note 103: ἐς codd.
106 Ch. 21 is transposed in Θ and V to after the first sentence
of ch. 25 (. . . τῶν ἐν γαστρὶ ἐχουσέων.).

20. If a woman does not accept (sc. male) seed although her menses are occurring according to nature, a membrane is in the way (sc. of her cervix); this can also happen from other causes. You will recognize this with a finger, if you palpate the obstacle. Make the following application: dissolve resin and flower of copper in honey, smear this on to a piece of linen, and, after tying a string from its end, insert this very deep. When you remove it, boil myrtle in wine, and have the woman douche herself with the warm wine. To remove the tunic is better.

21. There are women who become pregnant easily, but who are unable to carry their fetus to term, but abort it during the third or fourth month, although they have neither suffered any violence nor eaten anything inappropriate. The cause for this in such women will be one of the ones that have been mentioned, and especially when the uterus has allowed the material that should provide growth for the embryo to escape. These patients suffer diarrhea as well as weakness, violent fever, and loss of appetite at the time when the fetuses are aborted. The following can also cause this: if the uterus is smooth either naturally or because ulcers have formed in it—for when it is smooth, sometimes the membranes that envelop the fetus detach (sc. from the uterus) when the fetus begins to move, because they are less securely attached to the uterus than they should be due to its smoothness. A person

---

107 βίης ΘV: αἰτίης M.

108 ἀνεπι- ΘV: ἐπι- M.

109 καὶ ταύτῃσιν . . . εἰρημένων Θ: καὶ τοῦτ' ἐστὶ αἴτιον τὸ εἰρημένον M: ταύτῃσιν αἴτιον ἐστιν ἐν τῶν εἰρημένων V.

110 κοιλίη . . . καὶ ΘV: ἢ ἡ κοιλίη σφιν ταράσσεται M.

ἐουσέων. εἰδείη δ' ἄν τις τούτων ἕκαστα, εἰ ἐρωτῴη
ἀτρεκέως ταῦτα· περὶ δὲ τῆς λειότητος, εἰ ἑτέρη γυνὴ
ψαύσειε τῶν μητρέων κενεῶν ἐουσέων, οὐ γὰρ ἄλ-
62 λως[111] διάδηλον γίνεται. ἢν δὲ ἴωσι | τὰ καταμήνια
ταύτῃσιν, ἀλέα ἔρχεται. ἔστι δ' αὐτῶν ᾗσι γίνεται,
ὥστε ἐκφέρειν τὰ ἔμβρυα· μελεδαινομένῃσι δὲ ἐλπίδες εἰσὶ τόκου.

Ἀμφὶ δὲ τῶνδε ὧδε ἔχει.

22. Ἢν γυναῖκα μὴ δυναμένην τεκεῖν τοκήεσσαν
ἐθέλῃς γενέσθαι, χρὴ τὰ ἐπιμήνια σκέψασθαι, ἤντε
φλεγματώδεα ἤντε χολώδεα ᾖ. γνώσει δὲ τῷδε· ψάμμον ὑποβαλεῖν λεπτὴν καὶ ξηρήν, ὅταν οἱ τὰ ἐπιμήνια γίνηται, καὶ ἐν τῷ ἡλίῳ ἐπιχέαι τοῦ αἵματος,
καὶ ἐᾶν αὐανθῆναι· καὶ ἢν μὲν χολώδης ᾖ, ἐν τῇ
ψάμμῳ ξηραινόμενον τὸ αἷμα χλωρὸν ἔσται, ἢν δὲ
φλεγματώδης ᾖ, οἷον μύξαι. τούτων ὁπότερον ἂν ᾖ,
καθῆραι τὴν κοιλίην, ἤντε ἄνω ἤντε κάτω δέῃ· ἔπειτα
τὰς ὑστέρας καθαίρειν.

23. Ἢν δὲ θέλῃς συλλαβεῖν, τοῦ κισσοῦ ἑπτὰ κόκκους, ἢ τῶν φύλλων κατὰ μῆνα πίνειν ἐν οἴνῳ, παυομένων τῶν ἐπιμηνίων· ἢ σίδιον ἑψήσας ἐν οἴνῳ εὐώδει
ἀκρήτῳ, βαλάνιον ποιήσας, προστιθέναι ἔς τε μεσημ-
βρίην· ἢ στυπτηρίην Αἰγυπτίην τρίψας λείην, ἐς εἴριον ἐνδήσας δοῦναι[112] προστίθεσθαι, ἦμος ἥλιος δύ-
νει, εἶτα ἀφελομένη διανιζέσθω οἴνῳ εὐώδει· ποιέειν
δὲ ταῦτα, παυομένων τῶν ἐπιμηνίων.

24. Ἔχει δὲ καὶ τόδε οὕτως· ἐπὴν ἀποκαθαρθέωσιν
αἱ γυναῖκες, μάλιστα ἐν γαστρὶ λαμβάνουσιν ἱμερω-

can detect each of these things if he carefully inquires about the following: with regard to smoothness, if another woman were to palpate the uterus when it was empty, for otherwise it does not become obvious. If the menses pass in such women, they are copious. Some of these women do carry their fetus to term, and if they are cared for, they will give birth.

This is how the matter is.

22. If you want a woman who is unable to have children to become fertile, you must examine her menses to see whether they contain phlegm or bile. Determine this in the following way: spread out fine dry sand while her menses are passing, pour blood on to this in the sun, and let it dry. If the woman is bilious, when the blood dries in the sand it will be green, whereas if she is phlegmatic, it will be like mucus. Then, according to which of these she is, clean her cavity either upward or downward as needed; after that clean the uterus.

23. If you want (sc. a woman) to conceive, take seven ivy berries or some leaves, and have her drink them in wine, each month just when her menses are stopping. Or boil pomegranate peel in undiluted fragrant wine, make this into a small suppository, and insert it toward midday. Or grind Egyptian alum fine, bind it up in a piece of wool, and give it to be inserted when the sun is setting; then have the patient remove it and flush herself with fragrant wine. Do this when the menses are coming to an end.

24. The following also occurs. After their cleaning, women are most likely to become pregnant, since then

---

111 ἄλλως Linden after Calvus' *aliter*: ἁρμῷ Θ: ἁρμό- M: ἄλλω V.   112 δοῦναι om. MV.

θεῖσαι,[113] καὶ ἡ γονὴ σφίσι ῥώννυται, ἢν ὅτε χρὴ μι-
64 γέωσι, καὶ ἡ τοῦ ἀνδρὸς ῥηϊδίως μίσγεται, | καὶ ἢν
ἐπικρατήσῃ, τῷδε ἠδέλφισται· τότε γὰρ μάλιστα τὸ
στόμα τῶν μητρέων κέχηνε, καὶ[114] τετανόν ἐστι[115]
μετὰ τὰς καθάρσιας, καὶ αἱ φλέβες τὴν γονὴν σπῶ-
σιν· ἐν δὲ τῷ πρὶν χρόνῳ τό τε στόμα τῶν μητρέων
μέμυκε μᾶλλον, καὶ αἱ φλέβες πλέαι αἵματος ἐοῦσαι
οὐχ ὁμοίως σπῶσι τὴν γονήν. ἢν δὲ ἡ γονὴ ἀπορρέῃ
διειπετὴς, καὶ μὴ λήγῃ, οὐ μίσγεται ἀσπασίως τῷ
ἀνδρί, οὐδὲ κυΐσκεται, καὶ ἰξύες ἐπώδυνοι, καὶ πῦρ
ἔχει βληχρόν, καὶ ἀδυναμίη, καὶ ἀψυχίη· καὶ ἔστιν
ὅτε αἱ ὑστέραι σφῶν αὐτῶν ἕδρῃ ⟨μὴ⟩ εἰσίν.[116] ἢν μὲν
οὖν ὑπὸ πλησμονῆς ἴῃ, ἐὰν ἄριστον· ἢν δὲ ἡ ὑστέρη
χαλάσῃ, δίαιτα χόνδρος, κρέας ὕειον ἢ φάσσης, οἶνος μέλας, ποτήματα ὅσα πρὸς ῥόον γεγράψεται.

25. Νῦν δὲ ἐρέω ἀμφὶ νουσημάτων τῶν ἐν γαστρὶ
ἐχουσέων. φημὶ τῇ γυναικὶ ἐν γαστρὶ ἐχούσῃ δίμηνον ἢ τρίμηνον καὶ περαιτέρω, ἢν τὰ ἐπιμήνια χωρέῃ
αὐτῇ[117] κατὰ μῆνα ἕκαστον, ἀνάγκη λεπτήν τέ μιν
γενέσθαι καὶ ἀσθενέα· ἔστι δ' ὅτε καὶ πῦρ ἐπιλαμβάνει τὰς ἡμέρας ἕως ἂν χωρῇ τὰ ἐπιμήνια· καὶ ἐπειδὰν
χωρέῃ τὰ καταμήνια, καὶ μετὰ τὴν χώρησιν χλωρὴ
γίνεται, χωρέει δὲ ὀλίγα. ταύτῃσι κεχήνασιν αἱ μη-

---

113 ἱμερω- I: ἤμειρω- Θ: ἤμερω- MV.   114 κ., κ. om. Θ.
115 Add. καὶ MV.   116 σφῶν . . . εἰσίν Potter after
Calvus' suo loco non sedent and Froben σφ. αὐ. μὴ εἰσὶν ἕδρῃ:
ἕδρην εἰσὶν Θ: εἰσὶν ἕδρῃ MV.
117 χω. αὐ. Θ: παραχωρέῃ αὐτίκα Μ: παραχωρέῃ αὐτῇ V.

they feel desire and their seed is strong: if they have intercourse when they should, the man's seed is easily mixed
with theirs, and if his predominates, (sc. the child) will be
very like him. For at that time after the cleaning the mouth
of the uterus is most widely open and stretched, and the
vessels attract the seed. But during the time before that,
the mouth of the uterus is more closed and the vessels,
being filled with blood, do not attract the seed in the same
way. If the seed is expelled †in a continuous flow†[7] and
never stops, a woman is not eager to have intercourse with
her husband, nor will she become pregnant: her loins are
painful, and a mild fever arises along with weakness and
loss of consciousness; sometimes the uterus is ‹not› in its
proper position. If the movement (sc. of the flux) is the
result of fullness, it is best to leave it alone, whereas if the
uterus is gaping open, the diet followed should include
spelt groats, pork or some ringdove, dark wine, and the
potions prescribed for a flux.

25. Now I shall give an account of the diseases that
occur in women who are pregnant. Note that if, in a
woman who has been pregnant for two or three months or
even longer, her menses continue to pass each month, she
will inevitably become thin and weak; sometimes fever is
also present on the days the menses are passing, and then
and later the woman becomes greenish, although what
passes is little. In such women the uterus is gaping open

---

[7] Fuchs translates "in reinem Zustand" before conceding: "Da
der Sinn nichts Zwingendes ergiebt, bleibt auch der Text zwei-
felhaft."

τραι μᾶλλον τοῦ καιροῦ, καὶ παραμεθίασι τῆς αὔξης
τοῦ ἐμβρύου· κατέρχεται γάρ, ἐπὴν ἐν γαστρὶ ἔχῃ ἡ
γυνή, ἀπὸ παντὸς τοῦ σώματος αἷμα ἐπὶ τὰς μήτρας
66 κατ᾽ ὀλίγον, καὶ περιστάμενον | κύκλῳ περὶ τὸ ἐν
τῇσι μήτρῃσιν ἐὸν αὔξει κεῖνο· ἢν δὲ χάνωσιν αἱ
μῆτραι μᾶλλον τοῦ καιροῦ, παραμεθίασι τοῦ αἵματος
κατὰ μῆνα, ὥσπερ εἴωθεν χωρέειν, καὶ τὸ ἐν τῇσι
μήτρῃσιν ἐὸν λεπτόν τε καὶ ἀσθενὲς γίνεται.

Μελεδαινομένης δὲ τῆς γυναικός, ἄμεινόν τε τὸ
ἔμβρυον, καὶ αὐτὴ ἡ γυνὴ ὑγιαίνει· ἢν δὲ μὴ μελεδαίνηται, φθείρεται τὸ ἔμβρυον, κινδυνεύει δὲ καὶ αὐτὴ
τὸ νούσημα χρόνιον ἴσχειν, ἢν οἱ κάθαρσις πλείων
τοῦ δέοντος χωρέῃ μετὰ τὴν διαφθορήν, οἷα τῶν μητρέων μᾶλλον ἐστομωμένων. [καὶ κίνδυνος ἔσται.]¹¹⁸

Ἢν δὲ γυναικὶ ἐν γαστρὶ ἐχούσῃ ἡ κεφαλὴ φλεγματώδης, καταβαίνει τὸ φλέγμα δριμὺ ἀπὸ τῆς κεφαλῆς καὶ καταρρήσσει¹¹⁹ τὴν κοιλίην, καὶ πῦρ μιν
ἐπιλαμβάνει βληχρόν, καὶ παλμοὶ ἔστιν ᾗσιν ἀσθε-
νέες, ὑπεκλυόμενοι, ἐπαναδιδόντες, ὀξέες· καὶ ἀσιτίη
ἔχει καὶ ἀδυναμίη, κίνδυνός ἐστιν ἐν τάχει φθαρῆναι
τὸ ἔμβρυον, καὶ αὐτὴ ἐν κινδύνῳ ἔσται ἀπενεχθῆναι,
ἢν μὴ μελεδαίνηται· ἐπὴν ἀποφύγῃ, ἅτε τῆς κοιλίης
εὐρόου ἐούσης, ἀλλ᾽ αὐτίκα δεῖ καταλαμβάνειν.

Πολλοὶ δὲ καὶ ἄλλοι εἰσὶ κίνδυνοι, οἷς τὰ ἔμβρυα
φθείρονται· καὶ γὰρ ἢν ἡ γυνὴ ἐν γαστρὶ ἔχουσα

more than it should and allowing part of the nutriment destined for the fetus leak out: for when a woman is pregnant blood from her whole body moves downward toward her uterus, a little at a time, and standing all around what is in the uterus in a circle nourishes it. If the uterus gapes open more than it should, it lets some of this blood escape month by month in the way it is accustomed to pass, making what is in the uterus become thin and weak.

If this woman is cared for, the fetus improves and the woman recovers, but if she is not cared for, the fetus is aborted and the woman herself runs the risk of having a chronic disease, if her cleaning passes in a greater quantity than it should after the abortion, on account of the uterus being more dilated. [There will also be danger.]

If the head of a pregnant woman suffers from phlegm, the phlegm descends in an irritating state from her head and provokes a disturbance in her cavity, and a mild fever comes over her; in some women there are mild palpitations that sometimes intermit and sometimes increase more and more acutely. There are loss of appetite for food, weakness, and danger that the fetus will be soon aborted, and the woman herself will be in danger of being carried off unless she is cared for after she has lost the fetus, since her cavity is fluent, making immediate attention imperative.

There are also many other dangers causing embryos to be aborted, as may happen if a pregnant woman becomes

---

118 Del. Ermerins.   119 καταβαίνει . . . καταρρήσσει Ermerins: ἦ (ἔιη V) καὶ καταβαίνει (-βαίνῃ ΘV) τὸ φλέγμα δριμὺ ἐς τὴν κοιλίην, καὶ καταρήσσει (-ρήσσῃ ΘV) ἀπὸ τῆς κεφαλῆς (add. ἐς ΘV) codd.

νοσήση καὶ ἀσθενὴς ᾖ, καὶ ἄχθος βίῃ ἀείρῃ, ἢ
πληγῇ, ἢ πηδήσῃ, ἢ ἀσιτήσῃ ἢ λιποθυμίῃ ἴσχηται,
ἢ πλέονα ἢ ὀλίγην τροφὴν λαμβάνῃ, ἢ δειδίσσηται
καὶ πτύρηται, ἢ κεκράγῃ ἢ ἀκρατήσῃ·[120] καὶ τροφὴ δὲ
αἰτίη φθορῆς καὶ τὸ πόμα[121] πολύ. καὶ αὐταὶ δὲ αἱ
68 μῆτραι ἔχουσι φύσιας ᾗσιν ἐξαμβλέεται,[122] | πνευμα-
τώδεες, πυκναί, μαναί, μεγάλαι, μικραί, καὶ ἄλλα ὅσα
ἔοικεν. ἢν γυνὴ ἐν γαστρὶ ἔχουσα τὴν κοιλίην ἢ τὴν
ὀσφῦν πονέῃ, ὀρρωδέειν χρὴ τὸ ἔμβρυον ἀμβλῶσαι,
ῥαγέντων τῶν ὑμένων, οἳ περιέχουσιν. εἰσὶ δὲ αἱ φθεί-
ρουσι τὰ ἔμβρυα, [καὶ][123] ἢν δριμύ τι ἢ πικρὸν φάγωσι παρὰ τὸ ἔθος ἢ πίωσι, νηπίου τοῦ παιδίου ἐόν-
τος· ἐπὴν γὰρ τῷ παιδίῳ παρὰ τὸ ἔθος τι γένηται, ἢν
καὶ μικρὸν θνῄσκει, καὶ ἢν τοιαῦτα πίῃ ἢ φάγῃ ἡ
γυνή, ὥστε οἱ ἰσχυρῶς ταραχθῆναι τὴν κοιλίην, νηπίου ἐόντος τοῦ παιδίου· ἐπαΐουσι γὰρ αἱ μῆτραι τοῦ
ῥεύματος χωρέοντος ἐκ τῆς κοιλίης. καὶ ἢν ταλαιπωρήσῃ ἡ γυνὴ πλείονα τοῦ καιροῦ καὶ οἱ ἡ κοιλίη ἐρχθῇ ἢ καὶ μεγάλη γένηται, ἀπογίνεται καὶ οὕτω τὸ
παιδίον διαθερμανθὲν ὑπὸ τῆς ταλαιπωρίης καὶ πιε-
ζόμενον ὑπὸ τῆς κοιλίης· κάρτα γὰρ τὰ πολλά, μικρὰ
ἐόντα, ἔστιν ἄγυια. τὰ δὲ καὶ μεγάλα φθείρεται
παιδία· ὥστε οὐ χρὴ θαυμάζειν τὰς γυναῖκας, ὅτι
διαφθείρουσιν ἄκουσαι· φυλακῆς γὰρ καὶ ἐπιστήμης
πολλῆς δεῖ ἐς τὸ διενεγκεῖν καὶ ἐκθρέψαι τὸ παιδίον
ἐν τῇ μήτρῃ, καὶ ἀποφυγεῖν αὐτὸ ἐν τῷ τόκῳ.

26. Ἢν δὲ ἡ γυνὴ ἐν γαστρὶ ἔχουσα τὸ σῶμα

ill and weak, or uses force to lift some burden, or is struck, or jumps up, or goes without food, or loses consciousness, or takes too much or too little nutriment, or has a fright that makes her afraid, or shouts, or loses command over herself. Nutriment is also a cause of abortion, and excessive drink; furthermore the uterus itself has conditions causing miscarriage, such as being filled with air, or being too dense or loose in texture, too large or too small, and so on. If a pregnant woman has pain in her cavity or lower back it must be feared that her fetus will be aborted from the membranes that contain it being torn. There are also women who abort their fetus if they eat or drink something sharper or more bitter than usual while their fetus is still tender; for if anything out of the ordinary happens to a fetus when it is small, it dies. If a woman drinks or eats things that violently disturb her cavity while her fetus is still tender, her uterus reacts to the flux passing from the cavity. Also if a woman exerts herself more than she should, provoking her cavity to close in and swell, in this case too a child expires from becoming overheated by the exertion and crushed by the cavity. For many indeed are weak of limb due to their smallness, although large children, too, are aborted, so that it is no wonder that women have abortions without wanting to: for it requires much attention and knowledge to bring a child to term and provide for its nourishment in the uterus, and then to give birth to it.

26. If during pregnancy a woman's body is in an indif-

---

[120] κεκρ. ἢ ἀκρ. MV: κεκρακτήσῃ Θ.   [121] τὸ πόμα V: πόμα M: τὸ αἶμα Θ.   [122] Add. οὖσαι MV.
[123] Del. Parisinus Graecus 2144 (s. XIV).

φλαύρως ἔχοι, καὶ εἴη χολώδης καὶ ἐπίπονος, καὶ πυρεταίνοι ἄλλοτε καὶ ἄλλοτε, καὶ τὸ στόμα ἐκπικρά-
ζοιτο, γλῶσσα χλωρή, ὄμματα ἰκτερώδεα, ὄνυχες
χολώδεες, οὖρον δριμύ, ἄλλως τε καὶ εἰ πυρεταίνοι,[124]
70 ταύτῃ συμβήσεται, | ἐπὴν τέκῃ, τὴν κάθαρσιν χολώ-
δεα εἶναι, καὶ τὸ παιδίον[125] ἀσθενές.[126] καὶ εἰ χολώδεα
τὰ λοχεῖα ἢ μέλανά ἐστι κάρτα, καὶ ἐπιπολῆς λίπος
γίνεται, καὶ ἔρχεται κατ' ὀλίγον, καὶ οὐ ταχὺ πήγνυ-
ται· καὶ τὸν μὲν πρῶτον χρόνον ῥηϊτέρως οἴσει, ἔπειτα
χαλεπώτερον, καὶ ἐπικαθαίρεται ἐλάσσονα τοῦ δέ-
οντος· ἢν γὰρ τὸ σῶμα φλαύρως ἔχῃ, καὶ τὰ λοχεῖά οἱ
ἐλάσσονα χωρήσει καὶ πονηρότερα.

Πείσεται δὲ πάντα ταὐτὰ καὶ ᾗ τὰ καταμήνια ἐχώρει τὰ χολώδεα, ἐλάσσονα δὲ χρόνον νοσεῖ, καὶ κινδύνους τοὺς αὐτοὺς ἕξει ἡ νοῦσος, καὶ σημεῖα, καὶ
μεταλλαγάς· ἢ γὰρ ἔμετος ταύτῃ χολώδης ἢ κατὰ τὴν
κοιλίην κάθαρσις πρὸς αὐτῇσι γίνεται, καὶ ἑλκοῦνται
αἱ μῆτραι. φυλακῆς δὲ πολλῆς δεῖται ἡ γυνή, ὅταν τι
τοιοῦτον γένηται αὐτῇ, ὅπως μὴ θανεῖται ἢ ἄφορος
ἔσται. ἢν δὲ μηδὲν τούτων γένηται καὶ μὴ μελεδαίνηται, ἀλλὰ οἱ τὰ λοχεῖα κρυφθῇ, θνήσκει ἐν τριήκοντα
καὶ μιῇ ἡμέρῃ ὡς ἐπὶ τὸ πολύ. ταύτην φάρμακον πῖσαι χοληγόν, καὶ ἄννησον ἀρήγει, καὶ ὅσα ἐς οὔρη-
σιν· ἐμεῖν δέ, καὶ ἱδρῶτας ἄγειν, καὶ τὴν κοιλίην κλῦ-
σαι χυλῷ πτισάνης ἢ μέλιτι καὶ ᾠοῖς καὶ μαλάχης
ὕδατι.

[124] Add. ἄλλοτε καὶ ἄλλοτε MV.

ferent state, she is bilious and sensitive to fatigue, she has
fever from time to time, her mouth tastes bitter, her
tongue is greenish, her eyes are jaundiced, her fingernails
are stained with bile, her urine is irritating—but especially
if she is febrile—on giving birth her cleaning will contain
bile and her child will be weak. And if the lochia are especially bilious or dark, fat will form on their surface, they
will pass a little at a time, and they will not congeal quickly.
At first this patient will tolerate her condition more easily,
but then with greater difficulty, and less will be cleaned
out than should be: for if her body is in an indifferent state,
the lochia will pass both in an insufficient amount and accompanied with quite a lot of pain.

She will suffer all the same things as a woman whose
menstrual flux contains bile, and although she will be ill
for a shorter time her disease will have the same dangers,
signs and alterations, for such patients either vomit up
bilious material or have a cleaning through the cavity with
ulceration of the uterus. Such a woman requires much
care if she suffers in this way, to prevent her from dying
or becoming barren. If none of these signs appears and
she is not treated, but her lochia fail to appear, she will die
on the thirty-first day in most cases. Give this woman a
medication to drink that empties bile; anise would be of
benefit, and anything that promotes the passage of urine.
Stimulate her to vomit and to sweat, and flush her cavity
with barley gruel or honey, eggs and mallow in water.

---

125 καὶ τὸ παίδιον Θ: καὶ πᾶσα ἐλπὶς καὶ τὰ παίδια M:
καὶ τὸ αἰδοῖον V.

126 Add. ἦν (add. δὲ V) χολώδεα ἀποπατέῃ, κάρτα δέ γίνεται τοῦτο ῥήϊτερον διάξει MV.

27. Ὅσῃσιν ἐν γαστρὶ ἐχούσῃσι περὶ τὸν ἕβδομον
ἢ ὄγδοον μῆνα ἐξαπίνης τὸ πλήρωμα τῶν μαζῶν καὶ
τῆς γαστρὸς συμπίπτει, καὶ οἱ μαζοὶ συνισχναίνονται, καὶ τὸ γάλα οὐ φαίνεται, φάναι τὸ παιδίον ἢ
τεθνηκὸς εἶναι ἢ ζώειν τε καὶ εἶναι ἠπεδανόν. |

72   28. Ὅσῃσιν ἐχούσῃσιν[127] ἐν γαστρὶ ἐπιφαίνεται τὰ
ἐπιμήνια, τρωσμοὶ γίνονται, ἢν πλείονα ᾖ καὶ κάκοδμα, ἢ νοσώδεα τὰ ἔμβρυα γίνεται.

29. Ἢν γυνὴ ἐν γαστρὶ ἔχουσα φλεγματώδης ᾖ,
καὶ τὴν κεφαλὴν ἀλγέῃ, καὶ πυρεταίνῃ ἄλλοτε καὶ
ἄλλοτε, ἐν τῇ κεφαλῇ εἰλέεται τὸ φλέγμα, καὶ βάρος
ἔχει καὶ ψύξις, καὶ ἐς τὸ σῶμα διαχωρέει καὶ ἐς τὰς
φλέβας ὅταν ἡ κεφαλὴ ᾖ πλήρης· γίνεται δὲ μολύβδῳ ἡ χροιὴ ἰκέλη, καὶ ἐμεῖ φλέγμα, γλῶσσα λευκὴ
καὶ οὔρησις, κοιλίης ἔκλευκος ψυχρὴ τάραξις, δυσκι-
νησίη. ἐπὴν δὲ τέκῃ, χωρήσει οἱ ἡ κάθαρσις φλεγματώδης, καὶ φανεῖται ὑμενώδης, καὶ ὥσπερ ἀράχνια
διατεταμένα ἐν αὐτῇ ἔσται· καὶ πείσεται τὰ αὐτὰ
πάντα καὶ ᾖ τὰ καταμήνια ἐχώρει τὰ φλεγματώδεα,
ἐλάσσονα δὲ χρόνον νοσήσει, καὶ κινδύνους τοὺς
αὐτοὺς ἡ νοῦσος ἕξει, καὶ σημεῖα, καὶ μεταλλαγάς·
συμβήσεται γὰρ αὐτῇ, ἔμετον γενέσθαι φλεγματώδεα καὶ παθήματα ὅμοια ἐκείνῃ χρονίσαντα.[128] ἐξήρτηται γὰρ τῷ αὐτῷ τρόπῳ τὰ λοχεῖα καὶ τὰ καταμήνια
τὰ φλεγματώδεα, ἐλάσσονα δὲ χρόνον μένει τῶν
καταμηνίων.

---

[127] ὅσῃσιν ἐχούσῃσιν Littré: ὅσῃσιν ἔχουσιν Θ: εἰ γὰρ
ἔχουσιν MV.   [128] χρονί- ΘV: χωρή- Μ.

27. To women who in the seventh or eighth month of pregnancy suddenly have a rapid decline in the fullness of their breasts and belly, together with shriveling up of the breasts and the failure of milk to appear, announce that their child is either dead or barely alive.

28. Pregnant women in whom the menses reappear and are copious and ill-smelling will have an abortion, or their fetus will be sickly.

29. If a pregnant woman suffers from phlegm, has pain in her head, and is feverish from time to time, then phlegm is impacted in her head, heaviness and coldness will follow, and it (i.e., the phlegm) will run through into her body and vessels when her head has become full. The patient's skin becomes leaden, she vomits phlegm, her tongue and urine are white, and she has whitish diarrhea accompanied by a feeling of coldness and difficulty in moving. After she has given birth, her cleaning will be phlegmy and look like membranes, and it will contain a material like cobwebs spread out in it. This woman will suffer all the same signs as one whose menses are phlegmy, and although she will be ill for a shorter time her disease will have the same dangers, signs and alterations; for this woman too vomits material with phlegm in it and has the same chronic disorders as the other one. This is because lochia and menses that contain phlegm both arise in the same way,[8] although the lochia last for a shorter time than the menses.

[8] Cf. *Nature of the Child* 7, Loeb *Hippocrates* vol. 10, 47–51, where a direct relationship is drawn between the lochia and menstruation.

Καὶ ἢν μὴ ῥαγῇ ἡ κάθαρσις αὐτῇ χρονισθεῖσα,
θνήσκει ἐν πέντε καὶ τεσσεράκοντα ἡμέρῃσι· καὶ ἢν
οἱ φλεγματώδης ἡ[129] κάθαρσις χωρέῃ, ἐλάσσονα τῶν
ὑγιηρῶν χωρήσει· μελεδαινομένη δὲ ἡ γυνὴ ὑγιὴς
74 ἔσται, καὶ φυσηθήσεται | δὲ ἐξ ἀρχῆς μέχρι ὑγιανθῇ·
χαλεπὸν γὰρ τοῦτ' ἐστὶ τὸ νόσημα. ταύτῃ διδόναι
φάρμακον ὅ τι φλέγμα ἄγει, καὶ ἐπιπίνειν[130] γάλα
ἑφθὸν αἴγειον ἐν μέλιτι· ἢν δὲ μὴ ἐσακούῃ, κάρδαμον
ἢ κνῆκον[131] ἢ κνέωρον ἢ πουλυπόδιον ἢ ὀρὸν ἢ τὸ ἀπὸ
ἁλῶν συντιθέμενον διδόναι, καὶ ὅσα φλέγμα χαλᾷ τε
καὶ ἄγει.

30. Ἢν δὲ γυνὴ ἐν γαστρὶ ἔχουσα σπληνώδης ᾖ
ὑπὸ παθημάτων ὧν εἴρηται ἐν τῇ νούσῳ τῇ τὰ καταμήνια τὰ ὑδρωποειδέα καὶ φλεγματωδέα ἀφιείσῃ, τὰ
λοχεῖα χωρέει ὑδρωποειδέα, καὶ ἐλεύσεται ὁτὲ μὲν
πολλά, ὁτὲ δὲ ὀλίγα, καὶ γίνεται ὁτὲ μὲν ὥσπερ ἀπὸ
κρεῶν ὕδωρ, ὡς εἴ τις κρέα αἱματώδεα ἀποπλύνοι, ὁτὲ
δὲ ὀλίγῳ παχύτερα, καὶ οὐ πήγνυται. καὶ πείσεται
πάντα ταὐτὰ καὶ ᾗ τὰ καταμήνια τὰ ὑδαρέα ἐχώρεε,
καὶ κινδύνους τοὺς αὐτοὺς ἡ νοῦσος ἕξει καὶ μεταλ-
λαγάς· συμβήσεται γάρ οἱ ῥόον γενέσθαι ὑδαρέα, ἢ
κρυφθῆναι τὴν κάθαρσιν καὶ τραπέσθαι περὶ τὴν κοιλίην καὶ τὰ σκέλεα ἢ ἐς τὸ ἕτερον τούτων, καὶ[132] κίνδυνοι ἔσονται οἱ αὐτοί, οἳ καὶ εἴρηνται.

31. Ἢν κύουσα οἰδέῃ, κνίδης καρπὸν ὡς πλεῖστον
καὶ μέλι καὶ οἶνον κεκρημένον εὐώδεα διδόναι ποτὸν

If no lochial cleaning breaks out as time goes on, the woman dies in forty-five days; if her lochia pass containing phlegm, they will flow in a smaller amount than in a healthy woman. On being treated, this patient will recover, while suffering from wind from the beginning until she becomes healthy. The disease is difficult: prescribe a medication that draws phlegm, to be followed by a potion of boiled goat's milk with honey; if the patient fails to respond, give cress, safflower, spurge flax, polypody, whey or the compound made from salt, as well as whatever softens and evacuates phlegm.

30. If a pregnant woman has a disorder of her spleen due to the conditions referred to in my account of the disease that leads to the discharge of watery and phlegmy menses, her lochia will pass watery and more at one time than at another, and at other times resemble the water that runs out of meat—as if one were to rinse off bloody meat—and at still other times be a little thicker but fail to congeal. This woman will suffer all the same things as one in whom the menses pass off watery, and the disease will have the same dangers and alterations: the patient's flux will be watery, or her cleaning will fail to appear but be diverted around her cavity and legs, or toward one of these, and the same dangers will arise as have been described.

31. If a pregnant woman swells up with edema, give her a potion of a great number of stinging nettle seeds, honey,

---

129 Add. λοχείη MV.　　130 ἐπιπίνειν Θ: μίσγειν MV.

131 κνῆκον Froben: κνῆκος codd., Aldina.

132 ἕτερον τούτων, καὶ Grensemann: στέρνον τούτων Θ: στέρνον ἢ τι τούτων καὶ MV.

δὶς τῆς[133] ἡμέρης. ἢν κύουσαν χολὴ λυπέῃ, πτισάνην
δίδου, ῥόον ἐπιπάσσων τὸν ἐρυθρὸν ἢ τὸν ἐκ τῆς
συκαμίνου, ψυχρὸν δὲ ῥοφείτω, καὶ καταστήσεται. |

76   32. Ἢν δὲ πνὶξ προσπέσῃ ἐξαπίνης γυναικὶ ἐν
γαστρὶ ἐχούσῃ, γίνεται δὲ τοῦτο μάλιστα ἐπὴν ἡ
γυνὴ ταλαιπωρήσῃ καὶ ἀσιτήσῃ, θερμανθεισέων τῶν
μητρέων ὑπὸ τῆς ταλαιπωρίης καὶ ἐλάσσονος τῆς ἰκ-
μάδος γενομένης τῷ ἐμβρύῳ, καὶ ἅτε τῆς μητρὸς[134]
κενεωτέρης τὴν κοιλίην τοῦ καιροῦ ἐούσης, ἰθύει τὸ
ἔμβρυον πρὸς τὸ ἧπαρ καὶ τὰ ὑποχόνδρια, ἅτε ἰκμα-
λέα ἐόντα, καὶ πνίγα ποιέει ἰσχυρὴν ἐξαπίνης. ἐπιλαμβάνει γὰρ τὸν διάπνοον τὸν ἀμφὶ τὴν κοιλίην, καὶ
ἀναυδίη ἴσχει τὴν γυναῖκα, καὶ τὰ λευκὰ ἀναβάλλει
τοῖν ὀφθαλμοῖν, καὶ τἆλλα πάσχει πάντα ὅσαπερ
εἴρηται, ἤν τινα ἔφησα τὰς μήτρας πνίγειν.

Καὶ ἅμα ἄρχεταί τε ἡ πνὶξ γίνεσθαι τῇ ἐν γαστρὶ
ἐχούσῃ γυναικί, καὶ ἀπὸ τῆς κεφαλῆς καταρρεῖ
φλέγμα ἐς τὰ ὑποχόνδρια, οἷα τοῦ σώματος μὴ δυνα-
μένου τὴν ἀναπνοὴν ἕλκειν. καὶ ἢν μὲν ἅμα τοῦ φλέγματος τῇ κατελεύσει ἴῃ τὸ ἔμβρυον ἐς χώρην τὴν
ἑωυτοῦ, οἷα τὴν ἰκμάδα ἑλκύσαν καὶ κατενεχθὲν ὑπὸ
τοῦ φλέγματος, ὑγιὴς γίνεται ἡ γυνή· τρυλισμὸς δὲ
γίνεται, ἀπιόντος τοῦ ἐμβρύου ἐς χώρην τὴν ἑωυτοῦ,
καὶ ἡ γαστὴρ ὑγρὴ γίνεται ὡς ἐπὶ τὸ πλεῖον τῆς γυναικός. ἢν δὲ μὴ ἴῃ τὸ ἔμβρυον ἐν τάχει ἐς χώρην τὴν
ἑωυτοῦ, δύο ἤδη γίνεται τὰ πονέοντα τὸ ἔμβρυον, τὸ
φλέγμα τὸ καθελθὸν ἀπὸ τῆς κεφαλῆς, βαρύνει τε
γὰρ καὶ ψύχει ἐπιμένον, καὶ ἡ ἀηθείη τοῦ χωρίου· καὶ

and diluted fragrant wine twice a day. If bile is distressing a pregnant woman, give her barley gruel over which you have sprinkled red sumac or mulberry fruit, to be taken cold: the condition will settle down.

32. If suffocation suddenly befalls a pregnant woman, this usually occurs after she has exerted herself and gone without food, and is due to her uterus being warmed by the exertion and there being less nutriment for the fetus; and since the mother is emptier than normal in the cavity, the fetus pushes toward the liver and the hypochondrium, since they are full of fluid, which provokes an immediate violent suffocation. This blocks the breathing passage around the cavity causing the woman to become speechless, to turn up the whites of her eyes, and to suffer all the other things I described as happening in a woman whose uterus is causing suffocation.

At the same time that this suffocation sets in in a pregnant woman, phlegm also flows down from her head into her hypochondrium, since her body is unable to draw its inspiration. Now if, when this phlegm descends, the fetus returns to its proper place by attracting nutriment and being carried down by the phlegm, the woman recovers. A gurgling sound is heard as the fetus returns to its own space, and the woman's belly usually becomes fluid. If however the fetus does not move soon to its proper space, two conditions are then present that trouble it: phlegm coming down from the head which weighs down the fetus and cools it where it is residing, and the abnormality of its

---

133 δὶς τῆς MV: τρίτης Θ.

134 μητρὸς Potter, after γυναικὸς Grensemann and *mulierum* Lat.: μήτρης codd.

κινδυνεύσει, καὶ ἢν μή τις ἐν τάχει ἐπιτηδειοτέρως
διαιτῷη, ἀποπνιγείη ἂν ἡ γυνή.

Ἀμφὶ τούτων ὧδε ἔχει. |

78   33. Γυναικὶ ἐν γαστρὶ ἐχούσῃ, ἢν ὁ χρόνος ἤδη τοῦ
τόκου παρῇ, καὶ ὠδὶς ἔχῃ, καὶ ἐπὶ πολλὸν χρόνον
ἀποφυγεῖν ἡ γυνὴ τοῦ παιδίου μὴ οἵη τε ᾖ, ὡς ἐπίπαν
ἔρχεται πλάγιον ἢ ἐπὶ πόδας, χρειὼ δὲ ἐπὶ κεφαλὴν
χωρέειν. ὧδε δὲ γίνεται τὸ πάθημα· ὥσπερ εἴ τις ἐς
λήκυθον μικρόστομον πυρῆνα ἐμβάλοι, οὐκ εὐφυὲς
ἐξελεῖν πλαγιούμενον, οὕτω δὲ καὶ τῇ γυναικὶ χαλεπὸν πάθημα τὸ ἔμβρυον, ἐπειδὰν λοξωθῇ, οὐκ ἔξεισι
γάρ.[135] χαλεπώτερον δὲ καὶ ἢν ἐπὶ πόδας χωρήσῃ, καὶ
πολλάκις ἢ αὐταὶ ἀπώλοντο, ἢ τὰ παιδία, ἢ καὶ ἀμ-
φότερα. ἔστι δὲ καὶ τόδε μέγα αἴτιον τοῦ μὴ ῥηιδίως
ἀπιέναι, ἢν νεκρὸν ἢ ἀπόπληκτον ἢ διπλόον ᾖ.

34. Ἐπὴν ἐν γαστρὶ ἔχῃ γυνή, χλωρὴ γίνεται
πᾶσα, ὅτι αὐτῆς τοῦ αἵματος ἀεὶ τὸ ἀκραιφνὲς καθ᾽
ἡμέρην ὑπολείβεται ἐκ τοῦ σώματος, καὶ κατέρχεται
ἐπὶ τὸ ἔμβρυον, καὶ αὔξη οἱ γίνεται, καὶ ἐλάσσονος
τοῦ αἵματος ἐόντος ἐν τῷ σώματι ἀνάγκη εἶναι χλωρήν, καὶ ἱμείρεσθαι ἀτόπων ἀεὶ βρωμάτων, καὶ ἐπὶ
κοιλίην αἱματώδεα ἰέναι· καὶ ἀσθενεστέρη γίνεται, ὅτι
τὸ αἷμα μινύθει. φημὶ δὲ γυναῖκα, ἢν ἐπίτεξ ᾖ, πνεῦμα
πυκνὸν ἀφιέναι, καὶ ἢν κάθαρσις ἄρχηται, ἡ κοιλίη
80   πλήρης ἐστὶ καὶ θερμὴ πιεζομένη. | μάλιστα δ᾽ ἀνα-
πνεῖ πυκνόν, ἐπὴν τόκου πελάζῃ, καὶ τὴν ὀσφῦν τότε

135 οὐκ ἔξεισι γάρ V: οὐκ ἔξεισι Θ: χαλεπὸν ἐξελθεῖν Μ.

location. This woman will be in danger, and unless some-one quickly applies a more suitable regimen, she will die of suffocation.

So it is, regarding these matters.

33. If a pregnant woman is already approaching the time of delivery and having birth pangs, but she is unable for a long time to bring her child to birth, the child is usu-ally presenting sideways or in the direction of its feet, while it should be moving forward with its head. This oc-curs as follows: just as if someone were to place an olive stone into a narrow-mouthed oil flask he would be unable to extract it readily if the stone was turned sideways, so also in a woman a fetus presents a difficult problem when it has turned sideways, since it will not come out. Even more difficult is if the fetus moves in the direction of its feet, and often either mothers have perished from this, or children, or both. Other important causes of difficult birth are death of the fetus, paralysis, or its folding double.

34. When a woman is pregnant, she will take on a gen-eral green color because the pure part of her blood is regularly trickling down day by day from her body to her fetus giving it growth, so that as there is less blood in her body she must become green and always be desiring strange foods, and bloody material must move to her cav-ity; and because the woman's blood is diminishing in amount she becomes weaker. Note that if a woman is about to give birth, she will exhale rapidly, and that if her cleaning is beginning, her cavity feels full and warm when pressed. The rapid respiration occurs in particular as de-livery approaches, and then the woman suffers pain espe-

μάλιστα πονέεται· σπᾶται[136] γὰρ καὶ ἡ ὀσφῦς ὑπὸ τοῦ
ἐμβρύου· καρδιώσσει δὲ ἐν τῷ συνάπαντι χρόνῳ ἄλλοτε καὶ ἄλλοτε, τῆς κοιλίης περιστελλομένης ἀμφὶ
τὸ ἔμβρυον, ἢν τίκτῃ, μάλιστα δὲ τῆς ὑστέρης.[137]

Καὶ ἢν[138] ἐξανεμωθῇ, ἧπαρ οἶος ἢ αἰγὸς ἐς τέφρην
κρύψαι, καὶ μετέπειτα ἕψειν, καὶ λαμβάνειν, καὶ οἶνον,
ἢν μή τι κωλύῃ, πίνειν ζωρότερον παλαιὸν ἐπὶ τέσσα-
ρας ἡμέρας, ἢν πορρωτέρω ᾖ ἀπὸ τῆς τέξιος. ἢν δὲ
τὰς ἰξύας ἀλγέῃ, ἄννησον καὶ κύμινον Αἰθιοπικὸν
πινέτω, καὶ τῷ θερμῷ λουέσθω. ἢν δ᾽ ἄσθμα λάζηται,
θείου ὅσον κύαμον καὶ καρδαμώμου ἴσον καὶ πηγάνου καὶ κυμίνου Αἰθιοπικοῦ, ταῦτα τρῖψαι καὶ διεὶς
οἴνῳ πιεῖν δίδου νῆστι πυκνά·[139] καὶ σιτίων ἀπεχέσθω. ἢν δ᾽ ἐν τόκῳ κάθαρσις ἴῃ πολλή, ἡ ὑστέρη
ξυνέλκεται καὶ ἡ κύστις καὶ τὸ ἔντερον, καὶ οὔτε τὸ
κόπριον κατέχουσιν οὔτε τὸ οὖρον, προΐενται δέ· ᾠὰ
οὖν ῥυφεῖν διδόναι, καὶ ἄρτον ἐγκρυφίαν τρώγειν καὶ
ἄσσα γέγραπται. ἢν δὲ ᾖ ἐν τόκῳ ξηρὴ καὶ δύσικμος,
ἔλαιον πίνειν, καὶ καταιονᾶν τὰ χωρία ἐλαίῳ θερμῷ,
μαλάχης ὕδατι, κηρωτῇ τε ὑγρῇ διαχρίειν, καὶ ἔγχυτον χηνὸς ἄλειφα σὺν ἐλαίῳ. εἰ δὲ μὴ δύναιτο τίκτειν,
ὑποθυμία ῥητίνην ἢ κύμινον ἢ πίτυος φλοιόν· καὶ
τούτῳ ὑποθυμιᾶν.

Ἄσσα δὲ οἰδήματα γίνεται ὑστερικὰ ἐν τόκῳ ἢ ἐκ

---

[136] σπᾶται Θ: φλᾶται MV.

[137] ἢν τίκτῃ . . . ὑστέρης Θ: μάλιστα δὲ τῆς ὑστέρης M:
ἢν τεκούσῃ ἡ ὑστέρη V.

cially in her lower back from its being contused by the fetus. Over the whole period she also experiences intermittent heartburn, since the cavity—but in particular the uterus—is contracting around the fetus as birth takes place.

If it (sc. the uterus) becomes inflated, bury a sheep's or goat's liver in ashes, boil it, and give it to the patient to take, and if nothing speaks against it have her drink quite pure aged wine over four days after she is well beyond the delivery. If she suffers pains in her loins, have her take a potion of anise and Ethiopian cumin and bathe in hot water. If shortness of breath comes on, take sulfur to the amount of a bean, and the same amounts of cardamom, rue and Ethiopian cumin: grind and dissolve these in wine. Give this to the patient in the fasting state frequently to drink, and have her abstain from food. If at childbirth copious cleaning passes, the uterus will be contracted, and also the bladder and the intestine, and these will retain neither stools nor urine, but expel them. For this give eggs to eat, bread baked in ashes to chew on, and the things noted. If at the time of childbirth a woman is dry and bereft of moisture, have her drink olive oil and pour warm olive oil and water of mallow over her (sc. genital) parts, apply a moist wax salve, and make an injection with goose grease and olive oil. If a woman is unable to give birth, fumigate with resin, cumin or pine bark, applying this from below.

Edemas of the uterus that develop during or after

---

138 Add. τεκούσῃ ὑστέρη M: om. καὶ ἦν V.
139 After ἀπεχέσθω trans. πυκνά MV.

τόκου, οὐ χρὴ στύφειν, οἷα ἰητροὶ ποιέουσιν· φάρμακα δὲ τάδε ἄριστα,[140] κύμινον Αἰθιοπικόν, ὅσον
τοῖσι τρισὶ δακτύλοισι, καὶ ἀννήσου, καὶ τοῦ σεσέ-
82 λιος πέντε ἢ ἕξ, γλυκυσίδης χηραμύδος ἥμισυ | τῆς
ῥίζης, ἢ καὶ τοῦ σπέρματος, ταῦτα ἐν οἴνῳ λευκῷ
ἡδυόσμῳ μάλιστα νήστει διδόναι· ἢ δαύκου ῥίζαν
Αἰθιοπικοῦ, σέσελι, γλυκυσίδης ῥίζαν τὸν αὐτὸν τρό-
πον· ἢ ἱπποσελίνου καὶ δαύκου καρπὸν Αἰθιοπικοῦ[141]
ὡσαύτως· ἢ κρήθμου ῥίζαν, Αἰθιοπικὸν κύμινον,[142]
Ἀττικὸν τετρώβολον· ἢ πέπερι, ἄννησον, δαῦκος,
ἀκταίη, γλυκυσίδης ῥίζα· ταῦτα ἐν οἴνῳ τρίβειν καὶ
διδόναι· ἢ μυρτιδάνου κλωνία δύο ἢ τρία, καὶ κύμινον
Αἰθιοπικόν, γλυκυσίδης ῥίζαν, ἢ λίνου σπέρμα
ὁμοίως, ὃ τὰ παιδία βήσσοντα ψωμίζουσι ξὺν ᾠῷ
ὀπτῷ λεκίθῳ, σὺν σησάμῳ πεφρυγμένῳ.

Ἢν δὲ παιδίου ἀφθᾷ τὰ αἰδοῖα, ἀμύγδαλα τρίψας
καὶ βοὸς μυελὸν ἐν ὕδατι ἕψειν, καὶ ἄλητον ἐμβαλὼν
μικρόν, διαχρίειν τὰ αἰδοῖα, καὶ διακλύζειν τῷ ὕδατι
τῷ ἀπὸ τῶν μύρτων.

35. Νῦν δὲ ἐρέω ἀμφὶ λοχείων καὶ τῶν μετὰ τὸν
τοκετὸν ἰόντων. ὅταν γυνὴ ἢ τὰ λοχεῖα μὴ καθαρθῇ,
ἢ τὰ ἐπιμήνια μὴ ἴῃ, ἢ καὶ ἡ ὑστέρη σκληρὴ ᾖ, ὀδύνη
ἔχει τὴν ὀσφῦν, καὶ τοὺς κενεῶνας καὶ βουβῶνας καὶ
μηροὺς καὶ πόδας ἀλγέει πικρῶς,[143] καὶ ἡ γαστὴρ
ἐπαίρεται, καὶ φρῖκαι διὰ τοῦ σώματος διαΐσσουσιν,
ἐκ δὲ τῶν τοιούτων πυρετοὶ γίνονται ὀξέες. ταύτην, ἢν
μὲν ἀπύρετος ᾖ, διαιτᾶν λουτροῖσι, λιπαίνειν δὲ κε-

childbirth you should not treat with astringents, as physicians do. The best medications are: of Ethiopian cumin as much as you pick up with three fingers, of anise and hartwort five or six, and of peony root or also seed a half cheramys: give these in fragrant white wine, best to a fasting patient. Or root of Ethiopian dauke, hartwort, and peony root in the same way; or seed of alexanders and Ethiopian dauke in the same way; or four Attic obols of samphire root, or Ethiopian cumin; or pepper, anise, dauke, baneberry and peony root: grind these into wine and give; or two or three twigs of myrtidanon,[9] Ethiopian cumin and peony root. Or the linseed they feed to coughing children with a boiled egg yolk and parched sesame in the same way.

If aphthae develop on a child's genitalia, boil ground almonds with beef marrow in water, sprinkle in a little meal, and anoint this on the genitalia; also wash them well with water made from myrtle berries.

35. Now I will speak about the lochia and discharges that take place after childbirth. When a woman either fails to have a lochial cleaning, or her menses do not pass, or her uterus is indurated, an ache occupies her lower back; she has sharp pains in her flanks, groins, thighs and feet, her belly is raised, and shivering races through her body, from which acute fevers arise. When such a woman is without fever, treat her with baths and anoint her head

---

[9] Littré remarks: "plante indéterminée."

---

[140] Add. προσφέρειν MV.   [141] Aἰθιοπικοῦ Parisinius Graecus 2254 (s. XV): -κὸν ΘMV.   [142] Aἰθ. κύμ. ΘV: Aἰθιοπικὴν ἢ κύμ. M.   [143] ἀλγ. πικ. Θ: ἀλγέουσιν (-γεῦσι) MV.

φαλὴν ἐλαίῳ ἀνηθίνῳ· ἕψειν δὲ μαλάχην, ἢ ἔλαιον
κύπρινον ἐς ὕδωρ ἐγχέειν καὶ ἐγκαθίζεσθαι παρηγο-
ρικῶς· ἐν πάσῃσι δὲ τῇσι νούσοισιν, ἐφ᾿ ὧν πυρίη |
84 ἀρήγει, ἄμεινον ὕστερον χρίεσθαι λίπα· ἢν δὲ πῦρ
ἔχῃ, λουτρῶν ἀπέχεσθαι· πυριᾶν δὲ χλιάσμασι τὴν
νείαιραν γαστέρα καὶ τὴν ὀσφῦν θεραπεύειν· διδόναι
δὲ πίνειν τῶν φαρμάκων τῶν ὑστερικῶν, παραμίσ-
γοντα ἢ τῆς σηπίης τῶν ᾠῶν ἢ τοῦ κάστορος· μετὰ
δὲ ῥυφεῖν διδόναι ἄλητον σὺν πηγάνῳ ἑφθὸν ἢ πτισάνης χυλόν.

36. Ἢν δὲ γυναικὶ μετὰ τοῦ παιδίου ἐν τῷ τόκῳ μὴ
ἴῃ τὸ ὑγρὸν ὡς χρή, ἀλλὰ μεῖον, εἰ μὲν ἐν τῇ κεφαλῇ
ἔχοι τὸ ὑγρὸν ὑπὸ θέρμης εἰρυσθὲν ἐν τόκῳ καὶ πρὶν
ὀλίγῳ, κεφαλαλγήσει· ἢν δ᾿ ἐς τὴν κοιλίην ἔλθῃ ἁλές,
ἐπειδὰν συθῇ, διαταράξειεν ἂν αὐτὴν καὶ οὐχὶ πόρρω.
τιμωρέειν δὲ ὡς μὴ ἐκ τούτου διάρροια ἐπιγενομένη
σώματι φλαύρως ἔχοντι πονήσει μιν. ἢν δὲ ἀπὸ κεφαλῆς ἐλθὸν τὸ ῥεῦμα ἐς τὴν λοχείην κάθαρσιν τράπηται καὶ πολλὰ συθῇ, ῥαΐζει· ἢν δὲ πλέον τοῦ με-
τρίου, μελεδαίνειν· εἰ δ᾿ ἐς τὴν κοιλίην, ῥηΐτέρη ἂν ἡ
ἔξοδος τῷ παιδίῳ γένοιτο.

Εἰ δὲ ἡ κάθαρσις τῇ γυναικὶ ὀλίγη χωρέοι, πόνος
λάζεται ἰσχυρὸς ἰξύας τε καὶ τὸν ἀμφὶ τὰ αἰδοῖα
πάντα χῶρον, καὶ οἰδέει, καὶ οἱ μηροὶ πίμπρανται, καὶ
ἐκ τοῦ στόματος καὶ ἐκ τῶν ῥινέων ῥεῖ φλέγμα[144] ὑδα-
ρές, καὶ ἀλγέει κεφαλήν, καὶ πῦρ ἔχει καὶ φρίκη, καὶ
ἰδίει,[145] καὶ ὀδόντες βρύχουσι, καὶ ἀψυχέει,[146] καὶ ἡ

with olive oil seasoned with dill seed. Boil mallow or pour oil made from henna flowers into water and employ this in an emollient sitz bath. In all diseases for which fomentation is beneficial, it helps to anoint later with fat. If fever is present, prohibit baths, foment the lower abdomen with warm compresses, and attend to the lower back. Give medications favorable to the uterus to drink, to which you have added either cuttlefish eggs or castoreum; afterward give as gruel meal boiled with rue, or barley water.

36. If at the time of delivery a woman's water does not pass with the child as it should—but only in a smaller amount—and the fluid is held in her head after being drawn there by heat during childbirth or a little earlier, she will have a headache. If, however, it moves in a mass to the cavity, as it rushes down it will disturb the cavity, but have no effect beyond that: you must make sure that any diarrhea that arises in this way in the body when it is in an indifferent state does not harm the patient. If a flux coming from the head turns into the lochial cleaning and rushes down in a large amount, it will bring relief: if this is excessive, attend to it. If (sc. the flux turns) into the cavity, it will make the child's birth easier.

If a woman has little cleaning, severe pain befalls the flanks and all the area around the generative parts, they swell, the thighs become distended, and watery phlegm flows out of the patient's mouth and nostrils; she suffers pain in the head, fever and shivering are present, she sweats and grinds her teeth, and she loses consciousness;

---

144 Add. ἰσχυρῶς M.
145 καὶ ἰδίει om. Θ.
146 καὶ ἀψυχέει om. M.

γαστήρ οἱ στεγνὴ ἔσται καὶ ἡ κύστις, καὶ τώμματα ἀναδινέει, καὶ ζοφοειδὲς ὁρᾷ.

Γυναικὶ δὲ ἐκ τόκου ἐούσῃ ἡ κάθαρσις ἐπὴν ἴῃ, οὐκ εὐμαρῶς χωρέει, οἷα τῶν μητρέων ἐν φλογμῷ γενομένων, καὶ τοῦ στόματος σφέων μύσαντος· περιδινοῦται 86 γὰρ ὁ στόμαχος ὁ τοῦ | αἰδοίου μετὰ τὸ παιδίον τὴν ἐκχώρησιν ποιήσασθαι· ἢν γὰρ τούτων τι ᾖ, οὐ χωρήσει οἱ ἡ κάθαρσις· ἢν δὲ μὴ χωρέῃ οἱ ἡ κάθαρσις, συμβήσεται ὥστε μιν πυρεταίνειν, καὶ φρίκην ἔχειν, καὶ τὴν γαστέρα μεγάλην εἶναι· ἢν δὲ ψαύσῃ τις αὐτῆς ἀλγέει πᾶν τὸ σῶμα, μάλιστα ἤν τις τῆς γαστρὸς ψαύσῃ, καὶ καρδιώσσει ἄλλοτε καὶ ἄλλοτε, καὶ ὀσφῦν πονέει, καὶ ἀσιτίη καὶ ἀγρυπνίη καὶ νυγμός· ἔπειτα ἡμέρῃ πέμπτῃ ἢ ἑβδόμῃ ἔστιν ὅτε ἡ κοιλίη ταράσσεται, καὶ ὑποχωρέει μέλανα καὶ κάκοσμα κάρτα, ἄλλοτε καὶ ἄλλοτε καὶ ὡς ὄνειον οὖρον, καὶ ἢν ὑπέλθῃ, δοκέει οἱ ῥήϊτερον εἶναι, καὶ μελεδαινομένη ἐν τάχει ὑγιαίνεται· εἰ δὲ μή, κινδυνεύσει αὐτὴ διαρροίης ἰσχυρῆς ἐπιπεσούσης, καὶ τὰ λοχεῖά οἱ κεκρύψεται. ἢν δὲ ἡ κοιλίη μὴ ταράσσηται, μηδὲ ἡ κάθαρσις χωρέῃ αὐτομάτη, μηδέ οἱ προσφέρηται τὰ ἐπιτήδεια[147] ἐν τάχει, ὁ δὲ χρόνος προΐῃ, πονήσει τὰ προειρημένα μᾶλλον, καὶ ἐπὶ τούτοισι κινδυνεύσει πελιδνὴ γενέσθαι ὡς μόλιβδος, καὶ ὑδερωθῆναι, καὶ ὁ ὀμφαλὸς ἐκστήσεται αὐτῇ, ἀειρόμενος ὑπὸ τῶν μητρέων, καὶ ἔσται μελάντερος τῶν πέριξ. καὶ ἐπὴν ταῦτα γένηται, οὐχ οἵη τέ ἐσται περιγενέσθαι ἡ γυνή· θνήσκουσι δὲ ἄλλη ἄλλῳ χρόνῳ, ὅπως ἂν καὶ τὰ τοῦ

her belly and her bladder will be blocked, her eyes roll up, and her vision will be darkened.

When a woman's cleaning after childbirth is taking place, but it does not pass smoothly because her uterus has become inflamed and its mouth is closed, due to the opening of the vagina turning to the side after the child has left it, in this case the cleaning will not proceed. If the cleaning does not proceed, the woman will develop fever and chills, and her abdomen will swell up. If someone touches her, especially in her abdomen, she will have pain in her whole body, heartburn from time to time, an ache in her lower back, and anorexia, insomnia and a prickling sensation. Then on the fifth or seventh day the cavity is sometimes set in motion, and she passes dark very ill-smelling stools and also from time to time something like ass's urine. If these come down, the patient will seem to improve, and on being quickly attended to she will recover. If she is not treated, she will be in danger of an attack of violent diarrhea, and her lochia will stop. If the cavity is not set in motion, nor does the cleaning pass spontaneously, nor are the appropriate remedies expeditiously applied, but time goes by, the patient will suffer the things mentioned above more severely, and in addition she will be in danger of turning as livid as the color of lead and filling up with fluid; her navel will protrude, being raised due to the state of her uterus, and it will be darker than its surroundings. When these things happen, the woman will be incapable of surviving: one woman dies at one time, another at an-

---

147 ἐπιτήδεια ΘΜ: ἐπιμήνια V.

σώματος ἔχωσι καὶ τὰ τῆς πάθης· μιῆς δὲ καὶ εἴκοσιν
ἡμερέων οὐχ ὑπερβάλλουσιν, ὡς ἐπὶ τὸ πλέον οὕτω
συμβαίνει.

Ἢν δέ οἱ ῥαγῇ ἡ κάθαρσις εἴτε καὶ ὑπὸ φαρμάκων
εἴτε καὶ αὐτομάτη, γίνεται γὰρ καὶ τοῦτο, ἢν χαλάσωσι τὸ στόμα αἱ μῆτραι[148] βιασθεῖσαι ὑπὸ τοῦ
αἵματος ἀλέος ἐξαπίνης κατελθόντος, καὶ ἢν ῥαγῇ,
ἀποκαθαίρεται δύσοσμα καὶ πυώδεα, ἔστι δ' ὅτε καὶ
μέλανα, καὶ ῥηΐτερον ἔσται, καὶ μελεδανθεῖσα ὑγιαίνεται.

Γίνεται δὲ καὶ ἕλκεα ἐν τῇσι μήτρῃσιν οἷα τῶν
88   λοχείων σαπέντων· καὶ ἢν | γένηται, πλέονος μελεδώ-
νης δεήσεται,[149] ὅπως μή οἱ τὰ ἕλκεα μεγάλα καὶ ση-
πεδονώδεα ἔσται· κίνδυνος γὰρ ἀποθανεῖν ἢ ἄφορον
γενέσθαι. σημεῖα δὲ ταῦτα γίνεται ἢν ἕλκεα ἐνῇ· ἐπὴν
χωρέῃ ἡ κάθαρσις, δοκέει ὡς ἄκανθα διὰ τῶν μητρέων ἰέναι, καὶ πῦρ λάζεται[150] τὴν κοιλίην. φιλέει δὲ
ταῦτα ἐπιλαμβάνειν· ἀλγέει ἐπαφωμένη τὸ κάτω τοῦ
ὀμφαλοῦ, ὡς ἕλκεος καθαροῦ νευρώδεος εἰ θίγοις·
ἔπειτα ὀδύναι ἰσχυραὶ ἄλλοτε καὶ ἄλλοτε ἐμπίπτουσιν ἐς τὰς μήτρας, καὶ πυρετός, ἔστι δ' ὅτε πρὸς
χεῖρα βληχρός, καὶ ἄλλοτε[151] ὑποκακήθεα χωρέει τὰ
λοχεῖα, πυώδεα, δύσοσμα. ταῦτα σημεῖά ἐστιν, ἢν
ἕλκεα ᾖ ἐν τῇσι μήτρῃσι, καὶ δέεται πολλῆς μελεδώνης. ταύτης μέν νυν ἀμφὶ τῆς νούσου τόσαι τελευταί
εἰσιν.

Ἢν δὲ ἡ κάθαρσις ἡ λοχείη τὰς μὲν πρώτας ἡμέρας τρεῖς ἢ τέσσερας χωρήσῃ, ἔπειτα ἀπόληται ἐξ-

other time, according to the state of their body and their disease, but they do not make it beyond twenty-one days, as it usually turns out.

If cleaning occurs suddenly in a woman as the result of her medications or even spontaneously—for this too may happen, if her uterus opens up at its mouth on being over-powered by a mass of blood suddenly descending and breaking through—ill-smelling purulent material will be discharged, sometimes also dark, she will be relieved, and on being looked after she will recover.

Ulcers may also form in the uterus as a result of the lochia putrefying, and if this happens more attention will be required to prevent them from becoming large and putrid, for they threaten death and barrenness. The fol-lowing signs appear if ulcers are present (sc. in the uterus): when the cleaning is being discharged, something like a thorn seems to be piercing the uterus, and heat occupies the cavity. The following are also likely to ensue: on being touched in the region below the navel, the patient suffers pain as if you touched a fresh lesion in a cord. Then pains press intensely from time to time in the uterus, together with fever which is often mild to the touch, and sometimes the lochia that pass are slightly malignant, purulent, and ill-smelling. These are the signs if ulcers are present in the uterus, and they require active attention. Such are the outcomes of this disease.

If the lochial cleaning passes on the first three or four days, but then suddenly stops, this patient will have suf-

---

148 Add. ἢ M, μὴ V.   149 δεήσεται M: γενήσεται Θ:
δεηθήσεται V.   150 λάζ. Θ: μιν λάζ. μάλιστα MV.
151 Add. καὶ ἄλλοτε M.

ἀπίνης, αὕτη πάσχει παθήματα τῇ προτέρῃ ἀδελφά,
ἧσσον δέ· καὶ ἢν μεταπίπτῃ ἡ νοῦσος, ἐς τωὐτὸ
μεταπεσεῖται· χρονίη δὲ καὶ βληχροτέρη ἔσται τῆς
προτέρης, διαιτεομένη τε ἡ γυνὴ περιγίνεται, ἢν ἁρ-
μοῖ μελεδαίνηται.

Ἀμφὶ δὲ ταύτης τῆς νούσου ὧδε ἔχει.

37. Ἢν δ᾽ ἐκ τόκου μὴ καθαρθῇ, οἰδέει ἡ γαστὴρ
καὶ ὁ σπλὴν καὶ τὰ σκέλεα,[152] καὶ πῦρ ἔχει, καὶ ῥῖγος
λαμβάνει, καὶ ὀδύναι ἀΐσσουσι πρὸς τὰς ἰξύας, ἔστι
δ᾽ ὅτε καὶ πρὸς τὰ σπλάγχνα, καὶ ἀποψύχεται, καὶ τὸ
πῦρ ἔχει, σφυγμοὶ βληχροί, ἔστι δ᾽ ὅτε καὶ ὀξέες,
ἄλλοτ᾽ ἀειρόμενοι, ἄλλοτε ἐλλείποντες. ταῦτα πάσχει
ἀρχομένης τῆς νούσου, καὶ ὧδε ἔχει· ἢν δὲ προΐῃ τοῦ
90   χρόνου, τὰ κῦλα[153] τοῦ | προσώπου ἐρυθρὰ γίνεται.

Ὅταν ὧδε ἔχῃ, διδόναι κοῦφα σιτία· καὶ ἢν ὀργᾷ,
φάρμακον πῖσαι κάτω· ἢν μὲν χολώδης ᾖ, ὅ τι χολὴν
καθαίρει, ἢν δὲ φλεγματώδης, ὅ τι φλέγμα· μετὰ δὲ
πυριᾶσθαι τὰς ὑστέρας εὐώδεσι, καὶ προσθεῖναι μαλθακτήριον τὴν ἡμέρην. ἢν δὲ στερεὸν ᾖ τὸ στόμα,
πυριᾶν ἅπασαν ἡμέρην, καὶ τὰ μαλθακτήρια προστι-
θέναι· ἔπειτα λοῦσαι θερμῷ ὕδατι, καὶ ἐντιθέναι τοὺς
μολίβδους· μετέπειτα δὲ ἁλὸς χόνδρους καὶ σμύρναν
ἐς τρυχίον ἀποδήσας καὶ τὴν πίσσαν τὴν ἐφθὴν ἐν
εἰρίῳ, ἡδύσματα συμμίξας, ἴσον ἑκάστου, ποιέων
ἴσον κηκίδι μικρῇ, προσκεῖσθαι δὲ ἡμέρην καὶ εὐφρό-
νην· μετὰ δὲ διαλιπεῖν ἡμέρας τρεῖς, καὶ πυριᾶσαι
τοῖσιν αὐτοῖς.

ferings similar to the one before, but less intense, and if the disease has an alteration, it changes in the same way. This condition is chronic and weaker than the previous one, and if a woman follows the regimen prescribed, she will recover if she is attended to immediately.

So it is with this disease.

37. If after giving birth a woman is not cleaned, her belly, spleen, and legs will swell up, fever will set in, she will have a chill, and pains will dart to her loins and sometimes her viscera; she feels cold, fever sets in, and there is a gentle throbbing, which sometimes becomes violent, sometimes is raised, and sometimes intermits. These things the woman suffers at the beginning of the disease in the way described, but as time passes, the areas of her face below her eyes also become red.

When the case is such, give light foods, and if the patient becomes overfilled, administer a medicinal potion that acts downward; if she is bilious give a medication that cleans bile, if phlegmatic one that cleans phlegm. Afterward, foment her uterus with fragrant substances, and apply an emollient suppository during the day. If the mouth (sc. of the uterus) is hard, foment each day and apply emollient suppositories. Then bathe the patient in warm water and insert lead (sc. spatulas). After that tie up some lumps of salt with myrrh in a rag and boiled pitch in wool with aromatic substances—an equal amount of each—and form a suppository like a small oak gall and insert it for a day and a night. After that leave three days, and foment with the same things.

---

152 σκέλεα MV: ἕλκεα Θ.
153 κῦλα Θ: κοῖλα MV.

Προστιθέναι δὲ κόκκους ἐκλέψας ὅσον δύο πόσιας
καὶ πεπέριος, τρίψας λεῖα, παραμίξας ἔλαιον Αἰγύπτιον λευκὸν καὶ μέλι ὡς κάλλιστον, τοῦτο ἐμπλάσας
ἐς εἴριον, περὶ πτερὸν ἐλίξας, προστιθέναι ἡμέρην καὶ
εὐφρόνην, καὶ ἢν σοι δοκέῃ κεκαθάρθαι, ἄμεινον ἐὰν·
ἢν δὲ δοκέῃ ἔτι[154] δεῖσθαι καθάρσιος, δύο ἡμέρας διαλείπειν, καὶ αὖτις τὸ σὺν τῇ σικύῃ ἡμέρην καὶ νύκτα
προστιθέναι· κἄπειτα νέτωπον καὶ τὸ ῥόδινον ἔλαιον
εὐωδέστατον καὶ ἐλάφου στέαρ τήξας, ἐν εἰρίῳ προστιθέναι μίαν ἡμέρην, καὶ λούειν πολλῷ τῷ θερμῷ, ὡς
οἷόν τε ᾖ εὐμενές.

Καθαρτηρίοισι δὲ αὐτίκα καὶ θερμῷ καθηραμένη
τὰ πονεύμενα[155] χωρία, ἐναλειφέσθω τὸ στόμα τῶν
μητρέων χηνείῳ στέατι καὶ σμύρνῃ καὶ ῥητίνῃ χλι-
αρῇ, καὶ θάλπειν· κλυζέσθω δὲ τὰς ὑστέρας τῷ οἴνῳ
92 καὶ τῷ ναρκισσίνῳ ἐλαίῳ τῇ | ὑστεραίῃ· ἢν δὲ μὴ ᾖ
ναρκίσσινον, οἴνῳ· ταῦτα ὅπως σοι πρὸ τῶν ἐπιμηνίων ἡμέρῃ μιῇ πρόσθεν πεποιήσεται. ἐπὴν δὲ τὰ
ἐπιμήνια γενήται, τὰς μὲν ἐν ἀρχῇ ἡμέρας τρεῖς, τρίβουσα μέλαν[156] τὸ κύπριον, καὶ ἁλὸς χόνδρον ἐπιχέασα, εἰρίῳ ἀναφορύξαι· τοῦτο ἐν τῷ σώματι ἐχέτω ἐπ'
ὀλίγον καὶ νῆστις, καὶ οἶνον ἄκρητον εὐώδεα ἐπιρρο-
φεέτω. ἐπὴν δὲ παύσηται τὰ ἐπιμήνια, τὴν ἡμέρην τὸ
σὺν τῇ γλήχωνι προστιθεῖς, πρὸς τὸν ἄνδρα ἴτω· καὶ
ἢν ἐν γαστρὶ ἔχῃ, ὑγιὴς γίνεται.

Σιτίοισι δὲ χρήσθω ἐν τῇ καθάρσει. πρὸς τάδε
ἀρήγει[157] ἕψειν τὴν λινόζωστιν, καὶ ξυμμίσγειν πράσα

Application: take peeled (sc. Cnidian) berries to the amount of two portions and some pepper, grind finely, mix these into white Egyptian oil and the finest honey, anoint on to a piece of wool, wind this around a feather, and apply it for a day and a night. Once you think cleaning has taken place, it is better to leave off, but if more cleaning seems to be required, leave a period of two days, and apply another suppository made with a bottle gourd, for a day and a night. Then take oil of bitter almonds and very fragrant rose oil, melt deer's fat, apply these on wool for one day, and flush the patient with copious warm (sc. water) in the gentlest manner.

Immediately after the woman has been cleaned with suitable agents and warm water applied to her painful areas, have her anoint the mouth of her uterus with goose grease, myrrh and warm resin, and foment. On the next day she should inject her uterus with wine and narcissus oil—if there is no narcissus oil available, then just with wine. You should arrange these things so that they are over one day before the menses begin. When these arrive, during the first three days have the woman grind black copper, sprinkle coarse salt over this, and mix them up well in a piece of wool: have her hold this in her body for a short time and then drink unmixed fragrant wine in the fasting state. After the menses stop, apply a suppository with pennyroyal during the day, and have the woman go to her husband: if she becomes pregnant, she will recover.

During the cleaning, the patient should take food; it also helps to boil mercury herb, mix it with leeks, garlic,

---

154 ἔτι om. MV.   155 πονεύμενα Θ: ἀμφιπονεύμενα MV.
156 μέλαν Θ: μελάνθιον MV.   157 τάδε ἀρήγει Ermerins in note: τὰ ῥίγει Θ: τε τὰ ἀρήγοι M: τε τὰ ῥίγη V.

καὶ σκόροδα καὶ κράμβην κόκκωνά τε, καὶ τὸν χυλὸν
ῥυφείτω· τοῖσι δ᾽ ἄλλοισι, θαλασσίοισι μᾶλλον χρήσθω ἢ κρέασι· τῶν δὲ γλυκέων εἰργέσθω καὶ ἐλαιη-
ρῶν· πίνειν δὲ αἰεὶ νῆστιν τὸ ἀπὸ τῆς δᾷδος, ἔστ᾽ ἂν
καθαίρηται· ἐν δὲ τοῖσιν ἐπιμηνίοισι γάλα[158] πινέτω.

38. Ἢν δὲ ὀλίγῳ ἐλάσσονα τοῦ δέοντος χωρέῃ
γυναικὶ τὰ λοχεῖα, οἷα τῶν μητρέων στενοστόμων
ἐουσέων καὶ παρεστραμμένων, ἢ τοῦ αἰδοίου ἦ τι μεμυκὸς κάρτα ὑπὸ φλεγμασίης, ἡ γυνὴ πυρεταίνει
ὀξέως, καὶ καρδιώσσει, καὶ ἀλγέει τὸ σῶμα πᾶν, καὶ
σφαδάζει, καὶ ἐς τὰ ἄρθρα τῶν χειρῶν καὶ τῶν σκελέων καὶ τῆς ὀσφύος ἡ ὀδύνη φοιτᾷ, καὶ τὸν ἀμφὶ τὴν
δειρὴν χῶρον καὶ ῥάχιν καὶ βουβῶνας ἀλγήσει, καὶ
ἀκρατέα τινὰ τῶν μελέων[159] τοῦ σώματος γενήσεται |

94 καὶ ἔπειτα πῦρ ἠρεμαῖον γενήσεται καὶ[160] φρίκη πάνυ
φανερή· ἐμέουσι δὲ φλεγματώδεα, πικρά, δριμέα. καὶ
ἀμφὶ τῆσδε ὧδε ἔχει· καί οἱ συμβήσεται, ἢν μελε-
δανθῇ, ὑγιέα γενέσθαι· εἰ δὲ μή, χωλὴν[161] καὶ ἀκρατέα
τῶν μελέων τοῦ σώματος γενέσθαι. ἄφορος δὲ ἡ νοῦσος οὐ πάνυ.

Ἢν δὲ αἱ[162] μῆτραι ἑλκωθῶσι καὶ τὰ λοχεῖα μὴ
παρῇ, ὡς χρή, πάντα μιν πονήσει, καὶ ἢν μὴ μεγάλα
ᾖ τὰ ἕλκεα, μελεδαινομένη ἐν τάχει ὑγιαίνεται. χρὴ
δὲ τὴν μελέτην ἀτρεκέως ποιέεσθαι ἑλκέων τῶν ἐν
τῆσι μήτρησιν· ἅτε γὰρ ἐν ἁπαλῇ ἐόντα καὶ εὐεπαι-

---

[158] γάλα Θ: μάλα MV.   [159] μελέων ΘΜ: σκελέων V.
[160] γεν. κ. om. M.   [161] χωλὴν V: χολὴν ΘΜ.

cabbage and pomegranate seeds, and have her drink this juice. Otherwise, she should employ seafoods in preference to meats, and abstain from sweet and oily foods. While continuing to fast, have her drink a potion made from pinewood until she is cleaned; during her menses have her drink milk.

38. If a woman's lochia flow slightly less than they should due to the mouth of her uterus being narrowed and displaced to one side, or some part of her vagina is narrowly closed due to inflammation, she will have acute fever, heartburn, pain in her whole body, and restlessness; pain darts to the joints of her arms and legs and to her lower back, and she aches in the regions about her throat, spine and groins; various debilities in the limbs of her body also occur. Then a mild fever with evident shivering follows, and patients vomit material that is phlegmy, bitter, and sharp. So it is with this patient: if she happens to receive treatment, recovery will follow, but if she is not treated, she will become lame and disabled in the limbs of her body; the disease rarely[10] leads to barrenness.

If the uterus ulcerates and the lochia do not begin as they should, all the sufferings will be present, but if the ulcers are not large, on being treated promptly the patient will recover. The care of ulcers in the uterus must be carried out with precision, for since they are in a soft, very

[10] Grensemann suggests deleting the negation since: (a) the condition's severe symptoms suggest that it would result in barrenness; (b) ἄφορος is otherwise not used in a negative statement.

---

162 δὲ αἱ Grensemann: δὲ Θ: μὴ αἱ MV.

σθήτῳ καὶ νευρώδει κοιλίῃ,[163] πολλὰ δὲ τὰ κοινωνέο-
ντα, βρέγμα, στόμαχος, γνώμη, αὔξεται καὶ κακοτρο-
πέει, καὶ οὐ ῥηϊδίως ἐθέλει ξυνιέναι.

Ἢν δέ οἱ αἱ μῆτραι στενόστομοι γενοίατο, καὶ μὴ
παραχαλάσωσι τὴν λοχείην κάθαρσιν, καὶ φλεγμήνωσιν, ἢν μὴ μελεδαίνηται ἐν τάχει,[164] πάντα μιν
μᾶλλον πονήσει, καὶ ὀδμὴ πονηρή, καὶ οἰδίσκεται ἡ
ἔξοδος· καὶ ἢν μὴ φλεγμήνωσιν αἱ μῆτραι, αὐτόματον
ἔξεισι κακὸν ὀζόμενον καὶ πελιδνὸν ἐὸν ἢ ὑπομέλαν
ἐὸν θρομβοειδές, καὶ ἡ γυνὴ καθαίρεται τὰ λοχεῖα·
ἔστι δὲ ὅτε οὐκ ἔξεισιν, ἀλλὰ τῇ γυναικὶ θάνατον σημαίνει, ἢν μή τις ἐν τάχει φλέβα τάμοι ἢ κοιλίην
μαλθάξειεν, ἄμεινον δὲ καὶ κλυσμάτεσιν· ἢν δ' εὐή-
μετος ᾖ, καὶ ἐς ἔμετον ἄγειν· κρέσσον δὲ διουρέειν καὶ
ἰδίειν· τούτων δὲ καιρός, ὅτε δέοι, ἄριστος.

39. Ἢν δὲ ἐκ τόκου γυνὴ καθαρθῇ ὀλίγῳ πλέονα
ὧν χρή, γίνεται γὰρ καὶ τοῦτο, εἰ αἵ τε μῆτραι εὐρύστομοι γενοίατο καὶ τῶν φλεβῶν τινες καταρραγέ-
96 ωσιν, αἳ τείνουσιν ὑπὸ τὰς μήτρας, ὑπὸ | βίης τῆς
ἐξόδου τοῦ ἐμβρύου, πυρετὸς καὶ ῥῖγος ἕξει αὐτὴν
λεπτός, θέρμη τε ἀνὰ πᾶν τὸ σῶμα, ἔστι δ' ὅτε καὶ
φρίκη καὶ ἀσιτίη, καὶ βδελύξεται πάμπαν, καὶ λεπτὴ
ἔσται καὶ ἀσθενὴς καὶ χλωρὴ καὶ οἰδαλέος· ἢν δέ τι
φάγῃ ἢ πίῃ, οὐ πέσσεται· ἐνίῃσι δὲ καὶ κοιλίη καὶ
κύστις καταρρήγνυται, καὶ φρίκη ἔχει μᾶλλον.

Ἀμφὶ δὲ τῆσδε ὧδε ἔχει.

sensitive, sinewy hollow, they are in communication with many other parts such as the bregma, the cardia, and the mind, they expand, turn malignant, and tend not to unite.

If the mouth of a woman's uterus becomes narrow, fails to release the lochial discharge, and becomes inflamed, unless she is quickly treated her sufferings will all be more severe, including an unhealthy odor and swelling of the exit. If the uterus does not become inflamed, a spontaneous flux with clots will be discharged—ill-smelling and livid or darkish—and the woman will clean her lochia. But sometimes the lochia are not expelled; this presages death for the woman, unless someone quickly cuts a vessel or relieves the cavity—or even better applies enemas. If the woman vomits easily, induce her to vomit, too; very good are also diuresis and sweating: the optimal time for these is when they are necessary.

39. If a woman is cleaned slightly more than is necessary after childbirth—for this too happens, when her cervix is wide open and some of the vessels distributed to the uterus break open as a result of the violence of the fetus' passage—she will have a mild fever with chills, heat will spread through her whole body, and sometimes she also suffers from shivering and loss of appetite; she has a general loathing for food, and she becomes emaciated, weak, greenish in color, and swollen: if she eats or drinks anything, it cannot be digested. In some of these women, the cavity and the bladder have violent discharges, and more intense shivering is felt.

Such is this disease.

---

163 In V the text breaks off here and recommences immediately with ἢν ἑλκεωθῶσι in the first sentence of ch. 40.

164 Add. ἑκταίῃ ἢ ἑβδομαίῃ ἐούσαν Θ.

40. Ἢν δ' ἐκ τόκου ἐούσῃ συμφραχθῇ τι τοῦ αἰδοίου—ἤδη δὲ καὶ τοῦτο εἶδον, ἢν[165] ἑλκωθῇ τὸ στόμα τοῦ αἰδοίου· καὶ ἐπειδὰν ἑλκωθῇ ἐν τῷ τόκῳ βιηθὲν ὑπὸ τῆς ἐξόδου τοῦ παιδίου, ἐγένετο ἴκελόν τι ἄφθῃ, καὶ ἐφλέγμηνε κάρτα, καὶ τὰ χείλεα ὑπὸ τῆς φλεγμάντος ξυνέπεσε πρὸς ἄλληλα καὶ ἐλάβετο ἀλλήλων, ἄτε εἰλκωμένα ἐόντα. καὶ μύκης[166] ἄμφω τὰ χείλεα ἔχει συνδήσας, ἄτε τῆς καθάρσιος ἀπολελαμμένης· εἰ δὲ ἐχώρεεν ἡ κάθαρσις, οὐκ ἂν ἐμυκώθη τὰ ἕλκεα· νῦν δὲ ἐπιρρεῖ, καὶ παχύνεται ἀλλοκότῳ σαρκί. ἰῆσθαι οὖν ὡς τὰ ἐν τῷ ἄλλῳ σώματι, καὶ ἐς ὠτειλὰς ἄγειν· τὸ δὲ χωρίον λεῖον ἔστω καὶ ὁμόχροον.

Ἡ Φρόντις·[167] ἔπασχε δὲ ἡ γυνὴ πάντα ἃ πάσχουσιν αἱ μὴ[168] ἀποκαθαιρόμεναι τὰ λοχεῖα, καὶ ἐπὶ 98 τούτοισιν | ἤλγει τὸ αἰδοῖον, καὶ ψηλαφῶσα ἔγνω ὅτι οἱ συνεπέφρακτο, καὶ ἔφρασε, καὶ μελεδαινομένη ἀπεκαθήρατό τε καὶ ὑγιὴς ἐγένετο καὶ φορός· εἰ δὲ μὴ ἐμελεδάνθη, μηδέ οἱ ἡ κάθαρσις ἐρράγη αὐτομάτη, τὸ ἕλκος μέζον ἐποίησεν ἄν, καὶ ἐκινδύνευσεν, εἰ μὴ ἐμελεδάνθη, καρκινωθῆναι τὰ ἕλκεα.

41. Εἰ δ' ὁρμηθείη γυναικὶ λοχείη κάθαρσις ὡς ἐς κεφαλήν, θώρηκά τε καὶ πλεύμονα, γίνεται γὰρ καὶ τοῦτο, θνήσκουσιν[169] ἐν ταχεῖ, ἢν ἴσχηται· εἰ δὲ χωρέοι κατὰ στόμα ἢ ῥῖνας καλῶς, ἐξάντης γίνεται· εἰ

---

165 The text in V takes up at this point.

166 μύκης Littré: μυκηι Θ: ψύξις δὲ γίνεται καὶ μύκη M: θίξις γίγνεται καὶ μυκησσός V.

40. If a woman has a blockage in her vagina after child-birth—I have seen this too in the past, if the mouth of the vagina was ulcerated; and when lesions were caused by the fetus' forceful passage in childbirth, a kind of aphthae broke out, there was violent inflammation, and the labia fell together against one another from the inflammation, and adhered to one another due to their ulceration. A fungoid excrescence holds the labia joined together since cleaning is prevented: if cleaning does take place, the ulcers will not become fungous: but the further afflux causes the labia to become thick with extraneous tissue. This is to be treated like an excrescence in any other part of the body, by being made to scar: the site should become soft and uniform in color.

Phrontis: this patient suffered all the things women do who are not cleaned of their lochia, as well as having pain in her genitalia; she discovered by examining herself that she was obstructed and reported this, and on being treated she was cleaned, she recovered, and she was fertile. If she had not been treated or her cleaning had not broken out spontaneously, her ulcer would have become larger and brought her into the danger of her ulcers becoming cancerous, if left untreated.

41. If a woman's lochial cleaning rushes in the direction of her head, thorax and lung—for this too happens—such women quickly succumb if the flux is retained, but if it passes well through her mouth or nostrils, such a woman

---

167 Φρόντις Grensemann, after Frisk: φροντίς ΘV: φροντις
M.   168 μὴ Θ: μῆτραι MV.   169 Add. αὐτίκα MV.

δὲ ὀλίγον ἡ νοῦσος χρονιωτέρη γένοιτο καὶ πάσχοι
ἂν ἡ γυνὴ ὁποῖα εἴρηται ἀμφὶ τῆς παρθένου, ᾗ τὰ
ἐπιφαινόμενα πρῶτα ὥρουσεν ἄνω· ἡ δὲ γυνὴ πλέονα
χρόνον περιέσται τῆς παρθένου, καὶ βληχρότερα τὰ
παθήματα ἔσται οἱ, μέχρι ὁ πλεύμων διάπυος γένηται.

Εἰ δὲ μὴ χωρέοι οἱ ἡ λοχείη κάθαρσις κατὰ στόμα,
ἀλλ᾽ ἄνω ὁρμηθεῖσα τράπηται, κεκρύψεται τὰ λοχεῖα
καὶ οὐ χωρήσει κατά γε δίκην, καὶ βὴξ ὑπολήψεται
καὶ ἄσθμα, πληρεομένου τοῦ πλεύμονος ὑπὸ τοῦ
αἵματος· καὶ πονήσει τὸ πλευρὸν κάρτα καὶ μετάφρενον, καὶ ὅταν βήξῃ, ξηρὸν ἀποβήξεται, ἄλλοτε δὲ
ἀφρῶδες πτύσεται· τοῦ δὲ χρόνου προϊόντος, πτύαλον
ἐπιφαίνεται μέλαν[170] ἐὸν καὶ θολερόν, καὶ τὰ στήθεα
πῦρ ἔχει τοῦ ἄλλου σώματος πλέον,[171] οἷα τοῦ αἵματος[172] θερμήναντος αὐτά·[173] καὶ πυρεταίνει ἡ γυνή, καὶ
ἡ γαστήρ οἱ στεγνὴ ἔσται, καὶ ἀσιτήσει καὶ ἀγρυπνήσει, καὶ βδελύξεται, καὶ οὐ περιγίνεται, ἀλλὰ θα-
100   νεῖται ἐν | μιῇ καὶ εἴκοσιν ἡμέρῃσιν ὡς τὰ πολλά.

Ἢν δέ οἱ ἡ κάθαρσις ἄνω ὁρμηθεῖσα μὴ κατὰ
στόμα ἔλθῃ, μηδ᾽ ἐς τὸν πλεύμονα τράπηται, τρέψε-
ταί οἱ ἐς τὸ πρόσωπον τὰ λοχεῖα, καί οἱ ἐρυθρὸν
κάρτα ἔσται, καὶ ἡ κεφαλὴ βαρέη, οὐδὲ κινῆσαι ἄτερ
πόνου οἵη τε. καὶ οἱ ὀφθαλμοὶ ἐρυθροὶ κάρτα ἔσονται,
καὶ ἐκ σφέων αἷμα ῥεύσεται λεπτόν· καὶ ἐκ τῶν ῥινῶν
ἔστιν ᾗσιν αἷμα ῥέει, καὶ ἢν τοῦτο ἴῃ, ὧδε χρονιωτέρη

[170] μελ- Θ: ὑπομελ- MV.

escapes danger. If the disease is extended for a little longer time, a woman will suffer what is described as happening in a girl (sc. whose menses) rush upward for the first time: the woman will survive for a longer time than the girl, however, and her disease will be milder until her lung begins to suppurate.

If the lochial flow does not pass through the patient's mouth, but nonetheless turns upward and moves in that direction, the lochia will disappear and fail to pass as they should, and coughing and difficult breathing will result from the lung being filled with blood. This patient will also suffer great pain in her side and back, and when she coughs her cough will be dry: sometimes her sputum will also be frothy. As time passes, dark, turbid sputum will appear, and heat will occupy the chest more than the rest of the body since the blood there is heating it. The woman will have fever, her belly will be constipated, and she will lose her appetite for food, be unable to sleep, and feel loathing; she will be incapable of surviving, but generally die in twenty-one days.

If on rushing upward the patient's lochial cleaning neither passes through her mouth, nor is diverted to her lung, it will turn toward her face, which will become very red; her head will feel heavy and she will not be able to move it without pain; her eyes will be very red, and out of them will flow thin blood. In some women, blood also runs from the nostrils, and if this happens the disease becomes

---

171 πλέον Cordaeus: πολλόν codd.

172 αἵματος MV: ῥεύματος Θ.

173 αὐτά Cordaeus after Cornarius' *ipsa* (sc. *pectora*): αὐτό codd.

ἡ νοῦσος γίνεται· τοῖσί τε οὔασιν οὐκ ὠκέως εἰσ-
ακούει ἐκ τῆς νούσου· καὶ καρδιώξει, καὶ ἐρεύξεται,
καὶ ἀλλοφάσσει,[174] καὶ παράνοιαι γίνονται μανιώδεες·
ἔστι δ᾽ ᾗσι θράσος ὀμμάτων ἰλλωδέων· καὶ τἆλλα
πάντα πονήσει, ὅπως καὶ ἐς τὸν πλεύμονα, ὡς εἴρηται, ἢν ἡ κάθαρσις τράπηται, πλὴν οὐ βήξει οὐδὲ
πτύσεται τοιαῦτα, οὐδὲ τὸ μετάφρενον ἀλγήσει
ὁμοίως. καὶ μελεδαινομένη ὑγιαίνει· οὐ πολλαὶ δ᾽ ἐλπίδες εἰσὶ περιγενέσθαι· ἢν δ᾽ ἄρα καὶ περιγένηται,
κώφωσις ἔσται ὀφθαλμῶν ἢ ἀκοῆς τὸ ἐπίπαν. ἀμφὶ
τῆσδε τῆς νούσου ὧδε τελευτᾷ.

42. Ἢν δ᾽ ἐκ τόκου ῥόος λαμβάνῃ καὶ τὰ σιτία ἐν
τῇ γαστρὶ μὴ ἐμμένῃ,[175] ἀσταφίδα μέλαιναν καὶ ῥοιῆς
γλυκείης τὰ ἔνδον τρίψας, οἴνῳ διεὶς μελιχρῷ, τυρὸν
ἐπιξύσας αἴγειον, καὶ ἄλφιτα πύρινα πεφρυγμένα
ἐπιπάσας, εὔκρητον δίδου.

43. Ἢν δὲ αἷμα ἐκ τόκου ἐμέῃ, τοῦ ἥπατος θρὶξ[176]
τέτρωται, καὶ ὀδύνη πρὸς τὰ σπλάγχνα φοιτᾷ, καὶ
τὴν καρδίην σπᾶται. ταύτην χρὴ λούειν πολλῷ
θερμῷ, καὶ τῶν χλιασμάτων ἃ μάλιστα προσδέχεται |
προστιθέναι, καὶ πιπίσκειν ὄνου γάλα ἑπτὰ ἡμέρας ἢ
πέντε· μετὰ δὲ πιπίσκειν βοὸς μελαίνης ἄσιτον ἐοῦσαν,[177] εἰ οἵη τε εἴη, ἐφ᾽ ἡμέρας τεσσεράκοντα· ἐς δὲ
τὴν ἑσπέρην σήσαμον τριπτὸν πιπίσκειν. ἡ δὲ νοῦσος κινδυνώδης.

102

174 ἀλλοφάσσει Θ: ἀλλοφρονήσει MV.
175 μὴ ἐμ. Θ: μείνῃ MV.

longer. Because of her disease, the patient does not hear clearly with her ears, she suffers heartburn and belching, she talks nonsense, and her derangement becomes that of a madwoman; in some of these women, there is also a violent distortion of the eyes. Such a patient also suffers all the same things that have been described as occurring when the cleaning turns to the lung, except that she does not cough or expectorate the same kind of things, or have pains in her back in the same way; on being treated, she too may survive, although the hope of survival is not great. Now if she does survive, she will generally become blind in her eyes or lose her hearing. This is how the disease ends.

42. If after the birth of a child, a flux develops and food does not remain in the mother's belly: pound dark raisins and the insides of a sweet pomegranate, dissolve in dark-colored wine, grate over this goat's cheese, and sprinkle it with toasted wheat meal: dilute and give.

43. If a woman vomits blood after the birth of a child, a pipe on her liver has been injured: pain lancinates toward her abdominal viscera, and she has spasms in her heart. You should bathe this patient in copious hot water, apply the compresses to her that you know by experience she will best accept, and give her ass's milk to drink for seven days, or five. After that have her drink (sc. milk) of a black cow, and at the same time go without food—if she is able—for forty days. Toward evening, have her drink a potion with ground sesame. The disease is dangerous.

---

176 θρὶξ Θ: ἡ συρίγξ MV.
177 ἄσ. ἐοῦσαν Θ: γάλα ἀσιτεύσασα MV.

44. Τὸ δὲ γάλα ὅπως γίνεται, εἴρηταί μοι ἐν τῇ Γενέσει τοῦ Παιδίου τῇ ἐν Τόκῳ καὶ τἆλλα. ἢν δὲ γάλα σβεσθῇ, πράσα τρίψας, διεὶς ὕδατι, δίδου πίνειν· καὶ τῷ θερμῷ λουέσθω, καὶ πράσα καὶ κράμβην ἐσθιέτω· συνέψειν δὲ κυτίσου φύλλα, καὶ τὸν χυλὸν ῥοφεῖν· πιπίσκειν δὲ τοῦ μαράθου τὸν καρπὸν[178] καὶ τὰς ῥίζας, καὶ κριθὰς ἐπτισμένας καὶ βούτυρον ἑψήσας ὁμοῦ ψύξας, δίδου πιεῖν. ἀγαθὸν δὲ καὶ τὸ ἱππομάραθον καὶ τὸ ἱπποσέλινον καὶ κύτισος, ὁμοῦ ταῦτα πάντα γάλα πολὺ ποιέει, καὶ †ἐξισκυριαι† αἶγες,[179] τυροὶ δὲ μάλιστα· ἀγαθὸν καὶ τὸν ἐλελίσφακον ἕψειν ἢ[180] ἀρκευθίδων ἢ κεδρίνων[181] ἀποχέουσα τὸν χυλόν,[182] οἶνον ἐπιχέουσα, πινέτω· καὶ ἐς τὰ λοιπὰ ἔλαιον ἐπιχέουσα ἐσθιέτω. καὶ τῶν δριμέων καὶ τῶν ἁλμυρῶν καὶ ὀξέων καὶ ὠμῶν λαχάνων πάντων εἰργέσθω. τὸ δὲ κάρδαμον ἐν οἴνῳ πινόμενον ἀγαθόν· καὶ γὰρ τὸ γάλα καθαίρει· καὶ τῷ θερμῷ λουέσθω, καὶ ἀπὸ θερμῶν πινέτω· καὶ ἄγνου καρπὸν ἐν οἴνῳ διδόναι πίνειν, καὶ γάλα πολὺ ποιέει καὶ τεύτλου χυλός.

104 καὶ | σησάμου ἀπλύτου καὶ κριθέων τριμηνιαίων ἐμβαλὼν ἐς θυίαν, τρίψας πάντα, δι᾽ ὀθονίου ἐκχυλίσας, παραμίξας μέλι ἢ ἀμαμηλίδας, εἶτα ἐπ᾽ οἴνῳ μέλανι διδόναι πίνειν.

45. Ὅταν γυνὴ τέκῃ καὶ τοῦ ὑστέρου[183] ἀπαλλαγῇ, διδόναι ἄμεινον, ὑφ᾽ ὧν μάλιστα καθαίρεται τὰ λο-

---

[178] καρπὸν Θ: χυλὸν MV.
[179] ἐξισκυριαι αιγες Θ: ἄξει σκυρίαι MV.

44. How milk forms I have described in *Generation of the Child in Childbirth*,[11] along with other matters. If a woman's milk dries up, crush leeks, dissolve them in water, and give to drink; also have her bathe in warm water, eat leeks and cabbage, and boil tree medick leaves and drink the juice. Give a potion of fennel seed and its roots, winnowed barley (sc. flour), and the butter plant: boil all this together, cool and give to drink. Also good are horse fennel, alexanders, and tree medick; all these together produce much milk, and . . . goats,[12] especially cheese. It is also good for a woman to boil salvia, or, decanting juice from common or Phoenician juniper berries, to add wine and drink; also have her pour olive oil into what is left over, and eat this. She should avoid foods that are bitter, salty, or sour, as well as all raw vegetables. Cress drunk in wine is good, for this too cleans the milk. Also have the patient bathe in warm water, and afterward take a drink. Also give chaste tree fruit in wine to drink, and beet juice too produces much milk. Also pour unwashed sesame and three-month barley into a mortar, crush it completely, and make it into juice by straining it through a fine linen cloth; mix in honey or medlars, and then give in dark wine to drink.

45. After a woman has given birth and expelled the afterbirth, it is very good to give agents that clean the lo-

---

[11] Cf. *Nature of the Child* 10, Loeb *Hippocrates* vol. 10, 59–61.   [12] Littré suggests Σκυρίαι αἶγες (goats of Scyros) but concludes: "Phrase probablement alterée, mais où je n'ai rien pu trouver qui me satisfit."

---

[180] τῶν add. ΘΜ: ἢ om. V.   [181] ἢ κεδρίνων om. Θ.
[182] ἢ add. MV.   [183] Add. μὴ MV.

χεία, σκόροδα ἑφθὰ ἢ ὀπτὰ ἐν οἴνῳ καὶ ἐλαίῳ μετὰ
πουλυποδίων καὶ σηπίων ἀπ᾽ ἀνθράκων, ὅ τι ἂν βούληται τούτων· ἢ κάστορα ἢ νάρδον πίνειν ἢ[184] πήγανον ἐν οἴνῳ μέλανι γλυκεῖ νῆστιν, ἢ ἄνευ οἴνου· ἢν δὲ
μὴ ᾖ γλυκύς, ἄμεινον μέλι παραμίσγειν· καὶ κράμβην ἑφθὴν ὁμοῦ πηγάνῳ καὶ λινοζώστει, καὶ τῶν
σπερμάτων τι πίνειν τῶν ὑστερικῶν.

Ἢν δὲ θρομβωθῇ καὶ πόνος ἐν νειαίρῃ τῇ γαστρὶ
γένηται, διδόναι πράσα ἑφθά, καὶ ὅσα ἄγρια καὶ
ἥμερα· λιπαρὰ δὲ ποιέειν ἅπαντα· λούεσθαι δὲ διὰ
τρίτης ἡμέρης ἐν εὐδίῃ, ψῦχος γὰρ ταύτῃσιν ἐναν-
τίον· καὶ μετὰ τὸ λουτρὸν ἐπαλείφειν· ἄμεινον μὴ
πολλῷ θερμῷ.

46. Ὅταν τὸ ὕστερον μὴ αὐτίκα ἀπίῃ μετὰ τὸν
τόκον, τῆς νειαίρης γαστρὸς γίνονται πόνοι καὶ ἐν
κενεῶνι, καὶ ῥίγεα καὶ πυρετοί, καὶ ἀπαλλάσσεται[185]
τὸ ὕστερον, καὶ ὑγιαίνει ἡ γυνή· σήπεται δὲ ἐπὶ τὸ
πολύ· ἀπαλλάσσεται δὲ ἑκταίη ἢ ἑβδομαίη καὶ ἀνωτέρω ἔτι. τῇ τοιαύτῃ διδόναι χρὴ φάρμακα, ὧν ἂν ἐγὼ
γράψω, καὶ πνεῦμα κατέχειν· | ἄριστον δὲ πάντων ἀρ-
τεμισίη βοτάνη, καὶ δίκταμνον, καὶ λευκοΐου ἄνθος·
καὶ ὀπὸς σιλφίου, κράτιστον ἐν ὕδατι πινόμενος ὅσον
κύαμος Ἑλληνικός.

Ἢν τὰ ὕστερα μὴ δύνηται ἀποφεύγειν, ἀσιτεῖν
κἄπειτα πέταλα τῆς ἄγνου τρίψαι ἐν οἴνῳ καὶ μέλιτι
καὶ ἔλαιον ἐπιχεῖν, καὶ χλιήναντα διδόναι πιεῖν ὅσον
κοτύλην, καὶ ἔξεισιν. ἢν γυναικὶ τὸ χορίον ἐλλειφθῇ
ἐν τῇ μήτρῃ, τοῦτο δὲ γίνεται ὧδε, ἢν ῥαγῇ βίῃ[186] ὁ

chia: well boiled or baked garlic in wine and olive oil, together with small octopuses and cuttlefish prepared over hot coals; give her whichever of these she wishes to drink, or some castoreum, spikenard, or rue in sweet dark wine in the fasting state, or the same without wine; if the wine is not sweet, it is better to add honey. Also have her drink boiled cabbage together with rue and mercury herb, and some of the little seeds prescribed for the uterus.

If clots form and there is pain in the lower belly, give boiled leeks and agents both wild and cultivated. Prepare all these in fat. Bathe the patient every three days in good weather, for cold is hostile to such women; after the bath, anoint; it is better not to use very much hot water.

46. If the afterbirth is not expelled immediately after birth, pains will arise in the lower belly and the flanks, along with chills and fever, the afterbirth will be expelled—usually with suppuration—and the woman will recover; the expulsion takes place on the sixth or eighth day, or even later. Such a woman requires medications from among the ones I have recorded, and she must restrain her breathing: best of all are the artemisia plant, dittany, the white violet flower; also silphium juice, best drunk in water to the amount of a green bean.

If a woman is unable to expel her afterbirth, she must fast; then crush chaste tree petals in wine and honey, pour in olive oil, warm, and give a cotyle of this to drink: the afterbirth will come out. If a woman's placenta is left in her uterus, this happens in the following way: if the navel

---

184 πίνειν ἢ Θ: πινέτω· πίνειν δὲ καὶ Μ: πινέτω δὲ V.
185 καὶ ἀπ. Θ: κἢν ἀπαλλάσσηται MV.
186 βίη om. Θ.

ὀμφαλὸς ἢ ἀμαθίῃ ὑποτάμῃ ἡ ὀμφαλητόμος τὸν ὀμφαλὸν τοῦ παιδίου πρόσθεν ἢ τὸ χορίον ἐξιέναι ἐκ
τῶν μητρέων, αἱ μῆτραι ἀνασπῶσι τὸ ὕστερον ἄνω,
ἅτε ὀλισθηρὸν ἐὸν καὶ χεόμενον, καὶ κατίσχουσιν ἐν
ἑωυτῇσι· τέταται γὰρ τὸ χορίον ἐκ τοῦ ὀμφαλοῦ τοῦ
παιδίου, καὶ ὕστερος ἔξεισιν ὁ ὀμφαλὸς ἐκ τῶν μη-
τρέων· ἢν γὰρ πρότερος ἐξίῃ, δι᾽ αὐτοῦ οὐκ ἂν διέλθοι
ἡ τροφὴ τῷ παιδίῳ, κατότι[187] ἐξήρτηται ἐξ αὐτοῦ.

47. Ὅταν δ᾽ ἐν γαστρὶ ἔχουσα διαφθείρῃ τὸ ἔμβρυον μηνιαῖον ἢ διμηνιαῖον, καὶ[188] ἐξιέναι μὴ δύνηται, ᾗ δὲ λεπτή, ταύτης χρὴ αὐτίκα καθῆραι τὸ σῶμα
καὶ πιᾶναι· οὐ γὰρ πρότερον ἔξεισι τὰ ὕστερα[189] σα-
πέντα, ἢν μὴ ἰσχυραὶ αἱ μῆτραι ἔωσι καὶ εὐπαγέες.

48. Ἢν τὸ χορίον τῇ γυναικὶ ἐλλειφθῇ, ἢν ⟨μὴ⟩[190]
αἱ μῆτραι εὐρύστομοι ἔωσιν, χωρέει ἡ κάθαρσις
ἔλασσον τοῦ καιροῦ, καὶ ἡ γαστὴρ σκληρὴ γίνεται
καὶ μεγάλη, καὶ περίψυξις γίνεται, καὶ πυρετός μιν
ὀξύς, καὶ πόνος κατὰ πᾶν τὸ σῶμα, γαστρὸς δὲ τὸ
κατώτερον τοῦ ὀμφαλοῦ, καὶ βρῖθος γίνεται ἐν τῇσι
μήτρῃσι καὶ στροφὴ ὡς ἐμβρύου | ἐόντος, καὶ μελε-
δανθεῖσα ἐκβάλλει τὸ χορίον ἐν τάχει σεσηπός, καὶ
ὑγιαίνεται.

49. Ἢν δ᾽ ἐκ τόκου ἡ μήτρη ἑλκωθῇ, ῥόδων ἄνθει
ἰήσασθαι· διακλυζέσθω δὲ στρυφνοῖσιν. ἢν δ᾽ ἑλκωθῇ
τὸ στόμα καὶ φλεγμήνῃ, σμύρναν καὶ στέαρ χηνὸς
καὶ κηρὸν λευκὸν καὶ λιβανωτὸν λαγῴης θριξὶ τῇσιν ὑπὸ τὴν γαστέρα μίσγειν, καὶ προστιθέναι ἐν
εἰρίῳ λεῖα ποιέοντα.

108

is torn as the result of violence, or the midwife through ignorance trims the child's umbilical cord before the placenta has been expelled from the uterus, the uterus takes the afterbirth, which is slippery and moist, up, and retains it inside itself. For the placenta is suspended from the child's navel, and the cord separates later from the uterus, since if it were to separate earlier, no nutriment would come through it to the child suspended from it.

47. When a pregnant woman aborts her fetus at one or two months but cannot expel it and she becomes thin, you must immediately clean her body and build up her flesh, for the putrefying afterbirth will not be expelled unless the uterus is strong and robust.

48. If the placenta is left behind in a woman and her cervix is ‹not› wide open, less cleaning will pass than should, the belly will become constipated and enlarged, chills and acute fever will set in along with pain through her whole body but especially in the belly below the navel, there will be a sensation of heaviness in the uterus, and colic will be felt as if a fetus was present. If such a woman is cared for immediately, she soon expels the putrefying placenta and recovers.

49. If the uterus ulcerates after childbirth, heal it with flower of roses and have the woman flush herself with astringent substances. If the mouth of the uterus ulcerates and becomes inflamed, mix myrrh, goose grease, white wax, and frankincense with hairs from the belly of a hare, and after making this smooth apply it in wool as a suppository.

---

187 κατότι Grensemann: καὶ ὅτι codd.
188 Add. τὰ ὕστερα M.   189 ὕστερα Θ: ἔμβρυα MV.
190 μὴ Froben: μὲν codd.

50. Ἢν δ' ἐκ τόκου φλεγμήνωσιν αἱ ὑστέραι, πυρετὸς ἔχει βληχρὸς τὸ σῶμα, καὶ ἀχλύς· ἐκ δὲ τῆς κοιλίης οὐδέκοτε ἐκλείπει·[191] καὶ διψῇ, καὶ τὰ ἰσχία ἀλγέει, καὶ οἰδέει τὴν γαστέρα τὴν νείαιραν ἰσχυρῶς, καὶ ἡ κοιλίη ταράσσεται· ὑποχώρημα δὲ κακὸν ὀζόμενον, καὶ λάζεται τὸ πῦρ σφοδρόν, καὶ ἀποσιτίη,[192] καὶ κατὰ τὸ βρέγμα ὀδύνη, καὶ οὐ δύναται εἰρύσαι τῆς κοιλίης ὁ στόμαχος ποτὰ καὶ σιτία, καὶ ἀδυνατέει πέσσειν· καὶ ἢν μὴ θεραπεύωνται εὐθέως, αἱ πλεῖσται θνήσκουσιν, ἡ κοιλίη δὲ αἰτίη.

Τῆς ἀκτῆς οὖν[193] τὰ φύλλα ὡς ἁπαλώτατα ἐν πυρῶν κριμνοῖσιν ἑψήσας σητανίοισιν, ἀκροχλίερον ῥοφεῖν· καὶ διδόναι μελίκρητον καὶ οἶνον ὑδαρέα, καὶ τὸ ἦτρον καταπλάσαι τοῖσι ψυκτικοῖσι, σῖτον δὲ ὡς ἐλάχιστον προσφέρειν· καὶ τὴν γαστέρα ἵστασθαι, καὶ τὴν κεφαλὴν ἰᾶσθαι, ὑποχόνδριον δὲ καταπλάσσειν.

51. Μητρέων ποτόν, ἢν ἀλγέῃ ἐκ τόκου· ὅταν ἀλγέῃ τὴν ἕδρην ἢ ἄλλο τι, ἀρκεύθου καρπὸν ἢ λίνου σπέρμα καὶ κνίδης τρίβειν καὶ διδόναι. πινέτω, ἢν ἐκ τόκου ἀλγέῃ, ῥητίνην τερμινθίνην καὶ μέλι | καὶ οἶνον χλιερὸν διδόναι ῥυφεῖν, καὶ ἢν φλεγμήνωσιν αἱ μῆτραι, τοῦτο παύσει. ἢν[194] ἀλγέῃ τι τῶν μητρέων τόπον,[195] ἀμυγδάλας τρίψας πικρὰς καὶ ἐλαίης τὰ ἁπαλὰ φύλλα, καὶ κύμινον καὶ δάφνης καρπὸν ἢ τὰ φύλλα, καὶ ἄννησον καὶ ἐρύσιμον καὶ ὀρίγανον καὶ

50. If the uterus becomes inflamed after childbirth, a mild fever occupies the body and there is dimness of vision; from the cavity the fever never remits. There are also thirst, pains in the hips, severe swelling in the lower belly, and disturbance of the lower cavity. The stools are evil-smelling, the fever becomes violent, the patient loses her appetite for food, pain occupies the bregma, and the orifice of the cavity is unable to retain drinks and foods, nor can it digest them. If these women are not treated at once, most of them succumb from a condition of the cavity.

Now boil the tenderest of elder leaves in meal of spring wheat and have the woman ingest this lukewarm; also give melicrat and diluted wine, and apply poultices of cooling agents to the lower abdomen; administer very little food. Also bring the belly to a halt, treat the head, and apply poultices to the hypochondrium.

51. Potion for the uterus, if a woman suffers pain after giving birth. When she suffers pain in her seat or anywhere else, grind berries of the Phoenician juniper or seeds of linen and the stinging nettle, and give this to her to drink. If she suffers pain after giving birth, give resin of the terebinth tree, honey and warm wine to drink: if the uterus becomes inflamed, this will relieve it. If she suffers pain of some kind in the region of the uterus, grind bitter almonds, tender olive leaves, cumin, bay berries or leaves, anise, hedge mustard, marjoram, and soda, mix this

---

191 Add. τὸ πῦρ MV.   192 ἀποσιτίηι Θ: ἀσιτίη M: ἀσιτιεῖ V.   193 οὖν om. MV.

194 Add. δὲ μὴ δὲ MV.

195 τόπον Cordaeus after Calvus' *vulvarum locus* and Cornarius' *uterorum locum*: ποτὸν codd.

λίτρον, μίξας ταῦτα πάντα[196] λεῖα, κολλύρια ποιέειν
μητρέων. ἢν φλεγμήνωσι καὶ ὀδύνη ἔχῃ, ῥόδων
φύλλα καὶ κιννάμωμον καὶ κασίην τρίψας ἐν τῷ αὐτῷ
λεῖα, ἐπιχεῖν νέτωπον, καὶ ποιήσας φθοΐσκους ὅσον
δραχμιαίους, ὀστράκινον κυθρίδιον καινὸν διάπυρον
ποιήσας, περικαθίζειν, καὶ περιστείλας ἱματίοις, θυμιᾶσθαι ἐς τὰς μήτρας· τοῦτο ὀδύνας παύσει.

52. Ἢν δ' ἐκ τόκου αἱ ὑστέραι πονήσωσι, βληχρὸν
ἔχει τὸ πῦρ, ἔνδοθεν δὲ ἡ κοιλίη ἡ νείαιρα πυριφλε-
γέθης ἐστί, καὶ ἐς τὸ ἰσχίον ἐνίοτε ἀποιδέει, καὶ
ὀδύνη ἴσχει τὴν νείαιραν γαστέρα καὶ τοὺς κενεῶνας,
καὶ τὰ ὑποχωρέοντα χολώδεα καὶ κάκοδμα· καὶ ἢν
μὴ[197] σταθῇ ἡ κοιλίη, ἐξαίφνης θνήσκει.

Ὅταν ὧδε ἔχῃ, ψύχειν τὴν κοιλίην, φυλασσόμενον
ὅπως μὴ φρίξῃ· πινέτω δέ, ἢν μὴ ἵστηται, τὸ ἀπὸ τοῦ
κρίμνου ἢ ἄρτου, ἢ ἄλητον· ῥοφεῖν δέ, σίδης οἰνώδεος
τὸν χυλὸν κεράσας ὕδατι, ἐπίπασσε δὲ λέκιθον φακῶν, τοῦτο ἕψειν, καὶ μίσγων φακὸν καὶ κύμινον καὶ
ἅλα καὶ ἔλαιον καὶ ὄξος, τοῦτο διδόναι ῥόφημα ψυ-
112 χρόν, καὶ φακῆν ὀξέην, καὶ ἐπιπίνειν | οἶνον εὐώδεα
μέλανα[198] Πράμνιον· τῶν δὲ ἄλλων σιτίων ἀπέχεσθαι
χρὴ ἔστ' ἂν ᾖ ὁ πυρετός· ἢν δὲ δοκέῃ, καὶ λούσθω·
ἢν δ' ἀσθενὴς ᾖ, πίνειν πάλην ἀλφίτων·[199] ἢν δ' ἀκιδ-
νοτέρη ᾖ, ἐν ὕδατι ψυχρῷ· σιτίον δὲ προσφέρεσθαι

---

[196] μίξας τ. πάντα Θ: ταῦτα μίξας καὶ τρίψας MV.
[197] μὴ om. ΘV.
[198] εὐώδ. μέλ. ΘV: οἰνώδεα M.

smooth, and make a pessary for the uterus. If the uterus becomes inflamed and pains set in, grind rose leaves, cinnamon and cassia fine together, pour on oil of bitter almonds, and make pastilles weighing one drachma; heat a small terra cotta pot very hot, have the patient sit down over this, wrap her in blankets, and fumigate into her uterus. This will relieve pains.

52. If the uterus suffers pain after childbirth, there will be a mild fever but the lower cavity will be blazing hot inside; sometimes there will be a swelling toward a hip. Pains occupy the lower belly and the flanks, and the excreta are bilious and ill-smelling. If such a woman's belly does not stand still, she will suddenly die.

When the case is such, cool the cavity, while at the same time making sure that the patient does not suffer a chill. If the diarrhea does not stop, have her drink a potion made with coarse barley meal, bread or flour. As a soup mix juice of the vinous pomegranate with water, sprinkle lentil flour over this, boil with lentils, cumin, salt, olive oil and vinegar, and give this cold; also give acidic lentil soup, and after that a fragrant dark Pramnian wine. Other foods must be avoided as long as the fever persists. If it seems advisable, have the patient bathe herself. If she is weak, have her take a potion made with the finest barley flour[13]— if she is very feeble, made in cold water. When the fever

[13] Fuchs adds "(in warm water)" here to provide the necessary contrast with the following "in cold water."

199 πάλην ἀλφίτων is attributed by Foes in note 228 to *exemplaria Vaticana*: πάλιν ἀλφίτων ΘΜ: πάλιν ἀμφίτων V.

κοῦφον, ὅ τι μὴ ἰνήσεται,²⁰⁰ ὅταν τὸ πῦρ μεθῇ. ἡ δὲ
νοῦσος ὀξέη τε καὶ θανατώδης.

53. Ἢν δ' φλεγμήνωσιν αἱ ὑστέραι λεχοῖ, πίμπρα-
ται ἡ κοιλίη καὶ μεγάλη γίνεται, καὶ πρὸς τὰ ὑποχόνδρια πνὶξ ἔχει. ὅταν ὧδε ἔχῃ, καταπλάσσειν βρύῳ τῷ
θαλασσίῳ, ὃ ἐπὶ τοὺς ἰχθύας ἐπιβάλλουσι, κόψαι δὲ
ἐν ὅλμῳ καὶ συμμίσγειν ὠμήλυσιν καὶ σποδιὴν
κληματίνην καὶ λίνον φώξαντα, ἀλεῖν δὲ ταῦτα καὶ
ἀναφορύξαι ὄξει καὶ ἐλαίῳ, ποιέειν δὲ οἷον κυκεῶνα
παχύν· ταῦτα ἕψειν ἕως οἷον στέαρ γένηται· τούτῳ
καταπλάσσειν ὡς θερμοτάτῳ, καί, ἢν χρήζῃ, ἐγ-
καθιννύσθω.

54. Ἢν δὲ λεχοῖ φλεγμήνωσιν αἱ ὑστέραι, οἰδέουσι, καὶ ὅταν τὰ λοχεῖα ἐμμείνῃ, κρύβδην γίνονται,²⁰¹ γίνονται δὲ ἐπὴν πυκνωθέωσιν ἀπὸ ψύχεος.
ταύτῃσιν ἢν μὲν ἐπιψύχωνται, ἀλεαίνειν· ἢν δὲ πυρι-
φλεγέες ἔωσι καὶ τὸ ψῦχος ἀφῇ, προσθετὰ ποιέειν ἃ
τῇ φλεγμασίῃ ἐναντιοῦνται, καὶ λούειν, καὶ πυριᾶν,
καὶ φάρμακα προσάγειν, ὧν ἂν ἐγὼ γράψω, ἕλκειν τε
ἀτμίδα ἐς τὸ στόμα καὶ ἐς τὰς ῥῖνας. |

114   55. Ἢν δὲ πνίγωσι, φακοὺς ἕψειν ἐν ὄξει καὶ ἁλὶ
καὶ ὀριγάνῳ πολλῷ, καὶ ὁλκὴν ποιέεσθαι, καὶ τὴν λινόζωστιν ἐσθιέτω, καὶ ἐν τῷ χυλῷ λεπτὸν ἄλητον ῥυφεῖν.

56. Ὅταν δὲ τάχιστα τέκῃ, πρὶν τὴν ὀδύνην ἔχειν,
πρότερον διδόναι τῶν φαρμάκων, ὅσα τὰς ὑστέρας

remits, administer light foods that do not provoke evacuations. The disease is both acute and dangerous.

53. If in a woman who has just given birth the uterus becomes inflamed, her cavity becomes distended and large, and suffocation attacks her hypochondrium. When this happens, apply a poultice of the kind of seaweed they lay over fish, made by crushing it in a mortar. Mix with this bruised meal of raw grain, ashes of vine twigs, and roasted soda: mill these and knead them in vinegar and oil, giving them the consistency of a thick cyceon: boil this until it has the consistency of fat, and then apply it as a poultice, as hot as possible. If necessary, also have the patient take a sitz bath.

54. If in a woman who has just given birth the uterus becomes inflamed, it will swell up, and when the lochia remain inside, they are hidden; this happens when they congeal from the cold. If such women later have a chill, warm them; if they become blazing hot and the cold recedes, make suppositories that will counter the inflammation. Wash, foment, and apply the medications I have prescribed; have patient draw the vapor into her mouth and nostrils.

55. If the uterus provokes suffocation, boil lentils in vinegar, salt, and copious marjoram, and make an inhalation; have the patient eat mercury herb and drink some thin meal in its juice.

56. When a woman has given birth very quickly, before she has pain you must first give medications that prevent

---

200 ἰνήσεται Cordaeus after Cornarius' *qui non evacuet*: ἠνή-σεται Θ: εἰνήσεται M: σινήσεται V.

201 γίνονται Θ: τείνονται MV.

παύει τῆς ὀδύνης· καὶ σιτία διαχωρητικὰ προσφερέσθω. ἢν δὲ ἡ γαστὴρ θερμαίνηται, ὑποκλύσαι ὅτι τάχος.

57. Ἢν δ' αἱ μῆτραι φλέγματος ἐμπλησθῶσι, φῦσα ἐγγίνεται ἐν τῆσιν ὑστέρῃσι, καὶ τὰ ἐπιμήνια προέρχεται ἐλάσσονα, λευκά, φλεγματώδεα· ἔστι δ' ὅτε αἷμα λεπτόν, ἀκραιφνές, ὑμένων ἀνάπλεον, καὶ ἔστιν ᾗσιν[202] κυρκανᾶται, καὶ τρὶς τοῦ μηνὸς ἐπιφαίνεται. καὶ τῷ ἀνδρὶ ὑπὸ τῆς ὑγρότητος οὐκ ἐθέλει μίσγεσθαι, οὐδ'[203] ὀργᾷ τοῦτο δρᾶν, καὶ λεπτὴ γίνεται· ὀδυνᾶται δὲ τὴν νείαιραν γαστέρα καὶ τὰς ἰξύας καὶ τοὺς βουβῶνας· καὶ εἰ δάκνοι τὸ ῥέον καὶ εἰ ἑλκοῖ τὸ ἀμφιδήιον, χρόνιον φάναι τὸ ῥεῦμα. καὶ ἢν[204] πολὺ ἴῃ, φάκιον σὺν ἐλλεβόρῳ δοῦναι ἐμέσαι· ἔπειτα ἐς τὰς ῥῖνας ἐγχέαι, καὶ φάρμακον πῖσαι κάτω· σιτίων δὲ εἰργέσθω δριμέων· ἢν δὲ βαρύνηται καὶ ψύχηται καὶ νάρκα ἔχῃ, γάλα διδόναι καὶ οἶνον εὐώδεα· πινέτω δὲ νῆστις ὑπερικὸν, λίνου σπέρμα, ἐλελίσφακον ἐν οἴνῳ εὐώδει ὑδαρεῖ· καὶ κλύσαι τὰς ὑστέρας τῷ σὺν τῇ
116 τρυγί· καὶ ἢν μὴ εἱλκωμέναι | ἔωσι, διαλείποντα ἡμέρας δύο ἢ τρεῖς, κλύσαι τῷ σὺν τῷ κόκκῳ· μετὰ δὲ τοῦτο, στρυφνοῖσιν· ἢν δὲ εἱλκωμέναι ἔωσι, νίπτεσθαι τῷ ἀπὸ τῆς μυρσίνης καὶ δάφνης ἀφεψήματι, καὶ ἐγχρίεσθω τῷ σὺν τῷ ἀργυρέῳ ἄνθει. ἡ δὲ νοῦσος χαλεπή, καὶ ὀλίγαι ἐκφεύγουσιν.

58. Ἢν δ' αἱ κοτυληδόνες φλέγματος περίπλεαι ἔωσι, τὰ ἐπιμήνια γίνονται ἐλάσσονα, καὶ ἐν γαστρὶ

the uterus from having any; the patient should also take laxative foods. If her belly becomes warm, apply a douche as soon as possible.

57. If the uterus becomes filled with phlegm, wind collects in it and the menses pass in a reduced amount, white and phlegmatic, and also sometimes thin, pure, membranous blood; in some women the blood is mixed, and appears thrice a month. The woman does not want to have intercourse with her husband, on account of the moistness, nor does she have any desire to, and she becomes thin. She suffers pains in her lower belly, loins and groins. If the flux is irritating and ulcerates the ring of the uterus, indicate that the flux will be chronic. If much passes, give a lentil decoction with hellebore as an emetic. Then make an infusion into her nostrils, and give a purgative potion. Have the patient avoid sharp foods. If she feels weighed down and is chilled and numb, give milk and fragrant wine. Have her drink in the fasting state hypericum, linseed, and sage in fragrant, dilute wine; also wash out her uterus with an injection of wine lees. If the uterus is not ulcerated, leave two or three days and douche with an injection of (sc. Cnidian) berries, and after that with astringent substances. If it is ulcerated, wash with a decoction of myrtle and laurel (sc. berries), and have the woman anoint herself internally with flower of silver. The disease is troublesome, and few escape it.

58. If the cotyledons (sc. in a woman's uterus) fill up with phlegm, her menses will decrease in amount and she

---

202 Add. ἀνα- Θ.
203 οὐδ' MV: καὶ Θ.
204 Add. μὴ MV.

ἴσχει,[205] διαφθείρει δέ, ἐπὴν ἰσχυρότερον τὸ ἔμβρυον
γένηται· οὐ γὰρ ῥώννυται, ἀλλ᾽ ὑπορρεῖ. γνοίης δ᾽ ἂν
τῷδε· ὑγρὴ γίνεται, καὶ τὸ ἀπορρέον μυξῶδες καὶ γλί-
σχρον οἷα ἀπὸ κοιλίης φέρεται, καὶ οὐ δάκνει, καὶ
ἐν τοῖσιν ἐπιμηνίοισιν, ἐπὴν παύσηται τοῦ αἵματος
καθαιρομένη, καὶ δύο ἡμέρας καὶ τρεῖς βλένναι ἴασιν
ἐκ τῶν ὑστερέων, καὶ φρίκη ἔχει, καὶ θέρμη οὐκ ὀξέη,
καὶ οὐκ ἐκλείπει.

Ταύτην κλύσαι τῷ ἀπὸ τῶν ὀλύνθων καὶ ὑφ᾽ ὧν
ὕδωρ καθαίρεται, καὶ δὶς καὶ τρίς· ἐπὴν δὲ καθαρθῇ,
στρυφνοῖσι τὸ λοιπόν· προστιθέναι δὲ τὰ μαλθακά,
ὑφ᾽ ὧν καθαίρεται φλέγμα, καὶ πυριᾶν τὰς ὑστέρας
τῷ σὺν τῇ δάφνῃ, καὶ κλύζειν τῷ σὺν τῷ ὄξει, καὶ
θυμιᾶν, ἐπὴν παύσηται τὰ ἐπιμήνια, τοῖς ἀρώμασι·
κἄπειτα ἀσιτέειν χρή, καὶ ἀλουτείτω δὲ καὶ συνευδέτω[206] τῷ ἀνδρί, καὶ σιτία καὶ οἶνον ὀλίγον λαμβάνειν, καὶ ἀλεαίνειν, καὶ ῥήνικας ἀμφὶ τὰ σκέλεα ἑλίσσειν, καὶ ἐλαίῳ ἀλείφειν.

59. Ἢν δὲ ὕδερος ἐν τῇσι μήτρῃσιν γένηται, τὰ
ἐπιμήνια ἐλάσσονα καὶ πονηρὰ γίνεται καὶ προαπο-
λείπει, καὶ ἡ νείαιρα γαστὴρ ἐπανοιδέει, καὶ οἱ μαζοὶ
οἱ στερροί,[207] καὶ τὸ γάλα | πονηρόν, καὶ δοκέει ἐν
γαστρὶ ἔχειν, καὶ τούτοισι γνώσῃ ὅτι ὕδερός ἐστι·
σημαίνει δὲ καὶ ἐν τῷ στόματι τῇσιν ὑστέρῃσι, ψαυ-
ούσῃ γὰρ ἰσχνὸν[208] φαίνεται· καὶ ῥῖγος καὶ πῦρ λαμβάνει. ὅσῳ δ᾽ ἂν ὁ χρόνος πλείων γίνηται, ὀδύνη ἔχει
τὴν νείαιραν γαστέρα καὶ τὰς ἰξύας καὶ τοὺς κενεώ-

118

will become pregnant, but there will be a miscarriage when the fetus becomes larger, since it fails to thrive, and slips away. You will recognize this case as follows: the woman becomes moist and has fluxes which are mucous and sticky like what is carried off from the cavity—although not irritating. In the course of her menses, mucus comes out of her uterus for two or three days after blood is no longer being cleaned; she has chills and experiences a warming that is not acute but also does not remit.

Flush this patient two or three times with a douche made from wild figs to clean water, and follow this by cleaning with astringent medications. Apply a gentle agent to remove phlegm, foment the uterus with laurel, administer a douche made with vinegar, and when the menses end fumigate with aromatic agents. Then have the woman sleep with her husband in the fasting state and without bathing herself. She should also take food and a little wine, warm herself, wrap a sheepskin around her legs, and anoint herself with oil.

59. If dropsy arises in a woman's uterus, her menses decrease in amount, become troublesome, and prematurely cease; her lower belly swells up, her breasts become hard, and her lactation is difficult; she seems to be pregnant. From these (sc. signs) you will recognize that the woman has dropsy; the same is also indicated by the mouth of the uterus, for when she touches it, it seems to be dried up. Chills and fever set in, and as time passes pain occupies the lower belly, loins, flanks, and groins. This disease

---

205 ἐν γαστρὶ ἴσχει Θ: ἢν ἐν γ. ἴσχῃ MV.
206 δὲ κ. συνευδέτω Θ: ξυνίτω δὲ MV.
207 Add. μαλθακοὶ MV. 208 Add. καὶ ὑγρὸν MV.

νας καὶ τοὺς βουβῶνας. αὕτη ἡ νοῦσος ἐκ τρωσμοῦ
γίνεται, καὶ ἐξ ἄλλων δὲ προφασίων, καὶ ὅταν τὰ
ἐπιμήνια κρυφθῇ.

Λούειν πολλῷ καὶ θερμῷ, καὶ χλιάσματα προστιθέναι, ἢν ἡ ὀδύνη ἔχῃ· ἐπὴν δὲ παύσηται, φάρμακον
χρὴ πῖσαι κάτω, καὶ πυριᾶσαι τῷ σὺν τῷ βολίτῳ τὰς
ὑστέρας· ἔπειτα προσθεῖναι τὸ σὺν τῇ κανθαρίδι, καὶ
διαλείπειν δύο ἡμέρας ἢ τρεῖς· καὶ ἢν ῥώμη ἔχῃ, νετώπῳ κλύσαι· καὶ ἢν γαστὴρ λαπαρὴ γίνηται, καὶ οἱ
πυρετοὶ παύωνται καὶ τὰ ἐπιμήνια ἢν ἴῃ[209] κατὰ λόγον, τῷ ἀνδρὶ συγκοιμάσθω, καὶ ἐν τοῖσι προσθέτοισι μεσηγὺ ἡμέρης πινέτω κρήθμου φολιόν, γλυκυσίδης τοὺς μέλανας κόκκους πέντε, ἀκτῆς καρπὸν ἐν
οἴνῳ νῆστις· καὶ τὴν λινόζωστιν ἐσθίειν ὡς πλείστην,
καὶ σκόροδα ὠμὰ καὶ ἐφθά· καὶ τοῖσι μαλθακοῖσι
χρῆσθαι,[210] καὶ πουλύποσι, καὶ τοῖσιν ἄλλοισι μαλ-
θακιοῖσι,[211] θαλασσίοισι μᾶλλον ἢ κρέασι· ἢν δὲ
τέκῃ, ὑγιὴς γίνεται.

60. Ἢν δ᾽ ὕδρωψ γένηται ἐν τῇσι μήτρῃσι, τὰ
ἐπιμήνια ἐλάσσω γίνεται καὶ κακίω καὶ διὰ πλέονος
χρόνου· καὶ κύει δίμηνον ἢ[212] μικρῷ πλείονα· καὶ οἰ-
δέει ἡ κοιλίη, καὶ ἐπικτένιον, καὶ αἱ κνῆμαι, | καὶ
ὀσφῦς· ἐπειδὰν δὲ συχνὸς χρόνος γένηται, καὶ ἔχῃ ἐν
τῇ γαστρί, διαφθείρει καὶ ἐκβάλλει, καὶ ὕδωρ σὺν
αὐτῷ ἐκχεῖται, καὶ αὕτη θνῄσκει ὡς ἐπὶ τὸ πολύ· τὸ
δὲ αἷμα φθείρεται, καὶ ὑδεροῦνται. ταύτην γαλακτο-
ποτέειν, καὶ τῶν μηκώνων πίνειν, ἔστ᾽ ἂν κινεῖσθαι τὸ
ἔμβρυον δύναται· ἔτι δὲ πρὸ τούτου ὡς ἐπὶ τὸ πολὺ

120

can arise subsequent to an abortion, although it can also have other causes such as suppression of the menses.

Wash with copious hot water and apply warm compresses whenever pain is present; when it remits, give a purgative medication to drink, and foment the uterus with cow's excrement; then apply a suppository employing the blister beetle, and wait two or three days. If the woman is strong, douche her with oil of bitter almonds. If her belly becomes soft, her fevers remit, and her menses pass normally, have her sleep with her husband. As a potion administered on the day between, have the patient drink in the fasting state samphire bark, five black peony seeds, and elderberries in wine. Also have her eat a great amount of mercury herb and garlic both raw and boiled; she should employ mild foods, and also octopus and other mollusks—sea foods more than meats. If she gives birth, she will recover.

60. If dropsy develops in a woman's uterus, her menses will decrease in amount and quality, and occur at longer intervals of time. She becomes pregnant for two months or a little longer: her cavity, pubes, calves and lower back swell. As more time passes in her pregnancy, she will have an abortion and expel (sc. the fetus), water will be discharged with it, and the mother generally succumbs, with her becoming dropsical and her blood being spoiled. Have this patient drink milk and a preparation of poppies until the fetus becomes able to move. Generally her fetus is

---

209 ἢν ἴῃ Θ: εἴη M: ἴῃ V.   210 χρῆσθαι Θ: χρήσθω MV.
211 Add. χρήσθω καὶ πουλύποσι καὶ τοῖσι Θ.   212 καὶ
κύει δίμηνον ἢ Littré after *Nature of Women* 35: καὶ ἐλάσσονος
Θ: καὶ καίειν ἐλασσονα ἢ M: καὶ ἢν ἐλάσσονα ἢ V.

διαφθείρεται καὶ ἐξαμβλίσκεται, καὶ αἵματος[213] καὶ
ὕδατος ῥέουσιν αἱ μῆτραι· ταῦτα δὲ πάσχει οὐδέν τι
μᾶλλον ἐκ πόνου ἢ ἄλλως. τούτῳ δ' ἂν γνοίης ὅτι
ὕδρωψ ἐστίν, εἰ ἀφάσσων τῷ δακτύλῳ ὄψει τὸ στόμα
αὐτῶν ἰσχνὸν καὶ περίπλεον ὑγρασίης.

Ἢν δὲ αὕτη τὸ ἔμβρυον μὴ κατ' ἀρχάς, ἀλλ' ἤδη
δίμηνον, διαφθείρηται καὶ ἀποπνίγηται, ἥ τε γαστὴρ
ἡ νείαιρα ἐπανοιδέει, καὶ ἁπτομένη ἀλγέει ὡς ἕλκος,
καὶ πυρετὸς μέγας αὐτὴν καὶ βρυγμὸς λαμβάνει, καὶ
ὀδύνη ἰσχυρὴ τοῦ αἰδοίου, καὶ τὴν νείαιραν γαστέρα
καὶ τὰς ἰξύας, καὶ τοὺς κενεῶνας καὶ τὴν ὀσφὺν ὀξέη
τε καὶ σπερχνή. ὅταν οὕτως ἔχῃ, λούειν αὐτὴν θερμῷ,
ἢν ἡ ὀδύνη ἔχῃ, καὶ χλιάσματα προσάγειν, πειρώμενος ὅ τι ἂν μάλιστα προσδέχηται, καὶ φάρμακον
καθαρτήριον κάτω· διαλιπεῖν δὲ χρόνον, ὅσον αὐτῇ
δοκέει ἱκανὸς[214] εἶναι, καὶ κλύσαι, καὶ πυριᾶν· τῆς κυκλαμίνου ἐν ῥάκει μέλιτι δεύων προσθεῖναι πρὸς τὸ
στόμα τῶν μητρέων· καὶ τῆς κυπαρίσσου καταξύ-
σας[215] καὶ τέγξας ἐν ὕδατι, προσθεῖναι ὡσαύτως,
122 ἐλάσσω δὲ| χρόνον καὶ διὰ πλείονος τοῦτο, ὅσῳ μᾶλ-
λον δάκνει καὶ ξαίνει· καὶ μήλην ποιησάμενος κασ-
σιτερίνην ἐγκαθιέναι, καὶ τῷ δακτύλῳ ὡσαύτως, καὶ
τὰ ποτήματα ὅ τι ἂν μάλιστα προσίηται πιπίσκειν,
καὶ συγκοιμάσθω ἀνδρὶ ὡς μάλιστα τῶν καιρῶν παρ-
όντων· ἢν γὰρ συλλάβῃ τὴν γονὴν καὶ κυήσῃ, ὑπεκ-
καθαίρεται καὶ τὰ πρόσθεν ὑπόντα σὺν αὐτοῖς, καὶ
οὕτως ἂν μάλιστα ὑγιὴς γένοιτο.

124

aborted and miscarries before this, and blood and water run out of her uterus. A woman does not suffer this any more after exertion than otherwise. You will know dropsy is present, if on examining with a finger you find the mouth of the uterus to be shrunken and very full of moisture.

If such a patient aborts her fetus not at the beginning but after two months, and she suffers from suffocation, her lower belly will swell, and on being touched (sc. on the uterus) she will feel pain as if in an ulcer; great fever and chattering (sc. of the teeth) set in, and there is violent pain of the genitalia, and in the lower belly, loins, flanks, and sacrum—acute and violent. When the case is such, bathe the patient in hot water if the pain is present, apply warm compresses you know by experience to be the most acceptable, and administer a potion that will clean downward. Leave an interval of time you think is sufficient for the woman, and then give her a douche and a fomentation: smear cyclamen in a piece of cloth with honey, and apply this against the mouth of the uterus. Also shred some cypress wood, soak it in water, and apply in the same way, but for a shorter time and at longer intervals, since this might irritate more and cause lacerations. Fashion a tin probe and insert it, and also a finger in the same way. Also have the patient take the kind of drinks most suitable for her, and sleep very frequently with her husband at the propitious times. For if she retains his seed and becomes pregnant, what was present before will be cleaned out along with the birth, and in this way her chance of recovery is greatest.

---

213 καὶ αἵματος om. MV, add. M in marg.
214 ἱκανὸς om. Θ.   215 κ. MV: καταψύξας Θ.

125

61. Ἢν δὲ γυνὴ ὑδρωπιήσῃ, οἷα τοῦ σπληνὸς ὑδα
τώδεος καὶ μεγάλου οἱ ἐόντος, γίνεται δὲ ὁ σπλὴν
ὑδατώδης ἀπὸ τοῦδε τοῦ παθήματος, ἐπὴν πῦρ ἔχῃ
καὶ μὴ[216] ἀφίῃ τὴν ἄνθρωπον, καὶ δίψα μιν λαμβάνῃ
καρτερή, καὶ πίνῃ, καὶ μὴ ἀπεμῇ· τὸ μὲν[217] ἐς τὴν
κύστιν διελθὸν διουρεῖται, τὸ λοιπὸν δὲ ὁ σπλὴν ἕλκει
ἐς ἑωυτὸν ἀπὸ τῆς κοιλίης ἅτε ἀραιὸς ἐὼν καὶ σπογ
γοειδὴς κείμενός τε κατὰ τὴν κοιλίην· καὶ ἢν τούτων
οὕτω γινομένων μὴ ἱδρώῃ, μηδ᾽ οἱ ἡ κύστις διηθῇ,
μηδὲ ἡ κοιλίη χαλᾷ, διαίρεται[218] ὁ σπλὴν ὑπὸ τοῦ
ποτοῦ,[219] καὶ μᾶλλον ἢν ὕδωρ ᾖ τὸ ποτόν, καί μιν ἢν
τις ἐπαφήσαιτο, μαλθακὸς ὡς μνοῦς ἐστιν,[220] ἔστι δ᾽
ὅτε ἀντιτυπεύμενος· ἀερθεὶς δὲ καὶ ὑπερπιμπλάμενος
ἐκδιδοῖ κατὰ φλέβας τῷ σώματι, καὶ μάλιστα ἐς τὸ
ἐπίπλοιον καὶ τοῖσιν ἀμφὶ τὴν κοιλίην ἐοῦσι χωρίοισι
καὶ τοῖσι σκέλεσιν· ἕτερον γὰρ ἑτέρῳ διεκδιδοῖ ἐν τῷ
σώματι, καὶ μάλιστα[221] ἐπὴν πλέον ἑκάστῳ τοῦ και
ροῦ ᾖ καὶ μὴ δύναται κατέχειν. ἐπιγίνεται δὲ ἀπὸ τοῦ
<πο>τοῦ[222] ὕδρωψ αἰεί, ἐπὴν μάθῃ ὁ σπλὴν ἕλκειν ἐς
ἑωυτὸν [καὶ][223] φύσει ἀραιὸς ὢν[224] καὶ μανός.

124   Γίνεται δ᾽ ἔστιν ᾗσιν ἀρχὴ τῆς νούσου αὕτη | καὶ
ἄτερ πυρετοῦ, ἢν καῦμα ἐν τῇ κοιλίῃ ἐνστῇ οἷα[225]
φλέγματος ἐς αὐτὴν κατελθόντος, καὶ ἢν ἡ ἄνθρωπος
τὴν δίψαν μὴ κατέχῃ, μηδ᾽ ἡ κύστις μηδ᾽ ἡ κοιλίη
διηθέωσιν οὖρόν τε καὶ κόπριον κατά γε δίκην, μηδὲ
ἐπιτηδείῃ διαίτῃ χρέηται ἡ ἄνθρωπος.

---

[216] μὴ om. Θ.   [217] Add. γὰρ MV.

61. If a woman suffers from dropsy because her spleen is full of water and enlarged, the spleen fills with water in this disease when heat seizes her without remitting, she has violent thirst, she drinks, and she does not vomit. What passes through into the bladder is excreted as urine, but the rest of what comes from the cavity the spleen draws to itself on account of its porosity, sponginess, and location next to the cavity. If while this is taking place a woman fails either to sweat, or to excrete urine from her bladder, or to empty her cavity, her spleen will fill up from what she drinks—especially if it is water—and if someone palpates it, it will feel soft like down, although sometimes it will be firm. Because the spleen is raised and overfilled, it will empty itself through the vessels into the body, in particular to the omentum, the region of the cavity, and the legs, since any part of the body tends to pass fluid on to another one, especially when it has more than it should and it cannot retain it. Dropsy always arises after drinking when the spleen has become accustomed to attract fluid to itself on account of its porous and empty texture.

In some women, the beginning of the disease is even without fever, if heat is present in the cavity due to phlegm descending into it, if the patient fails to restrain her thirst and neither her bladder nor her cavity excretes urine and stools as they should, and if she follows an unsuitable regimen.

---

[218] χ. δ. Θ: χαλᾶται αἴρεται MV.   [219] τοῦ ποτοῦ Θ: τούτου MV.   [220] ἐστιν om. ΘV.   [221] καὶ μάλιστα om. MV.
[222] τοῦ ποτοῦ Linden after Cordaeus' ποτοῦ: τούτου Θ: τοῦ MV.   [223] Del. I.   [224] ὦν I: ἦ ΘMV.
[225] οἷα om. Θ.

Ἢν δὲ ὑδρωπιώδης ᾖ, ἔρχεται τὰ καταμήνια πολλὰ
ἐξαπίνης, ὁτὲ δὲ ὀλίγιστα, καὶ γίνονται ὁτὲ μὲν ὡς
ἀπὸ κρεῶν ὕδωρ, ὡς εἴ τις αἱματώδεα ἀποπλύνειεν,
ὁτὲ δὲ ὀλίγῳ παχύτερα,[226] καὶ οὐ πήγνυται, καὶ ἆσθμά
μιν λαμβάνει, πρὶν ἢ τὰ καταμήνια χωρέειν, καὶ
ὀδύνη ἐν τῷ σπληνί, καὶ μᾶλλον ἐπὴν τι γλυκὺ φάγῃ,
καὶ ἡ γαστὴρ ἐξαίρεται καὶ μεγάλη ἐστί· καὶ ἐπὴν
πλέονα τοῦ ἔθεος[227] φάγῃ, πονέεται τὴν γαστέρα, καὶ
τὴν ὀσφῦν ἀλγέει ἄλλοτε καὶ ἄλλοτε, καὶ πῦρ μιν
ἐπιλαμβάνει δι᾽ ὀλίγου. ἐπὴν δὲ ἀποκαθαρθῇ, ῥᾷον
δοκέει ἔχειν πρὸς τὰ πρόσθεν, ἔπειτα ἐς τωὐτὸ καθίσταται, καὶ ἐπειδὰν[228] μελεδαίνηται ὡς χρή, ὑγιὴς
ἔσται· ἢν δὲ μή, ὁ ῥόος ἐπιφανεῖται, καὶ διὰ παντὸς
τοῦ χρόνου ἀεὶ ῥεύσεται κατ᾽ ὀλίγον οἷον ἰχώρ, ἐπι-
μελείης δὲ δέεται.

Ἢν δὲ μὴ ὁ ῥόος ἐπιγένηται, ἀλλ᾽ αἱ μῆτραι ὑπὸ
τῶν πρόσθεν παθημάτων ἀερθεῖσαι[229] μὴ χαλάσωσι
τὰ ἐπιμήνια, ἥ τε γαστήρ οἱ μεγάλη ἔσται, καὶ βρῖθος ἐνέσται ὡς ἐν γαστρὶ ἐχούσῃ, καὶ δοκέει ὡσεὶ
παιδίον ἐν τῇ γαστρὶ κινέεσθαι, ἅτε τῶν μητρέων[230]
πλέων ἐουσέων καὶ τοῦ ὕδατος κινεομένου, ἄλλοτε
γὰρ καὶ ἄλλοτε κλυδάσσεται τὸ ὕδωρ ὡς ἐν ἀσκῷ·
καὶ ἀλγήσει ψαυομένη τὸ κάτω τοῦ ὀμφαλοῦ, καὶ αἱ
κληῖδες καὶ ὁ θώρηξ καὶ τὸ πρόσωπον καὶ τὰ ὄμματα
καταλεπτύνονται, καὶ αἱ θηλαὶ ἀείρονται. ἔστι δὲ ᾗσι
μὲν ἥ τε κοιλίη καὶ τὰ σκέλεα πλήθει τοῦ ὕδατος, ᾗσι
126   δὲ θάτερον τούτων· | καὶ ἢν μὲν ἄμφω πλησθῇ, οὐδε-

If a woman suffers from dropsy, she suddenly passes copious menses—although sometimes they are scanty and sometimes they are like the water that runs out of meat as if somewhere were flushing out sanguineous residues, although at other times they are slightly thicker and fail to clot—she has shortness of breath before they pass, she suffers pain in her spleen especially whenever she eats anything sweet, and her belly becomes raised and large. Whenever the woman eats more than she is accustomed to, her belly becomes painful, her lower back aches intermittently, and heat comes over her at brief intervals. After her cleaning has taken place, she seems to be easier than before, but then the same things return; when she receives the necessary treatment, she recovers. If untreated, her flux will continue to appear over the whole time, continually flowing a little at a time like serum, and requiring care.

If no flux appears, but the uterus fails to release the menses even though it is swollen by the diseases above, the woman's belly becomes large, she feels a weight in it like a woman who is pregnant, and a fetus seems to be moving there because the uterus is full of water and the water is moving, since it undulates from time to time like it would in a wineskin. If the patient is touched in the region below her navel, she feels pain; her clavicles, chest, face and eyes become emaciated, and her nipples protrude. In some women, both the cavity and the legs fill up with water, while in others only one or the other: if both

---

226 π. Θ: ἰσχυρότερα MV.   227 ἔθεος Θ: μάθεος MV.
228 ἐπειδὰν Θ: ἦν MV.
229 ἀερθεῖσαι Littré: ἀείρειται Θ: om. MV.
230 Add. ὕδατος MV.

μία ἐλπὶς περιγενέσθαι ἐστίν· ἢν δὲ[231] θάτερον τούτων,[232] ἐλπίδες ὀλίγαι, ἢν μελεδαίνηται καὶ μὴ λίην
τετρυχωμένη ᾖ. χρονίη δὲ ἡ νοῦσος αὕτη.

62. Γίνεται δὲ πάντα μᾶλλον μὲν τῇσιν ἀτόκοισι,
γίνεται δὲ καὶ τῇσι τετοκυίῃσιν· ἐπικίνδυνα δ' ἐστίν,
ὡς εἴρηται, καὶ τὸ πολὺ ὀξέα καὶ μεγάλα καὶ χαλεπὰ
συνιέναι, διὰ τοῦθ' ὅτι αἱ γυναῖκες μετέχουσι τῶν
νούσων, καὶ ἔσθ' ὅτε οὐδ' αὐταὶ ἴσασι τί νοσέουσι,
πρὶν ἢ ἔμπειροι νούσων γένωνται ἀπὸ καταμηνίων
καὶ ἔωσι γεραίτεραι· τότε δὲ σφέας ἥ τε ἀνάγκη καὶ
ὁ χρόνος διδάσκει τὸ αἴτιον τῶν νούσων, καὶ ἔστιν
ὅτε τῇσι μὴ γινωσκούσῃσιν ὑφ' ὅτευ νοσέουσι φθά-
νει τὰ νοσήματα ἀνίητα γινόμενα, πρὶν ἢ διδαχθῆναι
τὸν ἰητρὸν ὀρθῶς ὑπὸ τῆς νοσεούσης ὑφ' ὅτου νοσέει·
καὶ γὰρ αἰδέονται φράζειν, κἢν εἰδῶσι, καί σφιν δοκέουσιν αἰσχρὸν εἶναι ὑπὸ ἀπειρίης καὶ ἀνεπιστημο-
σύνης. ἅμα δὲ καὶ οἱ ἰητροὶ ἁμαρτάνουσιν, οὐκ ἀτρεκέως πυνθανόμενοι τὴν πρόφασιν τῆς νούσου, ἀλλ᾽
ὡς τὰ ἀνδρικὰ νοσήματα ἰώμενοι· καὶ πολλὰς εἶδον
διεφθαρμένας ἤδη ὑπὸ τοιούτων παθημάτων. ἀλλὰ
χρὴ ἀνερωτᾶν αὐτίκα ἀτρεκέως τὸ αἴτιον· διαφέρει
γὰρ ἡ ἴησις πολλὸν τῶν γυναικείων νοσημάτων καὶ
τῶν ἀνδρείων.

63. Ἢν δ' αἱ μῆτραι ἑλκωθέωσιν, αἷμα καὶ πῦα
καθαίρεται, καὶ ὀσμὴ βαρέη γίνεται, καὶ ὀδύνη ὀξέη
λαμβάνει ἐς τὰς ἰξύας καὶ ἐς τοὺς βουβῶνας καὶ ἐς
τὴν νείαιραν γαστέρα, καὶ ἄνω φοιτᾷ ἡ ὀδύνη ἐς τοὺς
128 κενεῶνας καὶ τὰς πλευρὰς καὶ ὠμοπλάτας, | ἐνίοτε δὲ

fill up, there is no hope of survival, while if only one there is a small hope if she is cared for and is not too severely wasted. The disease persists for a long time.

62. All these (sc. conditions) are more likely to occur in women who have not borne children, although they also happen in those who have. They are, as has been indicated, dangerous and in most cases acute, serious, and difficult to recognize, since they are occurring in women who sometimes only grasp themselves what their disease is when they have become familiar with the disorders that arise from menstruation, and are older: by then, both the necessary sequence of events and time itself have taught them the cause of these diseases. Sometimes in women who do not know the source of their illness, diseases have become incurable before the physician learned correctly from a patient the origin of her disease. Besides, women may be ashamed to speak out, even if they know, since the matter seems shameful to them, due their inexperience and ignorance. Furthermore, physicians too may err in not inquiring carefully about a disease's cause, and in treating them like diseases in men: indeed, I have seen many women perish in such cases. Rather you must question a patient immediately and in detail about the cause; for there is a great difference in the treatment of women's diseases and those of men.

63. If the uterus forms ulcers, blood and pus are discharged, an offensive smell develops, and acute pain occupies the loins, groins, and lower belly. Then pain shoots up to the flanks, sides, shoulder blades and sometimes the

---

231 ἢν δὲ Littré: om. Θ: δὲ ἢν MV.
232 καὶ ἢν μὲν ἄμφω . . . τούτων om. Θ.

καὶ ἐς τὰς κληῖδας ἀφικνεῖται, καὶ δάκνεται, καὶ κεφαλὴν ἀλγέει σφοδρῶς, καὶ παρανοεῖ· τῷ δὲ χρόνῳ
οἰδίσκεται πᾶσα, καὶ ἀσθενείη μιν λαμβάνει, καὶ
ἀψυχίη, καὶ πυρετὸς λεπτός, καὶ περίψυξις· οἰδίσκεται δὲ μάλιστα τὰ σκέλεα. ἡ δὲ νοῦσος λαμβάνει ἐκ
τρωσμοῦ, ἥτις ἂν διαφθείρασα τὸ παιδίον ἐγκατασα-
πὲν μὴ ἐκκαθαρθῇ, καὶ τὸ στόμα πῦρ ἔχει· λαμβάνει
δὲ καὶ ἐκ τῶν ῥόων, κἢν ἐπὶ σφέας δριμέα ᾖ καὶ χολώδεα, δάκνει.

Ἢν δὲ οὕτως ἐχούσῃ ἐπιτυγχάνῃς, ἐπὴν μὲν αἱ
ὀδύναι ἔχωσι, λούειν τε πολλῷ καὶ θερμῷ, καὶ χλιάσματα προστιθέναι, ὅπου ἂν ἡ ὀδύνη ἔχῃ· καὶ ἢν
μὲν ἄνω ὀδύναι ἔωσιν, ἢν μὲν ἰσχυρὴ ἡ γυνὴ ᾖ, πυ-
ριάσας ὅλην φάρμακον δοῦναι κάτω πιεῖν· καὶ ἐπὴν
ὥρη ᾖ τοῦ ἔτεος, ὀρὸν ἀφέψων, διδόναι ἐπιπίνειν ἡμέρας πέντε, ἢν δυνατὴ ᾖ· ἢν δὲ μὴ ᾖ ὁ ὀρός, ὄνου γάλα
ἀνέψειν, καὶ διδόναι πίνειν ἡμέρας τρεῖς ἢ τέσσερας·
μετὰ δὲ τὴν γαλακτοποσίην, ὕδατι ἀνακομίσαι αὐτὴν
καὶ σιτίοισιν ἐπιτηδείοισι, κρέασι μηλείοισιν, ἁπα-
λοῖσι, νέοισι, καὶ ὀρνιθείοισι, καὶ τευτλίῳ, καὶ κολο-
κύντῃ· ἀπεχέσθω δὲ καὶ ἁλμυρῶν καὶ δριμέων καὶ
τῶν θαλασσίων πάντων καὶ κρεῶν χοιρείων καὶ βοείων καὶ οἰείων·[233] ἄρτους δὲ ἐσθιέτω· ἢν δὲ ἀψυχίαι
ἔχωσι καὶ μὴ ἰσχύῃ καὶ περιψύχηται, ῥόφημα λαμβάνειν. εἰσὶ δέ τινες, οἳ ταύτῃσι κεφαλὴν ἀλγεούσῃσι
πιπίσκουσι γάλα, ὅτι κεφαλαλγέουσιν, οἱ δὲ ὕδωρ,
ὅτι λιποθυμέουσι· τἀναντία δὲ οἴομαι· εἰ κεφαλὴν ἀλ-

collarbones. The patient is irritated, she has an intense headache, and she becomes delirious. With time she swells through her whole body—but mainly in her legs—becomes weak, and has syncope, a mild fever, and violent chills. This disease arises after a miscarriage when a woman with a putrefying aborted fetus inside her is not cleaned and heat occupies the mouth (sc. of her uterus). It can also arise after fluxes; and if sharp and bilious matter is present, this irritates.

If you find a woman is this state, when the pains are present wash her in copious hot water, and apply warm compresses wherever the pain is located. If the pains are in the upper part of the woman's body and she is strong, foment her whole body and give her a purgative medication to drink. If it is the right season, then boil off whey and give it to the woman to drink for five days if she can accept it; if there is no whey, boil up ass's milk and give this to drink for three or four days. After the milk drink, build the patient up with water and appropriate foods: meat from tender young sheep or fowl, beets, or gourds. She should avoid salty and sharp foods, all sea foods, and pork, beef, and mutton, and eat breads. If she becomes faint and weak and has chills, choose a gruel. Some physicians have such women drink milk against any headache that is present, and others have them take water if they are fainting, but I think the opposite: if a patient has a head-

---

233 οειων Θ: αἰγείων MV.

γέοιεν καὶ φρενῶν εἴη ἄψις, ἁρμόσει ὕδωρ, ὅτε δὲ δάκνεται καὶ δριμέα ἐστί, γάλα ταύτῃσιν εὐμενές.

Ἐπὴν δέ σοι δοκέῃ ἰσχύειν, κλύζειν τὰς ὑστέρας, πρῶτα μὲν τῷ ἀπὸ τῆς τρυγός, μετὰ δὲ ταῦτα διαλεί130 πων | ἡμέρας τρεῖς ἢ τέσσερας κλύζειν τῷ σὺν τῷ κραμβίῳ χλιερῷ, καὶ αὖτις διαλείπων ἡμέρας τρεῖς κλῦσαι τῷ σὺν τῷ πικερίῳ· καὶ ἢν ταῦτα ποιήσαντι ὑγιάζωνται αἱ ὑστέραι, κλῦσαι τῷ σὺν τῷ σιδίῳ.

Ἐπὶ δὲ τὰ ἕλκεα ἐπιχρίειν ἀργύρεον ἄνθος, καὶ κηκίδα, καὶ σμύρναν, καὶ λιβανωτόν, καὶ τοῦ Αἰγυπτίου ἀκάνθου τὸν καρπόν, καὶ οἰνάνθην τὴν ἀγρίην, καὶ χρυσόκολλαν, καὶ λεπίδα, καὶ λωτοῦ πρίσμα, καὶ κρόκον, καὶ στυπτηρίην Αἰγυπτίην κατακεκαυμένην· τούτων ἐν ἔστω ἴσον ἑκάστου, τὴν δὲ στυπτηρίην καὶ τὴν κηκίδα καὶ τὸν κρόκον ποιῆσαι μίαν μοῖραν πάντων· τρίψας δὲ καὶ συμμίξας πάντα λεῖα, διεῖναι ἐν οἴνῳ λευκῷ γλυκεῖ· ἔπειτα ἕψειν, ἔστ᾽ ἂν παχὺ γένηται οἷόν περ μέλι· τούτῳ δὲ ἐπάλειφε δὶς τῆς ἡμέρης, νιψαμένην ὕδατι χλιερῷ· ἕψειν δ᾽ ἐν τῷ ὕδατι καὶ κισθὸν καὶ ἐλελίσφακον.

Καὶ ἐπήν σοι δοκέῃ ταῦτα ποιέοντι ῥάων εἶναι, προπιεῖν χρὴ[234] γάλα ἐφθὸν αἴγειον μίαν ἡμέρην, διδόναι δὲ καὶ[235] γάλα βόειον, τὸν αὐτὸν τρόπον οἷόνπερ ἐπὶ τῇσι προτέρῃσι· μετὰ δὲ τὴν γαλακτοποσίην παχῦναι σιτίοισιν ὡς μάλιστα, καὶ ποιέειν ὅπως ἐν γαστρὶ ἔχῃ· ὑγιὴς γὰρ γενήσεται. ὡς δὲ τὰ πολλὰ ἐκ τῆς νούσου ταύτης ἐκφεύγουσι, καὶ ἄτοκοι γίνονται· αἱ δὲ γεραίτεραι οὐ πάνυ τι. πίνειν δὲ μετὰ τὰ φάρ-

ache and a diminution of her consciousness, water is appropriate, but if she experiences irritation and a bitter taste, milk is beneficial for these.

When you think the woman is gaining strength, first irrigate her uterus with a douche made from wine lees, then leave her at peace for three or four days; next irrigate with warm cabbage juice and again leave three days; after that irrigate with butter. If, when you have done this, the uterus heals, irrigate with pomegranate peel.

On to the ulcers anoint flower of silver, oak gall, myrrh, frankincense, acacia fruit, wild grapevine blossoms, a dish of linseed and honey, copper scales, sawdust of nettle-tree wood, saffron, and burned Egyptian alum: let there be one measure of each of these except for making one sole measure of the alum, oak gall, and saffron together; grind and mix all these smoothly together and then dissolve them in sweet white wine. Boil this until it becomes thick like honey, and anoint it (sc. on to the ulcers) twice daily after washing the patient thoroughly with warm water. Also boil rockrose and salvia in water.

When your management seems to be leading to an improvement in the patient's condition, have her first drink boiled goat's milk for one day and then take cow's milk the same way as the patients above. After the milk drink, build her up as much as possible with foods, and have her become pregnant, which will lead to her recovery. In most cases women recover from this disease but become barren, although older women less so. After the

---

234 χρὴ om. MV.
235 διδόναι δὲ καὶ Θ: ἔπειτα διδόναι MV.

135

μακα λίνου σπέρμα πεφωγμένον, καὶ σήσαμον, καὶ
κνίδης καρπόν, καὶ γλυκυσίδης ῥίζαν τὴν πικρὴν τρίβων ἐν οἴνῳ οἰνώδει²³⁶ μέλανι κεκρημένῳ.

64. Ἢν δ' αἱ μῆτραι ἑλκωθῶσι, καὶ αἷμα καὶ πύος
ῥέῃ καὶ ἰχώρ, σηπομένων γὰρ τῶν μητρέων, νόσημα
ἀπ' αὐτῶν γίνεται, καὶ ἡ γαστὴρ ἡ νείαιρα ἐπαείρε-
132 ται, καὶ λεπτὴ γίνεται, καὶ ἀλγέει ψαυομένη, | ὡς ἕλ-
κος, καὶ πῦρ ἔχει καὶ βρυγμὸς αὐτήν, καὶ ὀδύνη ὀξέη
καὶ σπερχνὴ ἔς τε τὰ αἰδοῖα καὶ ἐς τὸ ἐπίσιον καὶ ἐς
τὴν γαστέρα τὴν νείαιραν καὶ ἐς τὸν κενεῶνα καὶ ἐς
τὰς ἰξύας· ἡ δὲ νοῦσος λαμβάνει μάλιστα μὲν ἐκ τόκου, ἢν αὐτῇ τι διακναισθὲν ἐνσαπῇ, ἢ καὶ ἐκ τρωσμοῦ, καὶ ἄλλως ἀπὸ ταὐτομάτου.

Ταύτης ἢν ἐπιτυγχάνῃς, λούειν ἐν ὕδατι θερμῷ καὶ
πολλῷ, καὶ ὅπου ἂν ἡ ὀδύνη λαμβάνῃ, τὰ²³⁷ χλιάσματα προστιθέναι καὶ σπόγγιον ἐξ ὕδατος,²³⁸ καὶ
κλύζειν, τὰ δριμέα καὶ τὰ στρυφνὰ ἀπεχομένη· τῶν
δὲ μαλακωτέρων διάμισγε ὅπως ἂν δοκέῃ σοι καιρὸς
εἶναι· τοῦ δὲ λίνου τὸν καρπὸν κόψας καὶ τῆς ἀκτῆς,
συμμίξον δὲ μέλιτι καὶ ποίησον φάρμακον· ἔπειτα
λοῦσον ὕδατι θερμῷ, καὶ σπόγγον λαβὼν ἢ εἴριον
μαλθακὸν βάπτων ἐς θερμὸν ὕδωρ διακάθαιρε τά τε
αἰδοῖα καὶ τὰ ἕλκεα· εἶτα ἐς οἶνον ἄκρητον ἐμβάπτων
πάλιν τὸν σπόγγον ἢ τὸ εἴριον τὸν αὐτὸν τρόπον·
ἔπειτα τῷ φαρμάκῳ τούτῳ ἐνάλειφε, ὁποσάκις ἂν δοκέῃ σοι καιρὸς εἶναι· ἔπειτα ῥητίνην καὶ στέαρ ὕειον
μίξας ὁμοῦ τῷ φαρμάκῳ, διαλείφειν τῷ δακτύλῳ πολλάκις μεθ' ἡμέρην καὶ τῆς νυκτός· μετὰ δὲ λίνου

medications, give a potion of roasted linseed, sesame, stinging nettle seed, and bitter peony root ground into diluted strong dark wine.

64. If the uterus forms ulcers, and blood, pus and serum are discharged due to the uterus's suppuration, a disease arises from this: the lower belly becomes raised, the woman becomes emaciated, on being touched she feels pain as if from an ulcer, fever and chattering of the teeth set in, and acute, violent pain invades her genitalia, pubes, lower belly, flanks and loins. The disease usually arises after childbirth if the patient suffers some laceration that suppurates in her uterus, or after an abortion, or otherwise spontaneously.

If you meet a case like this wash with copious hot water, and wherever there are pains apply warm compresses and a sponge soaked in water and squeezed out; apply an enema without sharp or astringent components, using instead a mixture of appropriate milder ones. Pound linseed and elderberries together, mix them with honey, and make into a medication. Then wash with hot water, and with a sponge or a piece of soft wool you have immersed in hot water carefully clean the genitalia and the ulcers; next immerse the sponge or piece of wool again in unmixed wine in the same way, and anoint with this as many times as you think right. Then mix resin and lard into the medication, and anoint it with a finger many times during the day and night. Afterward, toast, pound and sift linseed,

---

236 οἶν. Θ: εὐώδει MV.
237 τὰ Θ: τοιαῦτα τὰ MV.
238 Add. θερμοῦ MV.

σπέρμα φώξας καὶ κόψας καὶ σήσας, μήκωνα λευκὴν
κόψαι ἐν ἀλφίτοισι καὶ σῆσαι, καὶ τυρὸν αἴγειον
ὀπτᾶν περιξύσας τὸ ῥυπόεν καὶ τὴν ἄλμην, καὶ τὸ
πικέριον συμμίξαι καὶ πάλην ἀλφίτου, εἶτα ἐν τοῦ
φαρμάκου μέτρον καὶ τοῦ τυροῦ καὶ τοῦ ἀλφίτου
ποιῆσαι, καὶ[239] τοῦτο διδόναι πίνειν ἐξ ἠοῦς νήστει ἐπ'
οἴνῳ αὐστηρῷ κεκρημένῳ· ἐς δὲ τὴν ἑσπέρην μίσγων
καὶ παχὺν κυκεῶνα διδόναι, καὶ τῶν γυναικείων ὅ τι
ἂν μάλιστα προσδέχηται πιπίσκειν.

134   Καὶ μέχρι μὲν ἂν τὸ αἷμα πολὺ | ῥέῃ, καὶ ὀδύναι
ὀξέαι ἔχωσι καὶ ὀλίγον χρόνον διαλείπωσι, τοιαῦτα
ποιέειν· ἐπὴν δὲ τὸ ἕλκος ἔλασσον ᾖ ἄνω,[240] καὶ ὀδύναι βληχραὶ λαμβάνωσι κάτω[241] καὶ διὰ πλέονος
χρόνου, φάρμακα πιπίσκειν, ὑφ' ὧν μέλλει καθαίρεσθαι κάτω μᾶλλον ἢ ἄνω, διαλείπων χρόνον ὁπόσον
ἂν δοκέῃ καιρὸς εἶναι· καὶ πυριὴν βληχρῇσι πυρίη-
σιν, ἀνακαθίζοντα ὑψόθι, ἢν δοκέῃ ἑκάστοτε καιρὸς
εἶναι.

Ταῦτα ποιέουσα ὑγιὴς γίνεται· ἡ δὲ νοῦσος βληχροτέρη καὶ θανατώδης, καὶ διαφεύγουσιν αὐτὴν
παῦραι.

65. Ἢν ἑλκωθέωσι σφοδρῶς, αἷμα καὶ πῦον καθαίρεται, καὶ ὀδμὴ βαρέη, καὶ ὁπόταν ἡ ὀδύνη προσλάβῃ,[242] ὡς ὑπὸ ὠδῖνος τὸ πλῆθος[243] τὰ περὶ τὸν ῥόον
γίνεται, καὶ ὅταν ὁ χρόνος ᾖ, τὰ σκέλεα καὶ οἱ πόδες
οἰδέουσι, καὶ ἰῶνται οἱ ἰητροὶ ὡς ὕδρωπα· τὸ δὲ οὐ
τοῖον. ταύτην ἢν λάβῃς, λούειν θερμῷ, καὶ χλιαίνειν
καὶ κλύζειν δριμέσι καὶ μαλθακοῖσι καὶ στρυφνοῖσιν,

crush white poppy in barley meal and sift it, toast goat's cheese from which you have scraped off the parings and incrustation, and mix together butter and the finest barley meal: take one measure of the medication, of the cheese and of the meal, and give this to the patient to drink in the fasting state over diluted dry wine beginning at dawn. Toward evening mix a thick cyceon and give it, and also have the patient drink whichever of the medications for women's disorders you think is most acceptable.

As long as the blood continues to flow copiously and there are acute pains which briefly intermit, do this. But when the ulcer in the upper region recedes and mild pains invade the lower area for a longer time, give medications to drink that will clean the patient below rather than above, and then interrupt this treatment for as long as you think best. Also foment with mild agents by having the woman sit down over them whenever the right moment seems to come.

If the patient does these things, she will recover, but the disease is slowish and deadly, and few escape from it.

65. If (sc. the uterus) becomes very ulcerated, blood and pus will be discharged, an offensive smell develops, and when pain attacks, the flux is accompanied for the most part as if by the pangs of childbirth; with the passage of time the legs and feet swell, and physicians treat as if it were dropsy—which it is not. If you take on such a patient, wash her with warm water and keep her warm; apply an irrigation with acrid, emollient, astringent agents, water,

---

239 ποιῆσαι, καὶ om. MV.   240 ἄνω om. MV.
241 κάτω om. MV.   242 προσλάβῃ I: προσβάλῃ ΘMV.
243 τὸ πλῆθος om. MV.

ὕδατί τε καὶ οἴνῳ· καὶ πολύκαρπον καὶ πολύκνημον
καὶ μέλι ἕψων ὁμοῦ, εἶτα εἴριον ἐς τοῦτο βάπτων, διαχρίειν τὰ αἰδοῖα, καὶ ῥητίνην καὶ μέλι καὶ συὸς
ἔλαιον ἐγχρίειν· καὶ πιπίσκειν λίνου καρπόν, καὶ σήσαμον φώξας καὶ βούτυρον καὶ τυρὸν αἴγειον καὶ
ἄλφιτον ἐφ᾽ ἅπασι πιπίσκειν ἐν οἴνῳ νῆστιν, ἐς ἑσπέρην δὲ μέλι ἐπιχέων πολύ· καὶ ἕως μὲν ἂν τὸ αἷμα
καθαίρηται, καὶ ὀδύναι ὀξέαι ἔχωσι, καὶ ὀλίγον διαλείπωσι, τοῦτο ποιέειν· ὅταν δέ σοι ἔλασσον ἴῃ καὶ
ὀδύναι βληχρότεραι ἴσχωσι καὶ διὰ πλέονος χρόνου,
φάρμακον πιπίσκειν κάτω, καὶ διαλείπειν. ταῦτα
ποιέουσα, ὑγιὴς γίνεται· γενεὴ δὲ οὐκ ἔτι. |

136   66. Ὅσα δ᾽ ἑλκώματα γίνεται ἐν τῇσιν ὑστέρῃσιν
ἀπὸ τρωσμοῦ ἢ ὑπ᾽ ἄλλου του, ταῦτα δὲ χρὴ ἀποσκεπτόμενον ἐς τὸ ὅλον σῶμα θεραπεύειν πάντα, ὁποίης
ἂν δοκέῃ δεῖσθαι θεραπείης, ἤντέ σοι δοκέῃ ἐξ ἅπαντος τοῦ σώματος ἡ ἄνθρωπος θεραπευτέη εἶναι, ἤντε
ἀπ᾽ αὐτῶν εἴη τῶν ὑστερέων.[244] γνώσῃ δὲ εἰ ἀπ᾽ αὐτῶν
τῶν ὑστερέων ἐστὶν ὧδε· τὰ μὲν ἀπ᾽ αὐτῶν τῶν ἑλκω-
μάτων τὴν κάθαρσιν παρέχει πυοειδέα τε καὶ ξυν-
εστηκυῖαν, τὰ δὲ μὴ ἀπ᾽ αὐτῶν λεπτήν τε καὶ ἰχωροειδέα. ὅσα μὲν οὖν ἐστι λεπτὰ τῶν ῥευμάτων, ταύτας
μὲν χρὴ φαρμακεύειν πρῶτον καὶ ἄνω καὶ κάτω,
πρῶτον δὲ ἄνω· καὶ ἢν μὲν μετὰ τὴν φαρμακείην τὰ
ῥεύματα ἐλάσσω γίνηται καὶ εὐπετέστερα ᾖ, διαλείποντα αὖθις φαρμακεύειν τὸν αὐτὸν τρόπον· μετὰ δὲ
τὴν φαρμακείην διαιτᾶν διαίτῃ τοιαύτῃ, ἐν ὁποίῃ ἂν
εἴη μάλιστα ξηροτέρη[245] ἡ ἄνθρωπος· ἔσται δέ, ἢν

and wine. Boil willow weed and field basil together with
honey, dip a piece of wool into this, and anoint it on to the
genitalia; also smear on resin, honey, and lard. Give a drink
made with toasted linseed and sesame seed, over which
are added butter, goat's cheese, and barley meal for the
patient to take in wine in the fasting state, and toward
evening pour copious honey over it. As long as blood is
being cleaned and acute pains are present with little intermission, continue thus, but when you succeed in reducing
the patient's flux and her pains become milder and stretch
out over a longer time, give a purgative medication to
drink and then call a stop. If a woman does this, she will
recover, but no longer be fertile.

66. Ulcers that arise in the uterus as the result of an
abortion or anything else, you must treat by paying attention to the body as a whole for indications of how treatment should be conducted, whether you think it is the
whole body that requires attention, or the uterus alone.
You should recognize whether it is the uterus itself from
the following: conditions from (sc. uterine) ulcers themselves produce a discharge that is purulent and congealed,
while ones not from them themselves have a thin serous
cleaning. Now when the discharges are thin you must give
patients a medication to purge both upward and downward—upward first. If, after the medication, the fluxes
decrease and become more favorable, pause for a time and
then medicate again in the same manner. After the medication apply a regimen to make the woman very dry: this
you will achieve by fomenting her over her whole body

---

244 εἴη τῶν ὑστερέων om. MV.
245 ξηροτάτη MV.

αὐτὴν πυριῆς δι' ἡμέρης τρίτης ἢ τετάρτης ὅλον τὸ
σῶμα, καὶ ἐμέτους ποιέῃς ἐκ τῶν πυριῶν εὐθύς· μετὰ
δὲ τοὺς ἐμέτους καὶ τὰς πυρίας διαιτᾶν ἀλουσίῃσί τε
καὶ ὀλιγοποσίῃσι καὶ ἀρτοσιτίῃσιν· ὄψῳ μηδενί, ἀλλ'
ἢ οἴνῳ ἀκρήτῳ μέλανι, λαχάνῳ δὲ μηδενί· ὅταν δὲ τὸν
ἔμετον παρασκευάζῃς, τότε δὲ χρὴ λαχάνων πολλῶν
καὶ δριμέων ἐμπιπλάναι καὶ σιτίων πολλῶν καὶ ὄψου
ὁποίου ἂν βούλωνται, καὶ οἴνου πολλοῦ [ἐμπιπλά-
ναι]²⁴⁶ ὑδαρέος, καὶ λούειν ἐκ τῶν πυριῶν πολλῷ
θερμῷ. αὕτη μὲν ἡ θεραπείη τῶν τοιουτοτρόπων ῥευ-
μάτων· ἄμεινον δὲ ἄμφω φαρμακεύειν, καὶ ἐμέειν καὶ
ἄνω ἕλκειν· δίαιτα δὲ ξηραντικὴ κρέσσων, ἀλουσίη.²⁴⁷

138 τὰς δ' ὑστέρας χρὴ θεραπεύειν ὧδε· πρῶτον | μὲν πυ-
ριᾶν ὕδατι ἀκτῆς ἀφέψοντα τὰ φύλλα· ἔπειτα μετὰ
τὴν πυρίαν κλύζειν ἐκ τῆς ἰλύος τῷ σμήγματι· ἢν μὲν
σηπεδὼν ᾖ ἐν τοῖσιν ἕλκεσι καὶ τὰ ἀπιόντα δυσώδεα,
ἀκρητεστέρῳ τῷ σμήγματι· ἢν δὲ μηδὲν ᾖ τοιοῦτον,
ὑδαρεστέρῳ· μετὰ δὲ τὴν ἰλύν, ὕδατι· ἐν δὲ τῷ ὕδατι
ἐναφέψειν μυρσίνην καὶ δάφνην²⁴⁸ καὶ ἐλελίσφακον·
μετὰ δὲ τοῦτο οἴνῳ κλύζειν ἀκρήτῳ λευκῷ χλιαρῷ.
ὅταν δὲ δάκνηται²⁴⁹ ὑπὸ τῶν κλυσμάτων, τότε ἤδη
καθαρὰ τὰ ἕλκεά ἐστι· κλύζειν οὖν χρὴ ὑδαρεστέρῃ
τῇ ἰλύϊ καὶ οἴνῳ μέλανι· μετὰ δὲ τὸν οἶνον πιμελὴν
ὄϊος²⁵⁰ τήξαντα νεαράν, ἔλαιον παραμίξαντα, ἢν μὲν
ᾖ, χηνός, ἢν δὲ μή, ἄλλου του, μάλιστα ὄρνιθος, εἰ
δὲ μή, τὸ ἐκ τῶν ἐλαιῶν παλαιόν, τούτῳ χλιαρῷ κλύζειν, μετὰ δὲ τῷ οἴνῳ ἐς ἕτερον κλυστῆρα ἐγχέαντα·

every third or fourth day, and then making her vomit immediately after the fomentations. After the vomiting and fomentations, have her follow a regimen with no bathing and few drinks, but with wheaten bread; there should be no prepared dishes, but rather undiluted, dark wine and no vegetables. When you are preparing the emetic, at the same time fill patients with many sharp vegetables, much bread, and whichever prepared dishes they desire, as well as with generous dilute wine, and have them bathe after the fomentations in plenty of hot water. This is the treatment for such fluxes; it is better to evacuate by both paths, and for the patient to vomit and to attract upward. A drying regimen is preferable, and avoidance of the bath. The uterus you must treat as follows: first, foment by boiling elder leaves in water; then after the fomentation flush with an unguent made with wine lees—if there is mortification in the ulcers and the discharge is ill-smelling, the unguent should be very concentrated, but if there is nothing of that sort use a more watery one. After the wine lees inject water in which you have boiled myrtle, laurel and salvia; after that irrigate with undiluted warm white wine. When the patient is irritated by the irrigations, the ulcers are already clean, and so you must irrigate with more diluted wine lees and dark wine. After applying the wine, melt fresh sheep's fat and add olive oil and, if is available, goose oil; if this is not available, use the fat of something else, preferably a bird, or if not that, then aged oil of olives: irrigate with this warm, and after that with wine you have

---

246 Del. Gundert as dittography.   247 ἄμεινον δὲ ἄμφω
. . . ἀλουσίη om. Θ.   248 καὶ δάφνην om. MV.
249 Add. ἤδη MV.   250 ὄϊος Θ: ὑὸς MV.

ἐς δὲ τὸ στόμα, κἢν ᾖ εἱλκωμένον, κἢν μή, μοτοὺς
ποιεῦντας τῶν μαλθακτηρίων προστιθέναι, καὶ ἢν
καῦμα παρέχωσι προσκείμενοι, ἀφελομένην κελεύειν
νίψασθαι ὕδατι χλιαρῷ τοιούτῳ οἵῳ περ ἐκλύζετο. ἢν
δὲ πρὸς τὴν δίαιταν τὰ ῥεύματα μὴ ἀπαλλάσσηται,
ἐλάσσω δὲ γίνηται καὶ δάκνηται σφόδρα, καὶ τὰ
ἀπορρέοντα χολή τε εἴη καὶ ἅλμη, καὶ μὴ μόνον τὰ
ἔνδον, ἀλλὰ καὶ τὰ ἔξω ἑλκοῖ, μεταβάλλειν χρὴ τὴν
δίαιταν, ἐξυγραίνειν δὲ πᾶσαν, ὅπως τὰ ῥεύματα ὡς
ὑδαρέστατα ἔσται καὶ ἥκιστα δηκτικά, λουτροῖσι
θερμοῖσι πολλοῖσι, μάζῃ, λαχάνοισιν ἐφθοῖσι πᾶσι
λιπαροῖσιν, ἰχθύσι τοῖσι σελάχεσι, σὺν κρομμύοισι
καὶ σκορόδοισιν[251] ἕψοντα ἐν ἅλμῃ γλυκείῃ, ἕψειν δὲ
λιπαρῶς,[252] κρέασιν ἐφθοῖσι πᾶσι, πλὴν βοός,[253] δι-
140 έφθοισιν ἐν ἀνήθοισι καὶ μαράθοισιν, οἴνῳ | δὲ μελι-
χρόῳ, κιρρῷ, ὑδαρεῖ, πλέονι, γαλακτοποσίη μετ᾽ οἴνου γλυκέος· τὰ δ᾽ ἄλλα περὶ τῶν κλυσμάτων[254] κατὰ
τὸν ὑφηγημένον λόγον.

Αὕτη μὲν νῦν τῶν τοιῶνδε θεραπείη. ὅσα δὲ πυοειδέα τε καὶ συνεστεῶτα ἄπεισι, τούτων τὸ μὲν ὅλον
σῶμα οὐδὲν δεῖ κινέειν, κλύζειν δὲ καὶ ἀπὸ τούτων
τὴν θεραπείην πᾶσαν ποιέεσθαι, κλύζειν δὲ τοῖσιν
αὐτοῖσι κλύσμασιν οἷς πρόσθεν εἴρηται τὸν αὐτὸν
τρόπον. γεγράψεται δὲ καὶ ἄλλα κλύσματα.

Ἑλκέων ἴησις· ἐλάφου στέαρ χλωρὸν προσθετόν·
κλύζειν δὲ χρὴ αὐτίκα οἴνῳ σιραίῳ, πάντα κράτιστον

---

251 σκορ. Θ: κορίοισιν MV.

poured into another syringe. Into the mouth (sc. of the uterus), whether it is ulcerated or not, apply pessaries with emollient substances: if the application causes burning, instruct the patient to remove it and wash with the same kind of warm water she was using for irrigation before. If in response to this regimen, the fluxes are not averted, but they diminish in amount and cause violent irritation, the discharge is bilious or saline, and there is ulceration not only internally but also on the exterior, you must alter the regimen and saturate the woman's whole body, in order to make the fluxes as dilute as possible and the least irritating, by giving frequent warm baths and a diet of barley cake, vegetables all boiled with oil, cartilaginous fish with onions and garlic boiled in a sweet, salty sauce—boil in oil—all kinds of boiled meats except beef boiled with dill and fennel, generous diluted tawny, honey-colored wine, and a milk potion with sweet wine. Otherwise, the irrigations should conform with the method indicated.

Now this is the therapy for such conditions. If the discharge is purulent and congealed, you should not cause any movement anywhere within the patient's body, but use irrigations and base the whole treatment on these, using the same irrigations as mentioned above and in the same manner. Other irrigations will also be mentioned (sc. below).

Treatment of ulcers: a suppository made with fresh deer's fat: you must irrigate at once with new wine boiled down; most effective, if ulcers are present, is to irrigate

---

252 γλυκ. . . . λιπαρῶς M: γλυκίη δὲ λιπαρᾶ V: om. Θ.
253 Add. καὶ αἰγὸς MV.   254 Add. ποιέειν MV.

δὲ ψιμυθίῳ, ἢν ἕλκεα ᾖ, καὶ ἐλαίῳ ναρκισσίνῳ· σιτίοισι δὲ μαλθακωτάτοισι χρήσθω καὶ μὴ[255] δριμέσιν.
ἢν δὲ ῥερυπωμένα ᾖ καὶ νέμηται, καὶ τὸν πελαστάτω
χῶρον διακναίῃ, καθαίρειν, καὶ νέην σάρκα φύειν, καὶ
ἐς ὠτειλὰς ἄγειν τὴν σάρκα, ῥηϊδίως γὰρ ἀναχαλᾶται
καὶ κακήθεα οὐ[256] γίνεται, καὶ λούειν.

67. Ἢν δὲ γυνὴ ἐκ τρωσμοῦ τρῶμα λάβῃ μέγα, ἢ
προσθέτοισι δριμέσιν ἑλκώσῃ τὰς μήτρας, οἷα γυναῖκες δρῶσί τε καὶ ἰητρεύονται, καὶ τὸ ἔμβρυον
φθαρῇ, καὶ μὴ καθαίρηται ἡ γυνή, ἀλλά οἱ αἱ μῆτραι
φλεγμήνωσιν ἰσχυρῶς καὶ μεμύκωσι καὶ παραμεθιέ-
ναι τὴν κάθαρσιν μὴ οἷαί τε ἔωσιν, εἰ μὴ τὸ πρῶτον
ἅμα τῷ ἐμβρύῳ, αὕτη ἢν μὲν ἰητρεύηται ἐν τάχει,
ὑγιὴς ἔσται, ἄφορος δέ. ἢν δέ οἱ ῥαγῇ αὐτόματα τὰ
λοχεῖα καὶ τὰ ἕλκεα ὑγιανθῇ, καὶ ὧδε ἄφορος ἔσται.
ἢν δέ οἱ ἡ μὲν κάθαρσις γένηται, τὰ δὲ ἕλκεα μὴ
142 μελεδαίνηται, | κίνδυνος σηπεδονώδεα εἶναι. ἢν δέ οἱ
ἡ κάθαρσις ἴῃ τετρυχωμένη, θνήσκει. καὶ ἢν ἐν τόκῳ
κάρτα ἑλκωθέωσιν αἱ μῆτραι τοῦ ἐμβρύου μὴ κατὰ
φύσιν ἰόντος, πείσονται τὰ αὐτὰ τῇ ἐκ διαφθορῆς ἑλ-
κωθείσῃ τὰς μήτρας, καὶ μεταλλαγὰς τὰς αὐτὰς καὶ
τελευτὰς ἡ νοῦσος ἴσχει· ἤν δ' ἐκ διαφθορῆς ἢ ἐκ
τόκου αἱ μῆτραι ἑλκωθῶσι, καὶ τὰ λοχεῖα πάντα παρ-
ίωσιν, ἦσσον πονήσει, εἰ μὴ μεγάλα ἕλκεα εἴη, καὶ
μελεδαινομένη ἐν τάχει ὑγιαίνει.

[255] μὴ om. Θ.
[256] οὐ om. MV.

with white lead and narcissus oil. Have the patient employ the mildest of foods devoid of all sharpness. If the ulcers are putrid and spreading, corroding the adjacent territory, clean them and promote the growth of new tissue and the formation of scars, for such ulcers are quite easily relieved and they do not[14] become malignant; bathe the patient as well.

67. If a woman comes out of an abortion with a severe injury, or develops ulcers in her uterus as the result of irritating suppositories—given how many practices and treatments women apply—and the fetus is aborted and the woman is not cleaned, but her uterus becomes violently inflamed and closes and is unable to allow the cleaning to pass out unless it has occurred first along with the passage of the fetus, such a woman, if treated quickly, will recover but be barren. But if her lochia break forth spontaneously and her ulcers heal, even so she will be barren. If, however, her cleaning does take place but her ulcers do not receive attention, there is a danger they will putrefy; and if cleaning occurs but the woman is exhausted, she will die. Also if the uterus becomes very ulcerated at the time of childbirth, with the fetus not progressing as it should, such a woman will suffer the same things as one whose uterus ulcerates subsequent to a miscarriage, and the condition will develop and end in the same ways. If the uterus ulcerates after a miscarriage or birth and the lochia pass completely, a woman will suffer less, unless the ulcers are very large, and on receiving treatment she will quickly recover.

[14] This is the text of Θ. M and V omit the negative: "they become malignant."

Χρὴ δὲ τὴν μελέτην προσέχειν ἐν τάχει, ἢν ἕλκεα
ἐν τῇ μήτρῃ ἔνῃ· ἅτε γὰρ ἐόντα ἐν ἁπαλῷ αὔξεται,
καὶ σαπρὰ ταχὺ γίνεται. ἰᾶσθαι δὲ τὰ ἕλκεα, ὡς καὶ
τὰ ἐν τῷ ἄλλῳ σώματι, καὶ ἀφλέγμαντα χρὴ ποιέειν
καὶ ἀνακαθαίρειν καὶ ἀναπιμπλάναι καὶ ἐς ὠτειλὰς
ἄγειν· διδόναι δὲ ὕδωρ, οἶνον δὲ μή, σιτία ἀφαυρό-
τερα, πολλὰ δὲ μή.

68. Ὅσα τρωσμῶν γινομένων μὴ δύναται ἀπαλ-
λάσσεσθαι μεζόνων ὅλων τε ἢ τῶν μελέων τῶν ἐμβρύων ἐόντων, ἢ ἐλασσόνων καὶ πλαγίων τε καὶ ἀδυ-
νάτων, τὰ τοιαῦτα ἢν μὲν κατὰ φύσιν ἴῃ, διδόναι τῶν
φαρμάκων τι ὧν ἂν ἐγὼ γράψω, προλούσαντα θερμῷ
παμπόλλῳ· καὶ ἢν θέλοντα προϊέναι μὴ εὐλύτως ἀπίῃ
κατὰ φύσιν ἰόντα, τῇσι τοιαύτῃσι τοῦ πταρμικοῦ
προσφέρειν, ἐπιλαμβάνειν δὲ τὸν μυκτῆρα καὶ πτάρ-
νυσθαι, καὶ τὸ στόμα πιέζειν, ὅπως ὅτι μάλιστα ὁ
πταρμὸς ἐνεργὸς ᾖ. χρῆσθαι δὲ καὶ σεισμοῖσι· σείοις
δ' ἂν ὧδε· κλίνην λαβεῖν ὑψηλήν τε καὶ ῥωμαλέην,
144 ὑποστορέσαντα δὲ ἀνακλῖναι τὴν γυναῖκα | ὑπτίην, τὰ
δὲ στήθεα καὶ τὰς μασχάλας καὶ τὰς χεῖρας προσ-
καταλαβεῖν ταινίῃ ἢ ἱμάντι πλατεῖ[257] πρὸς τὴν κλί-
νην· καὶ ζωννύειν,[258] καὶ τὰ σκέλεα συγκάμψαι καὶ
κατέχειν ἐκ τῶν σφυρῶν· ὅταν δὲ εὐτρεπίσῃ, φρυγάνων φάκελον μαλθακῶν ἤ τι τῷδε ἐοικὸς εὐτρεπίζειν
ὅσον τὴν κλίνην οὐ περιόψεται ἐπὶ τὴν γῆν ῥιπτεομέ-
νην, ὥστε ψαῦσαι τοῖσι πρὸς κεφαλὴν ποσὶ τῆς γῆς·

257 Add. μαλθακῷ MV.   258 ζωννύειν Aldina: ζώνην ΘMV.

If ulcers develop in the uterus, treatment must be applied quickly, since they are growing in soft tissue and rapidly putrefy. Treat ulcers here the same as you would in any other part of the body: you must bring down their inflammation, clean them, and promote tissue regrowth and scar formation. Prescribe water but not wine, and weakish foods but not very many.

68. If what is in the process of being aborted cannot be expelled either because the whole fetus or its limbs are too large or because although smaller they are in a transverse position and powerless, if there is movement in the natural direction, first bathe the patient with copious hot water and then give her some of the medications I am about to record. Or, if the fetus is tending to come forth but is not free to pass out easily by a natural progression, give the mother a sternutatory and, while you are making her sneeze, hold her nose and compress her mouth in order that the sneezing will have the greatest possible effect. Also employ shaking in the following way: take a bed that is high and sturdy, make the bed, lay the patient down on it on her back, and fasten down her chest, axillae, and arms to the bed with a band or a broad piece of cloth; gird her loins,[15] fold her legs together, and attach them at the ankles. When she is ready, prepare a bundle of flexible sticks or something of the sort that will not allow the bed to touch the ground with its feet at the head end when it is

---

[15] The text's καὶ ζώνην (and waist) might make sense if it was after "tie down her chest, axillae, and arms" (cf. Countouris' translation, "lege den Brustkorb, die Achselhöhle, die Arme und die Taille in eine Binde"), but here seems lost.

καὶ κελεύειν αὐτὴν λαβέσθαι τῇσι χερσὶ τῆς κλίνης,
καὶ μετέωρον πρὸς κεφαλῆς τὴν κλίνην ἔχειν, ὡς
κατάρροπος ᾖ ἐπὶ πόδας, φυλασσόμενος ὅπως μὴ
προπετὴς ἔσται ἡ ἄνθρωπος· ὅταν δὲ ταῦτα ἐνεργῆται
καὶ μετάρσιος ᾖ ἡ κλίνη, ἐκ τῶν ὄπισθεν ὑποθεῖναι
τὰ φρύγανα, κατορθοῦσθαι δὲ ὡς μάλιστα, ὅπως οἱ
πόδες μὴ²⁵⁹ ψαύσωσι τῆς γῆς, ῥιπτεομένης τῆς κλίνης, καὶ τῶν φρυγάνων ἔξω ἔσονται, αἴρειν δ' ἐξ ἑκατέρου ποδὸς ἄνδρα τῇδε καὶ τῇδε, ὡς κατ' ἰθὺ ἡ κλίνη
πεσεῖται ὁμαλῶς καὶ ἴσως καὶ μὴ σπασμὸς ᾖ· σείειν
δὲ ἅμα τῇ ὠδῖνι μάλιστα· καὶ ἢν μὲν ἀπαλλάσσηται,
αὐτίκα πεπαῦσθαι, εἰ δὲ μή, διαλαμβάνοντα σείειν,
καὶ αἰωρέειν ἐπὶ κλίνης φερομένην. ταῦτα μὲν νῦν
οὕτω ποιέεται, ἢν ὀρθά τε καὶ κατὰ φύσιν ἀπαλλάσσηται. χρὴ δὲ κηρωτῇ ὑγρῇ προχρίειν, ἐπὶ πάντων δὲ
τῶνδε τῶν ἀμφὶ τὴν ὑστέρην τοιῶνδε παθημάτων ἄριστον τοῦτο, καὶ μαλάχης ὕδωρ καταιονᾶν, καὶ βουκέ-
ρας, ἢ πτισάνης πυρίνης μᾶλλον χυλός· χρὴ δὲ ἄχρι
βουβώνων ἕδρην τε καὶ αἰδοῖον πυριᾶν, καὶ ἐνίζεσθαι
δέ, ὅταν αἱ ὠδῖνες σφοδραὶ καὶ ὄχλοι ἔωσι μάλιστα,
καὶ μηδὲν ἐν νόῳ ἕτερον ἔχειν. τὴν δὲ ἰητρεύουσαν τὰ
στόματα μαλθακῶς ἐξανοίγειν, καὶ ἠρέμα τοῦτο δρᾶν,
ὁμαλὸν²⁶⁰ δὲ συνεφέλκεσθαι τὸ ἔμβρυον. |

146   69. Ὅσα δὲ δίπτυχα πτύσσεται καὶ ἔγκειται ἐν τῷ
στόματι τῶν ὑστερέων, ταῦτα δέ, ἤντε ζῶντα ἤντε
τεθνεῶτα ᾖ, προώσαντα ὀπίσω πάλιν στρέφειν, ὅπως
κατὰ φύσιν ἴῃ²⁶¹ ἐπὶ κεφαλήν. ὅταν δὲ ἀπωθέειν
βούλῃ ἢ στρέφειν, ἀνακλίναντα χρὴ ὑπτίην ὑπὸ τὰ

moved. Instruct the patient to take hold of the bed with her hands, set the bed higher toward the head end so she is inclined toward her feet, but at the same time be careful that she does not slip downward. When this has been set up and the bed has been lifted, lay down the sticks from behind straightening them as well as you can in order that its feet will not touch the earth when it is moved and they (i.e., the feet) will be outside of the sticks: have one man on each side lift the bed by one leg so that the bed will fall evenly in a level, straight line and the woman will not be stretched. Shake especially at the time of the (sc. uterine) contraction and if it (i.e., the fetus) moves, stop immediately, but if it does not, after taking a break shake again, and swing the woman held on the bed. This is how it is performed if the fetus is straight and moving according to nature. You must apply a moist cerate—this is best for all such conditions involving the uterus—foment with mallow water and fenugreek, or better gruel of peeled wheat; also foment the seat and the genitalia as far as the groins, and apply a sitz bath when the contractions are strong and especially troubling, but ignore everything else. The midwife should gently open the cervix, doing this carefully, and draw the fetus evenly with the contractions.

69. Fetuses that are folded double and become lodged in the mouth of the uterus, whether living or dead, are to be pushed back and turned around so that they will move head first in the natural direction. When you wish to push them back or turn them, lay the woman down on her back,

---

259 μὴ om. Θ.
260 ὁμαλὸν Θ: ὀμφαλὸν MV.
261 ἴη I: εἴη ΘMV.

ἰσχία ὑποστορέσαι τι μαλθακόν, καὶ ὑπὸ τοὺς πόδας
τῆς κλίνης, ὅπως ὑψηλότεροι ἔσονται οἱ πρὸς ποδῶν
πόδες συχνῷ, ὑποτιθέναι τι[262] χρή· καὶ ἀνωτέρω δὲ τὰ
ἰσχία τῆς κεφαλῆς ἔστω, προσκεφάλαιον δὲ μηδὲν
ὑπέστω τῇ κεφαλῇ· προμηθεομένοισι δὲ ταῦτα· ὅταν
δ᾽ ἀπώσῃ τὸ ἔμβρυον καὶ περιδινῆται τῇδε καὶ τῇδε,
κατὰ φύσιν καθίστασθαι καὶ τὴν κλίνην καὶ τὰ
ἰσχία, ὑπεξελὼν τὰ ὑπὸ τοὺς πόδας τῆς κλίνης[263] καὶ
τὰ ὑπὸ τῶν ἰσχίων· πρὸς κεφαλῆς δὲ ὑποθεῖναι προσ-
κεφάλαιον· τὰ τοιαῦτα τῷ τοιούτῳ τρόπῳ θεραπεύειν.

Ὅσα δὲ ζῶντα τῶν ἐμβρύων τὴν χεῖρα ἢ τὸ σκέλος
ἔξω προτείνει[264] ἢ ἄμφω, ταῦτα χρή, ὅταν τάχιστα
προσημήνῃ, ἔσω ἀπωθέειν τῷ προειρημένῳ τρόπῳ,
καὶ στρέφειν ἐπὶ κεφαλήν, καὶ ἐς ὁδὸν ἄγειν. καὶ ὅσα
πτύσσεται τῶν ἐμβρύων πεπτηῶτα ἢ ἐς τὸν κενεῶνα
ἢ ἐς ἰσχίον ἐν τῷ τόκῳ, χρὴ ταῦτα ἀπορθοῦσθαι, καὶ
στρέφειν, καὶ προσκαθίννυσθαι ἐς ὕδωρ θερμόν, ἄχρι
οὗ ἰαίνηται.

70. Ὅσα δὲ τεθνεῶτα τῶν ἐμβρύων ἢ τὸ σκέλος ἢ
τὴν χεῖρα ἔξω ἔχει, ταῦτα ἄριστον μέν, ἢν οἷόν τε,
ἀπώσαντα ἔσω ἐπὶ κεφαλὴν στρέφειν· εἰ δὲ μὴ οἷόν
τε ᾖ, ἀνοιδίσκηται δέ, τάμνειν τῷδε τῷ τρόπῳ· σχί-
σαντα τὴν κεφαλὴν μαχαιρίῳ ξυμφλάσαι, ἵνα μὴ
θράσῃ, τῷ πιέστρῳ, καὶ τὰ ὀστέα ἕλκειν ὀστεολόγῳ,
καὶ[265] τῷ | ἑλκυστῆρι, παρὰ τὴν κληῖδα καθέντα ὡς ἂν
ἔχηται, ἕλκειν, μὴ κατὰ πολύ, ἀλλὰ κατ᾽ ὀλίγον, ἐξ-
ανιέντα καὶ αὖτις βιώμενον. ὅταν δὲ ταῦτα μὲν εἰρύ-

148

spread out something soft below her hips, and also put something under the feet of her bed in order that it will be much higher at the foot end; the patient's hips should also be raised higher than her head, and under her head there should be no pillow: these are the preparations. When you have pressed the fetus back and rotated in this way and that, lower the bed and the woman's hips to the natural position by removing what was under the bed's feet and under her hips, and place a pillow under her head. Conduct the treatment in such cases this way.

In the case of living fetuses that protrude an arm or a leg or both, as soon as it appears you must push the limb back inside in the way described above, turn them in the direction of their head, and lead them on their path. Any fetuses that fold over after they have fallen toward the flanks or hip in the process of birth you must straighten out and turn, as well as having the mother sit down in hot water until she relaxes from the warmth.

70. Dead fetuses that project an arm or a leg it is best, if possible, to push back in and turn around so they will move toward their head. If this is not possible due to their swelling, you dissect them in the following way. Sever the head with a scalpel, break it up with a cranioclast in such a way as to avoid splintering it into fragments, and remove the bones with a bone forceps; pass a hook over the clavicle in order to catch hold of it, and then draw (sc. the remainder of the fetus) a little at a time rather than all at once, alternately slackening and applying force again.

---

262 τι om. ΘV.   263 Add. καὶ τοὺς λίθους MV.
264 προτίνει Θ: προσπίπτει M: προπίπτει V.
265 ὀστεολόγῳ καὶ Θ: ὥστε ὀλίγῳ ἢ MV.

σης, ἐν δὲ τοῖσιν ὤμοισιν ᾖ, τάμνειν τὰς χεῖρας
ἄμφω ἐν τοῖς ἄρθροισι μετὰ τῶν ὤμων· καὶ ὅταν
ταῦτα κομίσῃ, ἢν μὲν οἷόν τε ᾖ ἰέναι, καὶ τἆλλα εὐπετέως ἕλκειν· ἢν δὲ μὴ ἐνακούῃ, τὸ στῆθος πᾶν μέχρι τῶν σφαγέων σχίζειν, φυλάσσεσθαι δὲ ὡς μὴ
κατὰ τὴν γαστέρα τάμῃς, καὶ ψιλώσῃς τι τοῦ ἐμβρύου,²⁶⁶ ἔξεισι γὰρ ἡ γαστὴρ καὶ τὸ ἔντερον καὶ κό-
προς· ἢν δέ τι τούτων ἐκπέσῃ, πραγματωδέστερον
ἤδη γίνεται· ξυμφλάσαι δὲ καὶ τὰ πλευρά, καὶ τὰς
ὠμοπλάτας συναγαγεῖν, καὶ ῥηϊδίως χωρήσει τὸ λοιπὸν ἔμβρυον, ἢν μὴ ἤδη οἰδαλέον ᾖ τὴν κοιλίην· ἢν
δὲ ᾖ τι τοιοῦτον, ἄμεινον τὴν γαστέρα τοῦ ἐμβρύου
τρῆσαι πρηέως, ἔξεισι γὰρ φῦσα μοῦνον ἐκ τῆς γαστρός, καὶ εὐπετέως χωρήσει.

Ἢν δὲ ἐκπεπτώκῃ ἡ χεὶρ ἢ τὸ σκέλος τεθνεῶτος
τοῦ ἐμβρύου, ἢν μὲν δυνατὸν ᾖ, εἴσω ἀπῶσαι ἄμφω,
καὶ εὐτρεπίσαι τὸ ἔμβρυον, ταῦτα ἄριστα· ἢν δὲ μὴ
οἷόν τε ᾖ τοῦτο ποιῆσαι, ἀποτάμνειν ὅ τι ἂν ἔξω ᾖ ὡς
ἂν δύνηται ἀνωτάτω, καὶ τοὐπίλοιπον ἐσμασάμενος
προῶσαι καὶ στρέψαι τὸ ἔμβρυον ἐπὶ κεφαλήν· ὅταν
δὲ στρέφειν ἢ κατατάμνειν μέλλῃς τὸ παιδίον, τὰς
χεῖρας χρὴ ἀπονυχίσασθαι, τὸ δὲ μαχαίριον, ᾧ ἂν
κατατάμνῃς, καμπυλώτερον ἔστω ἢ ἰθύτερον,²⁶⁷ καὶ
τοῦτο κατὰ κεφαλὴν καλύπτειν ἀμφὶ²⁶⁸ τῷ λιχανῷ
δακτύλῳ, ἐσματευόμενον καὶ ὀδηγέοντα καὶ ὀρρω-
δέοντα, ὅπως μὴ ψαύσῃς τῆς ὑστέρης.

71. Περὶ δὲ μύλης κυήσιος τόδε αἴτιον· ἐπὴν πολλὰ
τὰ ἐπιμήνια ἐόντα γονὴν ὀλίγην καὶ νοσώδεα συλλά-

When you have drawn this out and the fetus is at the level of its shoulders, cut the two arms at their shoulder joints. After you have removed these, if it is possible to dislodge it, simply draw the rest out. But if it does not respond to your traction, sever the whole chest up as far as the jugulars, being careful not to cut into the belly or strip off any part of the fetus, since the belly, intestine and feces would then come out, and if any of these do the case becomes more troublesome. Also break up the sides and bring the shoulder blades together so that the rest of the fetus will pass more easily, unless its belly is already swollen. If something like this has happened, it is better gently to perforate the fetus' belly, for only wind will be expelled and the fetus will pass out easily.

If an arm or a leg protrudes from a dead fetus, if it is possible to reinsert them both and turn the fetus correctly, this is best. But if this is not possible, cut off the part outside as high up as possible, and in addition insert (sc. your hand) to push the fetus back and turn it around toward its head. Whenever you are about to turn or incise a child, you must cut your finger nails short, chose a blade that is more curved than straight, cover its extremity with your index finger, put your hand in to feel and guide the way, and take care not to come into contact with the uterus.

71. This is the cause of a molar pregnancy: when copious menstrual fluid takes up a small amount of morbid

---

266 ἐμβρύου MV: ὀστέου Θ.
267 ἰθύ- Θ: εὐθύ- MV.
268 κ. ἀ. Θ: ἀμφικαλύπτειν MV.

150 βωσιν, οὔτε κύημα ἰθαγενὲς | γίνεται, ἥ τε γαστὴρ
πλήρης ὥσπερ κυούσης, κινέεται δ᾽ οὐδὲν ἐν τῇ γαστρί, οὐδὲ γάλα ἐν τοῖς τιτθοῖς ἐγγίνεται,[269] σφριγᾷ
δὲ τοὺς τιτθούς. αὕτη οὖν δύο ἔτεα, πολλάκις δὲ καὶ
τρία οὕτως ἔχει. κἢν μὲν μία σὰρξ γένηται, ἡ γυνὴ
ἀπόλλυται· οὐ γὰρ οἵη τέ ἐστι περιγενέσθαι· ἢν δὲ
πολλαί, ῥήγνυται αὐτῇ κατὰ τὰ αἰδοῖα αἷμα πολὺ καὶ
σαρκῶδες· κἢν μὴν μετριάζῃ, σῴζεται· εἰ δὲ μή, ὑπὸ
ῥόου ἁλοῦσα ἀπώλετο.

Τὸ μὲν νόσημα τοιοῦτό ἐστι· κρίνεσθαι δὲ χρὴ τῷ
πληρώματι, καὶ ὅτι οὐ κινέεται ἐν τῇ γαστρὶ τὸ βρέφος. καὶ[270] τὸ μὲν ἄρσεν τρίμηνον, τὸ δὲ θῆλυ τετρά-
μηνον τὴν κίνησιν ἔχει· ἐπὴν οὖν τοῦ χρόνου παρελ-
θόντος μὴ κινέηται, δηλονότι τουτέστι· ἔστι δὲ καὶ
τοῦτο τεκμήριον μέγα· ἐν τοῖσι τιτθοῖσι γάλα οὐκ
ἐγγίνεται. ταύτην μάλιστα μὲν μὴ ἰῆσθαι· εἰ δὲ μή,
προειπόντα ἰῆσθαι.

Καὶ πρῶτον μὲν πυρίησον τὴν γυναῖκα ὅλον τὸ
σῶμα, ἔπειτα κατὰ τὴν ἕδρην κλύσον, ὅκως αἷμα
καταρραγῇ πολύ· καὶ γὰρ ἴσως κλύσας κινήσειας ἂν
τὸ ἔμβρυον τὸ δοκέον εἶναι τὸ συνεστηκός, διαθερ-
μανθείσης τῆς γυναικὸς ὑπὸ τοῦ φαρμάκου· κλύειν
δὲ καὶ κατὰ τὰς μήτρας, ὅκως αἷμα ἀπαγάγῃς· εἰ δὲ
μή, προσθέτοισι χρῆσθαι τοῖσιν ἀπὸ τῆς βουπρήστιος ὡς ἰσχυροτάτοισι, καὶ πιπίσκειν τὸ δίκταμνον
τὸ Κρητικὸν ἐν οἴνῳ· εἰ δὲ μή, τὸν καστόριον ὄρχιν·
καὶ ὄπισθεν αὐτῇ σικύην προσβάλλειν πρὸς τοὺς κε-

seed, no proper pregnancy occurs, but the belly fills up as it does in a woman who is pregnant. There is no movement in the belly, nor does milk form in the breasts, although the woman swells up in them. Now this (sc. disease) persists for two, often even three years in this state. If only one fleshy object is formed, the woman dies, since she is unable to prevail, but if many form, she will have a copious fleshy hemorrhage through her vagina: if the flux moderates, the patient will be saved, but if not, she will die as a result of the flux.

The disease is as follows: you must recognize it by the fullness, and by the fact that the fetus does not move in the belly—normally a male fetus first moves in three months and a female one in four months: thus, when this time passes without there being any movement, the condition is revealed. It is also an important sign when no milk forms in the breasts. Generally such a woman is not to be treated, or if she is, then after giving a warning.

First, apply a vapor bath to the woman's entire body, and then apply an enema to her seat in order that much blood will break out; for very probably when you apply the enema you will move the embryo that apparently exists, the agglomeration, as the woman is warmed by the medication. Also apply a douche to the uterus in order to draw blood. If not that, then apply very forceful suppositories made from the buprestis beetle, and have the patient drink Cretan dittany in wine; if not that, then castoreum. After that apply a cupping instrument to the woman's

---

269 ἐγγίνεται om. Θ.
270 τὸ βρέφος. καὶ om. MV.

νεῶνας, καὶ ἀφαιρέειν ὅτι πλεῖστον αἷμα· πρόσβαλλε
δὲ καὶ ὅτι μάλιστα τεκμαιρόμενος κατὰ τὰς μήτρας.

72. Καὶ τούτων μὲν περὶ τῶν νοσημάτων τῶν ἀπὸ
152 λοχείων | γινομένων οὕτως εἴρηται· εἰσὶ δὲ οἱ κίνδυνοι
οὐ μικροὶ ἐν αὐτοῖσιν· ὀξέα γάρ ἐστι καὶ ταχὺ μεταλ-
λάσσονται, καὶ μᾶλλον πονέονται αἱ πρωτότοκοι ἢ
αἵτινές εἰσιν ἔμπειροι τόκων. χωρέει δὲ τὰ λοχεῖα τῇ
ὑγιηρῇ γυναικὶ ἱκανὸν ὅσον Ἀττικὴ κοτύλη καὶ ἡμί-
σεια τὸ πρῶτον ἢ ὀλίγῳ πλέονα, ἔπειτα ἐπὶ ἐλάσσονα
κατὰ λόγον τούτου, μέχρι παυσῆται· χωρέει δὲ οἷον
αἷμα ἀπὸ ἱερείων,[271] ἢν ὑγιηρὴ ἡ γυνὴ καὶ μέλλῃ ὑγι-
αίνειν, καὶ ταχὺ[272] πήγνυται. καὶ καθαίρεσθαι [καὶ][273]
μετὰ τὸν τόκον ὡς ἐπὶ τὸ πλεῖον τὴν ὑγιηρὴν συμβαίνει, ἐπὶ μὲν τῇ κούρῃ ἡμέρας τεσσεράκοντα καὶ δύο
τὴν χρονιωτάτην κάθαρσιν, ἀκίνδυνος δέ ἐστι καὶ
εἴκοσι καὶ πέντε ἡμέρας καθαιρομένη· ἐπὶ δ' αὖ τοῦ
κούρου τριήκοντα ἡμέρας ἡ κάθαρσις γίνεται ἡ χρονιωτέρη, ἀκίνδυνος δέ ἐστι καὶ εἴκοσιν ἡμέρας γενομένη. καὶ τῶν διαφθαρεισέων τὰ ἔμβρυα κατὰ λόγον
ἡ κάθαρσις γίνεται τούτων τῶν ἡμερέων, καὶ ἐπὶ
τοῖσι νεωτέροισι φθαρεῖσιν ἐλάσσονας ἡμέρας, ἐπὶ δὲ
τοῖσι γεραιτέροισι πλέονας. παθήματα δὲ τὰ αὐτά
ἐστι περὶ λοχείων φθαρείσῃ τε τὸ ἔμβρυον καὶ τεκούσῃ, ἢν μὴ νήπιον φθείρῃ τὸ παιδίον· καὶ κινδυ-
νεύουσιν δὲ αἱ φθείρουσαι μᾶλλον· αἱ γὰρ φθοραὶ
τῶν τόκων χαλεπώτεραί εἰσιν· οὐ γάρ ἐστι μὴ οὐ βι-
αίως φθαρῆναι τὸ ἔμβρυον ἢ φαρμάκῳ ἢ ποτῷ ἢ
βρωτῷ ἢ προσθέτοισιν ἢ ἄλλῳ τινί· βίη δὲ πονηρὸν

flanks, and draw as much blood as you can—make the application as great as you judge possible according to the uterus.

72. This is my account of diseases arising from the lochia. There is no small danger in these, since they are acute and rapidly changing; women at their first delivery suffer more from them than those experienced in childbirth. In a healthy woman, the lochia are adequate if they first flow to the amount of one-and-a-half Attic cotyles, or a little more, and then regularly less until they cease. The flux is like the blood of a sacrificial animal, if the woman is healthy and is going to remain that way, and it congeals quickly. Also the cleaning (sc. of the lochia) after birth in a healthy woman usually occurs over 42 days at the longest after the birth of a female child: but danger would also be escaped if the cleaning lasted just 25 days. After the birth of a male child, the cleaning takes place over 30 days at the longest, but danger would also be escaped if the cleaning lasted just 20 days. After fetuses are aborted, the cleaning is usually for the same number of days—in the case of early abortions for fewer days, of later ones for more. The pathologies connected with the lochia are the same whether a woman has aborted a fetus or given birth to one, unless she has aborted a very small fetus, but women with abortions run more danger, since abortions are more difficult than births. For it is not possible to abort a fetus without violence—whether subsequent to a potion, a particular food, a suppository, or something else—and vio-

---

271 ἱερείων MV: κρέων Θ.
272 ταχὺ Θ: παχὺ MV.
273 Del. Littré.

ἐστιν· ἐν τῷ τοιούτῳ δὲ κίνδυνός ἐστι καὶ τὰς μήτρας ἑλκωθῆναι ἢ φλεγμῆναι· τοῦτο δὲ ἐπικίνδυνόν ἐστι.

73. Τὸ δὲ γάλα ὅκως γίνεται, εἴρηταί μοι ἐν τῇ
154 Φύσει τοῦ Παιδίου | τοῦ ἐν Τόκῳ· ἐπὴν δὲ κυΐσκηται ἡ γυνή, τὰ καταμήνια οὐ μάλα χωρέει, εἰ μὴ ἔστιν ᾗσιν ὀλίγα· τρέπεται γὰρ ἐς τοὺς μαστοὺς τὸ γλυκύτατον τοῦ ὑγροῦ ἀπὸ τῶν σιτίων καὶ ποτῶν, καὶ ἐκθηλάζεται· καὶ ἀνάγκη ἔσται τὸ ἄλλο σῶμα κεκενῶσθαι μᾶλλον, καὶ ἧσσον πλῆρες γίνεται τοῦ αἵματος·[274] τοῦτο οὕτω γίνεται. εἰσὶ δὲ αἵτινες φύσει ἀγάλακτοί εἰσι, καὶ σφέας ἐπιλείπει τὸ γάλα πρὸ τοῦ καιροῦ· αὗται φύσει στερεαί εἰσι καὶ πυκνόσαρκοι, καὶ οὐ διέρχεται ἐπὶ τοὺς μαστοὺς ἀρκέουσα ἰκμὰς ἀπὸ τῆς κοιλίης, πυκνῆς τῆς ὁδοῦ ἐούσης.

74. (1) Ἐπιμήνια κατασπάσαι· ἐλατηρίου δύο πόσιας, συμμίσγεται δὲ καὶ στέαρ ὄϊος ἀπὸ τῶν νεφρῶν, ὅσον τὸ ἐλατήριον, μὴ διαθρύπτεσθαι, ποιέειν δὲ δύο προσθετά· ἢ μελάνθιον τὸ ἐκ τῶν πυρῶν τρίψας ὕδατι φορύξαι καὶ προσθετὰ δύο ποιέειν· προστιθέναι δὲ πρὸ τῶν ἡμερέων ᾗσι μέλλει· ποιέει δὲ ῥίγεα καὶ πυρετούς.

(2) Μαλθακὰ ὑφ᾽ ὧν καθαίρεται ὕδωρ καὶ ψάμμος, καὶ ἄγει ἐπιμήνια, ἢν μὴ πολυχρόνια ᾖ, καὶ στόμα

[274] αἵματος MV: σώματος Θ.

lence does damage. In this case there is a danger that the uterus will become ulcerated or inflamed, and this entails danger.

73. How milk is formed I have explained in *Nature of the Child in Childbirth*.[16] When a woman becomes pregnant, her menses do not flow at all, except a little in some women, since the sweetest part of the moisture derived from foods and drinks turns toward her breasts and is sucked out (sc. of the body),[17] so that the remainder of the body must be emptier and contain less blood: this is how it comes about. Some women are naturally bereft of blood, and milk deserts these before the proper time (sc. for lactation to end): such women are by nature of a solid and dense flesh, so that sufficient moisture cannot pass through their cavity to their breasts because the passageway is compressed.

74. (1) To draw the menses. Two cups of squirting cucumber juice and sheep's fat taken from the kidneys: mix this to the same amount as the juice, but do not grind it; form into two suppositories. Or grind some ergot of wheat, mix it with water, and make two suppositories. Apply these before the days on which the menses are expected; they also bring on chills and fever.

(2) Emollients by which water and sand are expelled, and which also bring on the menses—unless they are chronically absent—and soften the mouth (sc. of the

---

[16] Cf. *Nature of the Child* 10, Loeb *Hippocrates* vol. 10, 59–61. [17] ἐκθηλάζεται implies "sucked out" (sc. of the breast by a child), but this passage is describing the situation during pregnancy, where nutriment is being drawn out of the body into the breasts, leaving the body with less.

μαλθάσσει· νάρκισσον, κύμινον, σμύρναν, λιβανωτόν, ἀψίνθιον, κύπαιρον, ἴσον ἑκάστου, ναρκίσσου
δὲ μοίρας τέσσερας, ἐπικτένιον ὠμόλινον συμμίξας,
ταῦτα τρίβειν ὀριγάνου ἡψημένου σὺν ὕδατι, καὶ
ποιέειν βάλανον, προσθεῖναι. ἢ κυκλαμίνου μίσγε ὡς
156 ἀστράγαλον· καὶ ἄνθος χαλκοῦ ὡς | κύαμον τρίψας,
μέλιτι δὲ δεῦσαι καὶ ποιέειν βάλανον, καὶ προστιθέ-
ναι·

ἢ γλήχωνα, σμύρναν, λιβανωτόν, ὑὸς χολὴν καὶ
βοὸς μέλιτι ἀναταράσσειν καὶ ἀναπλάσαι βάλανον.

(3) Ἢν τὰ ἐπιμήνια μὴ γίνηται, χηνὸς ἔλαιον καὶ
νέτωπον καὶ ῥητίνην ξυμμίσγουσα[275] προσθέσθω,
εἰρίῳ ἀναλαβοῦσα.

(4) Προσθετὸν καθαρτικὸν μαλθακόν·[276] ἰσχάδα
λαβὼν δίεφθον ποιέειν, καὶ ἀποπιέσας τρίβειν ὡς λει-
οτάτην, καὶ πρόσθες ἐν εἰρίῳ καὶ ῥοδίνῳ μύρῳ.

(5) Τὸ δριμύ· κράμβης, πηγάνου, ἑκατέρου ἥμισυ
τρίψας, τὸν αὐτὸν τρόπον χρῶ.

(6) Καθαρτικόν· χηνὸς μυελόν, ἢ βοός, ἢ ἐλάφου,
ὅσον κύαμον, παραχέοντα μύρου ῥοδίνου καὶ γάλα
γυναικός, τρίβειν ὡς φάρμακον τρίβεται, εἶτα τούτῳ
ἐναλείφειν τὸ στόμα τῆς μήτρης.

(7) Καθαρτικόν· χηνὸς μυελὸν ὅσον κάρυον, κηροῦ
ὅσον κύαμον, ῥητίνης σχινίης ἢ τερμινθίνης ὅσον
κύαμον, ταῦτα τῆξαι ἐν μύρῳ ῥοδίνῳ ἐπὶ πυρὸς μαλ-
θακοῦ, καὶ ποιῆσαι ὡς κηρωτήν· εἶτα τούτῳ χλιερῷ
ἐναλείφειν τὸ στόμα τῆς μήτρης, καὶ τὸν κτένα κατα-
βρέχειν.

uterus): narcissus, cumin, myrrh, frankincense, wormwood, galingale, an equal amount of each except four portions of narcissus, mix with the tow of raw flax that is left
on the carding comb, knead together in water in which
marjoram has been boiled, and form into a suppository:
apply. Or mix cyclamen to the amount of a vertebra; also
grind flower of copper to the amount of a bean, mix with
honey, form into a suppository, and apply.

Or knead pennyroyal, myrrh, frankincense, hog's gall
and bull's gall into honey and form into a suppository.

(3) If a woman's menses do not appear, have her mix
together goose oil, oil of bitter almonds and resin, take this
up in a piece of wool, and apply as a suppository.

(4) An emollient cleaning suppository: take a dried fig,
boil it well, squeeze it out, and grind it very fine; apply this
with rose oil in a piece of wool.

(5) A sharp suppository: grind cabbage and rue, half of
each, and apply in the same way.

(6) A cleaning agent: marrow of goose, cow, or deer to
the amount of a bean, add rose unguent and woman's milk,
knead it like a medication, and then anoint the mouth of
the uterus with this.

(7) A cleaning agent: goose's marrow to the amount of
a nut, wax to the amount of a bean, resin from mastic or
terebinth to the amount of a bean: melt these in rose unguent over a gentle fire, and make into a cerate: then
anoint this warm to the mouth of the uterus, and moisten
the genitalia.

---

275 ξυμμίσγουσα MV: μίσγουσα Θ.
276 ἰσχάδα λαβὼν ... (7) Καθαρτικόν om. V.

(8) Προσθετὸν καθαρτικόν· ἄλευρον σητάνιον, σμύρνης τριώβολον, κρόκου τὸ ἴσον, καστορίου ὀβολόν, ταῦτα τρίψας μύρῳ ἰρίνῳ προστιθέσθω· ἢ κνίδης καρπὸν καὶ μολόχης χυλὸν ἐν χηνὸς στέατι τρίψαντα προσθεῖναι.

(9) Προσθετὸν καθαρτικόν, ἢν τὰ γυναικεῖα μὴ φαίνηται· στύρακα καὶ ὀρίγανον τρίψας λεῖον καὶ ξυμμίξας, ἐπίχεον χηνὸς ἔλαιον, καὶ ὧδε προστίθεσθαι. |

158   (10) Προσθετὸν καθαρτικόν, ὥστε μήτρας ἐκκαθαίρειν καὶ αἷμα ἐκκενοῦν· ἀψινθίου ῥίζαν τρίψας λείην μέλιτι, μίξας ἐν ἐλαίῳ²⁷⁷ πρόσθες.

(11) Προσθετὸν καθαρτικόν· βουπρήστιος ἀφελεῖν κεφαλὴν καὶ πόδας καὶ τὰ πτερά· τὰ δ' ἄλλα τρίβειν, καὶ συμμίσγειν τοῦ σύκου τὸ ἔνδοθεν· διπλάσιον δὲ τὸ πῖον ἔστω· τοῦτο φυσᾷ τὰς ὑστέρας, τοῦτο καὶ τῆσιν ἀπαυδώσῃσιν ἄριστον. ἢ λινοζώστιος τὰ φύλλα λεῖα προσθετὰ ποιέειν· τοῦτο λεπτὴν ἄγει χολώδεα κάθαρσιν· ἀρτεμισίη ποιέει ὡς λινόζωστις, καὶ καθαίρει ἄμεινον. ἐλλέβορος μέλας λεῖος ἐν ὕδατι ἄγει οἷον ἀπὸ κρεῶν ὕδωρ. καὶ ἡ στυπτηρίη δὲ καὶ ἡ ῥητίνη τωὐτὸ δρᾷ.

Κύπαιρος, ἀψίνθιον, ἀριστολοχεία, κύμινον, ἅλες, μέλι, ταῦτα πάντα ἐν τωὐτῷ τρίβειν καὶ προστιθέναι.

Καὶ ἐλλέβορος ἐν οἴνῳ γλυκεῖ· αἰρῶν ἄλευρον καὶ πύρινον μέλιτι φυρήσας· ἐν εἰρίῳ προστιθέναι.

(12) Προσθετά, ἢν μὴ τὰ κατάποτα καθαίρῃ· λινό-

(8) A cleaning suppository: meal of spring wheat, three obols of myrrh, the same amount of saffron, and obol of castoreum, knead these into iris unguent and have the patient apply it; or knead nettle seed and mallow juice in goose grease, and apply.

(9) A cleaning suppository if the menses do not appear: crush storax and marjoram fine, mix together, add goose oil, and apply thus.

(10) A cleaning suppository, to clean and empty the uterus of blood: grind root of wormwood fine in honey, mix in olive oil, and apply.

(11) A cleaning suppository: remove the head, legs and wings from a buprestis beetle, crush what remains and mix this together with the inside of a fig—the fatty material should be double in amount—and blow this into the uterus; it is also very good for women who have lost their speech. Or make suppositories of finely ground leaves of mercury herb: this cleans out thin bile; artemisia acts like the mercury, and it cleans better. Thin black hellebore diluted in water draws off a fluid resembling what runs out of meat; alum and resin do the same.

Galingale, wormwood, aristolochia, cumin, salt, and honey: knead all these together and apply as a suppository.

Also hellebore in sweet wine; take darnel and wheat meal, and mix them in honey: apply in a piece of wool.

(12) Suppositories to be used in case pills do not clean:

---

277 μέλιτι . . . ἐλαίῳ Θ: μέλιτι τρίψας ἐν ἐλαίῳ M: καὶ
ξυμμίξας ἐπίχεον χηνὸς ἔλαιον ἐν μέλιτι καὶ V.

ζωστιν, σμύρναν, λευκόϊον, κρόμμυον ὡς δριμύτατον,
καὶ μελάνθιον τὸ ἡδύοσμον, ἢν ὑποφέρῃ.

(13) Αἷμα προσθετὰ δριμέα ἄγει· κανθαρίδας πέντε,
πλὴν τῶν ποδῶν καὶ πτερῶν καὶ κεφαλῆς, συμμίσγειν δὲ σμύρναν καὶ λιβανωτὸν καὶ μέλι, ἔπειτα βάψας ἐς ἄλειφα ῥόδινον ἢ Αἰγύπτιον προσθέσθω τὴν
160 ἡμέρην, καὶ ἐπειδὰν δάκνῃ, ἀφαιρεῖσθαι· καὶ | βά-
πτειν πάλιν εἰς γάλα γυναικὸς καὶ μύρον Αἰγύπτιον,
προσθίθεσθαι δὲ ἐς νύκτα, καὶ διανίζεσθαι ὕδατι εὐώδει, καὶ προσθιθέναι στέαρ.

Ἀρμόζοι δ᾽ ἂν καὶ βούπρηστις, ἢν σμικρὴ ᾖ, ἄνευ
πτερῶν καὶ κεφαλῆς καὶ ποδῶν· ἢν δὲ μεγάλη, ἥμισυ,
μίσγειν δὲ τὰ αὐτὰ ἃ καὶ τῆσι κανθαρίσι, καὶ προστιθέναι ὁμοίως· ἢν δὲ μαλθακωτέρου δέηται, τῇ βου-
πρήστει μῖξαι οἶνον καὶ κύμινον Αἰθιοπικόν, ἀσταφίδα καὶ πάλην σεσέλιος καὶ ἀννήσου, καὶ ἀναζέσαι
τὸν οἶνον· ἀποχέας δὲ τρῖψαι λεῖον, καὶ πλάσαι φθοΐσκους ὅσον δραχμιαίους· τούτων προσθιθέναι, σμύρναν καὶ λιβανωτὸν μίσγοντα, ποιέειν δὲ τὸν αὐτὸν
τρόπον, ὅνπερ ἐπὶ τῆσι κανθαρίσιν.

Μελάνθιον τὸ ἐκ τῶν πυρῶν τρίβοντα λεῖον μέλιτι
φυρῶντα προσθιθέναι· ἢ τὸ μελάνθιον τὸ ἐκ τῶν πυρῶν τρίβειν σὺν μέλιτι, καὶ ποιέειν οἷον βάλανον·
πτερῷ δὲ περίπλασσε.

(14) Προσθετὸν ἐνεργόν· ὀπὸν μανδραγόρου καὶ
κολοκυνθίδος ἀγρίης γάλακτι γυναικείῳ πρόσθες.

Τρύγα ξηρὴν ἐξ οἴνου παλαιοῦ λευκοῦ καίειν, καὶ
οἴνῳ σβέσαι.

mercury herb, myrrh, white violet, a very pungent onion, and fragrant black cumin, if the patient can stand it.

(13) Sharp suppositories that draw blood: take five blister beetles without their legs, wings and head, add myrrh, frankincense and honey, then dip in rose or Egyptian unguent, and have the woman insert this during the day; when the suppository causes irritation, remove it. Dip it again in woman's milk and Egyptian unguent, and apply overnight. Also wash thoroughly with fragrant water, and insert fat.

Also applicable would be a buprestis beetle without its wings, head, and legs—(sc. a whole one) if it is small, but if it is large, then a half—mixed with the same things as blister beetles are, and applied in the same way. If something milder is needed, mix wine and Ethiopian cumin with a buprestis beetle, a raisin, and powders of hartwort and anise; boil these ingredients in wine, pour off the fluid, knead smooth, and form a pastille of one drachma. Apply mixed with myrrh and frankincense, and in the same way as blister beetles.

Grind ergot from wheat fine, mix with honey, and apply as a suppository; or grind ergot from wheat with honey, make into a suppository, and plaster it around a feather.

(14) An effective suppository: apply mandrake juice and juice from a wild gourd together with woman's milk.

Burn dry lees from aged white wine, and quench this with wine.

Καὶ κολοκύντη ἀγρίη, λινόζωστις, λίτρον καὶ ἐρύ-
σιμον· θᾶσσον κατασπᾷ καὶ μανδραγόρου ῥίζα, κανθαρίς, ἕρπυλος, δάφνης καρπός, μύρον ἴρινον, δάφνι-
νον.

Τιθυμάλλου τὸν ὀπὸν καὶ ἀνακινέειν, καὶ τὸν ἰξὸν
ἀφαιρέειν· διδόναι δὲ ὅσον ὄροβον, προσθετὸν ἄρι-
στον· εἰ δὲ πλέον ἴοι, οἴνῳ προσκλυζέσθω.

Ἢ ὀπτοῦ χαλκοῦ ἐκδιείς, χλιερὸν ἀναλαβεῖν εἰρίῳ,
καὶ προστιθέναι. |

162   75. (1) Κυητήριον· κεδρίης ἐμβάφιον, στέατος βοείου δραχμὰς τέσσερας, λεῖα τρίψας καὶ ἐς τωὐτὸ
μίξας, πεσσοὺς ποιέων, προστίθει νήστει, καὶ προσ-
κειμένη ἐκνηστευέτω τὴν ἡμέρην· προστιθέσθω δὲ
δίς, πρωῒ καὶ δείλης, μετὰ τὰ καταμήνια, καὶ μετὰ τὸ
δεῖπνον λούσθω, καὶ[278] κοιμάσθω ξὺν ἀνδρί.

Ἢ μελάνθιον φλάσαι, καὶ ἐς ῥάκος ἐνδῆσαι· καὶ
χηνὸς ἔλαιον δοῦναι προσθέσθαι.

(2) Κυητήριον· γυναῖκα θεραπεῦσαι, ὥστε ἔχειν ἐν
γαστρί· οὖρον λαβὼν παλαιὸν καὶ σιδήρου σκωρίαν
ὅσην διπάλαιστα[279] θρύμματα· ἔπειτα—καθίσας ἐπὶ
δίφρου—καὶ συγκαλύψασα καὶ τὸ σῶμα καὶ τὴν κεφαλήν, ὑποθεῖσα ἐς πόδας—ἰητρὸν[280] ἐμβαλεῖν κατὰ
τρία διάπυρα τὰ θρύμματα· ἔστω δὲ τὸ οὖρον ὅσον
χοῦς· τοῖσι δὲ σύμπασι πυριᾶν αὐτὴν ὅσον τριήκοντα
πυριάς· ἐπὴν δὲ πυριάσῃς, σμῆχε τὴν κεφαλὴν τῷ

---

[278] Add. μὴ Θ.   [279] διπάλαιστα Ermerins: διπαλαστα
Θ: διπλὰ ἐς τὰ MV.

Also wild gourd, mercury herb, soda, and hedge mustard; to draw down more quickly, mandrake root, blister beetle, tufted thyme, bay berry, iris unguent, and laurel unguent.

Stir spurge juice, remove the sticky part, and give an amount equal to a vetch: excellent suppository. If too much passes, have the patient irrigate herself with wine.

Or dissolve some burned copper, take it up warm on a piece of wool, and apply it.

75. (1) An agent to promote pregnancy: a saucer (= oxybaphon) of cedar oil and four drachmas of cow's fat, mix together and knead smooth, form into suppositories, and apply with the patient in the fasting state; when it is in place, have her fast through the day. After her menses, she should insert it twice, at dawn and in the evening; after dinner have her wash herself and sleep with her husband.

Or crush black cumin and bind it in a rag; also give goose oil to apply as a suppository.

(2) An agent to promote pregnancy, i.e., to treat a woman so that she will become pregnant: take some old urine and pieces of iron slag two palms long; then—after sitting the patient down on a stool—have her cover both her body and her head, and place (sc. the fomentation) between her feet: the physician should throw the pieces of slag red hot three at a time into one chous of urine. Foment the patient with about thirty of these pieces altogether. After you have fomented (sc. the woman once),

---

280 ἐς πόδαν ἰητρὸν Θ: ἐς πόδανι πτῆρα M: ἐς πόδας νιπτῆρα V.

οὔρῳ ᾧ ἂν[281] πυριαθῇ, ἐναποσβεννύων πάλιν τοὺς λίθους καὶ τὴν πυρίην πάλιν θερμήνας· μετὰ δὲ[282] ταῦτα
λοῦε κατὰ κεφαλῆς ὡς πλείστῳ, ἔστω δ᾽ ἐν τῷ ὕδατι
πόλιον καὶ τῆς λύγου ὡς πλεῖστα· ταῦτα δὲ ποιεῖ ἡμέρας ἑπτά· τρὶς δὲ τούτων ὑποθυμιᾶν ἑκάστην πυρίην
164   πρὸ τοῦ σμήχεσθαι· χριέσθω δ᾽ ἐκ τοῦ | λουτροῦ
ἐλαίῳ δαφνίνῳ.

Μετὰ δὲ τὸ δεῖπνον φαγοῦσα κρόμμυα ἐμβαπτό-
μενα[283] ἐς μέλι, καὶ μελίκρητον χλιαρὸν ὅσον κοτύλας
τέσσερας πιοῦσα, ἔπειτα, ἐπὴν φάγῃ, σμικρὸν ἐπ-
ισχοῦσα, ἀπεμείτω· ἔπειτα πάλιν πιοῦσα τὸ ἴσον
ἐμείτω, καὶ ἀνακλιθεῖσα ὑπτίη, τοῦ πηγάνου ἐχέτω
καὶ ἐν τοῖσιν ὠσὶ καὶ ἐν τῇσι ῥισί· καὶ ἄρτον ζυμίτην,
ὅσον ἕκτον μέρος χοίνικος, ἐς ζωμὸν ἐνθρύψασα ὄρνιθος, ἔχοντα σελίνου ὅσον χήμην, προσφερέσθω·
καὶ πάλιν δίδου τὸ ἴσον ἐπὶ τῷ δείπνῳ· τωὐτὸ δὲ ποιεῖ
τὰς ἑπτὰ ἡμέρας. ἔπειτα τὴν κοιλίην κλύζε ἡμέρας
ἑπτά· ἔστω δὲ κλύσμα ῥητίνης δραχμαὶ τέσσερες,
μέλιτος ὀξύβαφον τῶν πλατέων, ἔλαιον ἴσον, πιτύρων
σητανίων χυλός, λίτρου ἀφρός, ᾠὰ ἑπτά· κοτύλαι δὲ
ὀκτὼ τοῦ κλύσματος, τούτων αἱ τρεῖς πτισάνης χυ-
λοῦ· κλυζέτω δὲ πλαγίην, καὶ λοῦε πολλῷ.[284] προστιθέσθω δὲ βαλάνους,[285] κατεχέτω δὲ ἔστ᾽ ἂν κατατακῇ·
ἔστωσαν δὲ λιβάνου, λίτρου, χαλβάνης, μέλιτος
ἑφθοῦ· σιτίῳ δὲ χρήσθω τῷ αὐτῷ.

[281] ᾧ ἂν Littré: ἕως Θ: ἕως ἂν MV.
[282] μετὰ δὲ Foes in note 334, with a reference to Cornarius᾽
et fomentum rursus calefacias. Postea vero: δὲ μέτα codd.

anoint her head with the same urine with which she has been fomented, quenching hot stones again to reheat the fomentation fluid. After that bathe the patient very copiously down over her head with water containing hulwort and much agnus castus; continue for seven days; fumigate with these three times for each fomentation before you anoint (sc. with the urine). The patient should anoint herself with laurel oil on coming from the bath.

After her dinner have the patient eat onions dipped in honey and drink four cotyles of warm melicrat, retain these for a short time after she has eaten them, and then vomit; then have her drink the same thing and vomit again, lie down on her back, and insert some rue into her ears and nose. Next she should crumble leavened bread to the amount of one sixth choinix into chicken soup containing a cheme of celery, and take that. Give the same amount again for dinner, and continue for seven days. Then flush the cavity for seven days with the following enema: four drachmas of resin, a level oxybaphon of honey, the same of olive oil, water (sc. extracted) from this year's wheat, aphronitrum, and seven eggs. The enema fluid should measure eight cotyles, three of which are barley gruel. The patient should apply the enema while lying on her side, wash herself with copious water, and then apply suppositories, retaining each until it melts: these should be of frankincense, soda, all-heal juice, and boiled honey; she should take the same foods.

---

283 -όμενα Potter: -ομένη ΘΜ: -ομένην V.
284 πολλῷ Θ: ὀλίγῳ MV.
285 Add. ἑπτὰ τῆς ἡμέρης MV.

ὑποθυμιᾶν δὲ πόλιον, ὄνου τρίχας, λύκου κόπρον, ἐπίβαλλε δὲ ὡς πλεῖστον ἐπ' ἀνθρακιήν, καὶ περικαθίσας αὐτὴν καὶ περιστείλας θυμία, φυλασσόμενος μὴ κατακαύσῃς.

(3) Ἢν δὲ γυνὴ μὴ δύνηται τίκτειν πρόσθεν τίκτουσα, λίτρον καὶ ῥητίνην καὶ σμύρναν καὶ κύμινον Αἰθιοπικὸν καὶ μύρον τρίβειν ἐν αὐτῷ, καὶ προστίθεσθαι.

Ἡ γλήχωνα ξηρὴν ἐν ὀθονίῳ προστιθέσθαι· πίνειν δὲ τὴν γλήχωνα, ἐπὴν εὕδειν μέλλῃ.

(4) Κυητήριον· διαιτᾶν δεῖ τὴν γυναῖκα ἥτις δεῖται κυήσιος, καὶ διδόναι αὐτῇ ἅπερ λεχοῖ καὶ ἐσθίειν καὶ πίνειν, τῷ δὲ ἀνδρὶ τῆς γυναικὸς τὰ ἄλλα | πλὴν σκορόδου, καὶ κρομμύου, καὶ ἔτνους, καὶ ὀπίου, καὶ[286] σιλφίου, καὶ ὅσα φυσητικά· τούτων δ' ἀπεχέσθω.

(5) Ἔγχυτον κυητήριον· γάλα γυναικὸς κουροτρόφου, σίδης νεαρῆς τοὺς κόκκους τρίψας, ἐκπιέσας τὸν χυλόν, καὶ χελώνης θαλασσίης τὸν περίνεον κατακαύσας,[287] τρίψας, ἐγχέαι ἐς τὰ αἰδοῖα.

(6) Ἔγχυτον κυητήριον, ὅταν μὴ κυίσκεται· γάλα καὶ ῥητίνην καὶ σίδης γλυκείης χυλόν, ταῦτα ξὺν μέλιτι μίξας, ἐγχεῖν πάντα.

(7) Κυητήριον· βολβοῦ τοῦ λευκοῦ καρπὸν ἢ τὸ ἄνθος τρίψας ξὺν[288] μέλιτι, ἐν εἰρίῳ ἑλίξας, προσθέσθω πρὸς τὴν μήτρην ἐπὶ τρεῖς ἡμέρας· τῇ δὲ τετάρτῃ, μαλάχην ἀγρίην τὴν πλατύφυλλον τρίψασα, μῖξαι γυναικὸς γάλακτι, καὶ ἐς εἴριον ἐνελίξασα,

166

Fumigate from below with hulwort, ass's hair, and wolf's excrement: pour good amounts of these on to burning charcoal and have the patient sit down over this: wrap her and carry out the fumigation carefully so as not to burn her.

(3) If a woman who has given birth before is no longer able to, knead soda, resin, myrrh, Ethiopian cumin and unguent together, and apply as a suppository.

Or have her apply dry pennyroyal in a piece of linen; she should also drink pennyroyal at bed time.

(4) To promote pregnancy: any woman who must become pregnant should follow a regimen, and you should give her the same things to eat and drink as you would to a parturient; to her husband give everything except garlic, onions, bean and pea soups, opium, silphium and flatulents, which he should avoid.

(5) An infusion to promote pregnancy: take milk from a woman who is nursing a male child, crush the seeds of a fresh pomegranate and squeeze out the juice, burn and triturate the perineum of a sea turtle: inject all this into the patient's genitalia.

(6) An infusion to promote pregnancy, when a woman is unable: milk, resin, and juice of the sweet pomegranate: mix with honey, and inject.

(7) An agent to promote pregnancy: grind seed of white tassel hyacinth or its flower into honey, wrap this in a piece of wool, and have the woman apply it against her cervix for three days. On the fourth day have her crush some broad-leafed mallow, mix it with woman's milk, wrap this

---

286 ὀπίου καὶ Θ: ὀποῦ MV.

287 -καύσας MV: -σκευάσας Θ.   288 ξὺν MV: οὖν Θ.

προσθέσθω· εἶτα κοιμάσθω μετ' ἀνδρός· προρροφείτω
δὲ γλήχωνα ἐν ἀλεύροισιν ἐφθοῖσι, καὶ πινέτω γλήχωνα ἐν οἴνῳ λεπτήν. ἢν δὲ τούτου μὴ ἐσακούῃ, λαβὼν κόνυζαν εὔοσμον, συγκόψας καὶ ἐκθλίψας τὸν
χυλὸν καὶ συμμίξας οἴνῳ, πινέτω νῆστις.

(8) Κυητήριον· ἀσπαράγου καρπὸν πινέτω ἐν οἴνῳ.

(9) Κυητήριον· χορίον γυναικὸς καὶ τῶν εὐλέων τὰς
κεφαλὰς τρίψαι, διεὶς στυπτηρίην Αἰγυπτίην ἐν χηνὸς στέατι, ἐν εἰρίῳ προσθέσθω πρὸς τὸ στόμα τῆς
μήτρης.

(10) Κυητήριον· ἰὸν χαλκοῦ, ἄνθος, ἡμιωβόλιον
ἑκατέρου, λιβανωτὸν ἄρσενα, στυπτηρίην σχιστήν,
οἰνάνθην²⁸⁹ ἀμπέλου, κηκίδα, σμύρναν, σίδιον, ῥητίνην, πόλιον, ὀβολὸν ἑκάστου, ἐν μέλιτι τρίψασα,
προσθέσθω ἐπὶ τρεῖς ἡμέρας δὶς τὴν ἡμέρην· ἢν δὲ
δριμύτερον ᾖ, παραμίσγειν χηνὸς στέαρ καὶ λίτρον
168 ὀπτόν· | διδόναι δὲ οἶνον, φεύγοντα τὸ ἐν αὐτῷ μένος.

(11) Κυητήριον· εὐλὴν ἢ τὴν κέρκον ἔχει, λαβὼν
αὐτῶν τρεῖς ἢ τέσσερας καὶ ὀρίγανον λεπτήν, τρίψας
ἐν ῥοδίνῳ μύρῳ, προσθέσθω πρὸς τὸν στόμαχον.

(12) Κυητήριον· ἀνδράχλην τρίψας ἐν χηνὸς στέατι
χηνὸς καὶ σμύρναν καὶ πράσου σπέρμα καὶ βοὸς χολήν, εἰρίῳ ἐνελίξας προσθεῖναι πρὸς τὸ στόμα²⁹⁰ τῆς
μήτρης.

---

²⁸⁹ λιβ. . . . οἰνάνθην Littré: λιβανωτοῦ ἄρσενος, στυπτηρίης, σχιστῆς, οἰνάνθης ΘΜ: λιβάνου ἄρσενος, στυπτηρίην,
σχιστῆς, οἰνάνθης V.

in a piece of wool, and apply it to herself. Then have her sleep with her husband: beforehand she should take some pennyroyal in boiled meal, and drink tender pennyroyal in wine. If she does not respond to this, take fragrant fleabane, crush it, and squeeze out the juice, mix this with wine, and have the patient drink it in the fasting state.

(8) An agent to promote pregnancy: have the patient drink seed of stone sperage in wine.

(9) An agent to promote pregnancy: crush a piece of a woman's placenta and some maggot heads, dissolve in Egyptian alum and goose grease, and have the patient apply this in a piece of wool against her cervix.

(10) Another to the same purpose: verdigris and flower of copper—a half obol of each—and male frankincense, powdered alum, grapevine blossom, oak gall, myrrh, pomegranate peel, resin and hulwort—an obol of each: have the woman knead these in honey and apply the suppository to herself twice daily for three days. If this is too sharp, mix in goose grease and roasted soda. Also give wine, but be careful to avoid its violence.

(11) An agent to promote pregnancy: take three or four of the worms that have a tail, together with some tender marjoram, knead these in rose unguent, and have the patient apply it against her cervix.

(12) An agent to promote pregnancy: knead purslane in goose grease along with myrrh, leek seed, and bull's gall, wrap in a piece of wool, and apply against the patient's cervix.

---

290 τὸ στόμα Θ: τὸν στόμαχον MV.

(13) *Ην[291] καταμήνια γίνηται πολλά, καὶ μὴ ξυλλαμβάνῃ, κυητήριον· χαλκοῦ ἄνθος, ὀβολοὺς δύο, καὶ στυπτηρίης σχιστῆς, τρίψας λεῖον ἐν μέλιτι, εἶτα εἰρίῳ ἀνασπογγίσας, ἐνδήσας τὸ εἴριον ἐς ὀθόνιον λίνῳ, προσθέσθω ὡς ἐσωτάτω· τὸ δὲ λίνον ὑπερεχέτω· εἶτα ὅταν καλῶς ἀποκαθαρθῇ, ἀφελέτω, καὶ οἶνον ἀναζέσας εὐώδεα, μυρσίνης φύλλα ἐμβάλων, τούτῳ προσκλυσάσθω, καὶ ἴτω πρὸς τὸν ἄνδρα.

(14) Κυητήριον προσθετόν· μέλι, σμύρναν, μυρίκης τὸν καρπόν, ῥητίνην ὑγρήν, χηνὸς ἄλειφα, τρίψας πάντα ἐν τῷ αὐτῷ, εἰρίῳ ἐνελίξας, προστιθέσθω.

(15) Κυητήριον προσθετὸν σφόδρα ἀναστομῶσαι μήτρην, ὅταν μεμύκῃ καὶ μὴ δύνηται κυῆσαι, ὕδωρ ἐκκαθῆραι· λαβὼν σχεδιάδα τὴν ὀλίγην, καὶ σχῖνον, καὶ κύμινον, καὶ κύπαιρον, καὶ ἀγρίην κολοκύντην, καὶ λίτρον ἐρυθρόν, καὶ ἅλας Αἰγύπτιον, καὶ σχεδιάδα τὴν μεγάλην, ταῦτα πάντα λεῖα ποιήσας, δι᾽ ὀθονίου διηθῆσαι· λαβὼν δὲ μέλι, ἕψησον ἐπὶ πυρὶ μαλθακῷ· ἐπειδὰν δὲ ζέσῃ, παράμιξον κηρόν, ῥητίνην· ἔπειτα συμμίξας πάντα, ἔλαιον ἐπιχέας, ἀφελών, χλιήνας, εἴριον ἀγκυλιδωτὸν[292] ἐνελίξας τὴν μήτρην
170 προστίθει, | μέχρι ἂν καθαρθῇ.

(16) Κυητήριον· ἢν γυναῖκα βούλῃ κυῆσαι, [καὶ][293] τοῖσι καθαρτηρίοισι νῆστις χρήσθω,[294] καὶ ἢν δέῃ[295] πρὸς τὸν ἄνδρα ἰέναι, δαφνίδας μελαίνας δέκα, λιβα-

---

291 Add τὰ MV.   292 ἀγκυλιδωτὸν (cf. Galen vol. 19, 69)
V: ἀγκαλίδωτον Θ: ἀγκαλίδοτον Μ.

(13) An agent to promote pregnancy if a woman's menses become excessive and she does not become pregnant: take two obols of flower of copper together with powdered alum, knead fine in honey, soak this up in a piece of wool, bind the wool in a piece of linen with a thread, and have the woman apply this as far inside as she can, while still leaving the thread to hang out. Then, after she has been well cleaned, have her remove the pessary. Boil some fragrant wine, add myrtle leaves, and have the patient flush herself with this and then go to her husband.

(14) A suppository to promote pregnancy: take honey, myrrh, tamarisk seed, moist resin and goose grease, knead these all together, wrap them in a piece of wool, and have the woman apply it.

(15) A suppository to promote pregnancy, to open the mouth of the uterus forcefully when it is closed and a woman cannot become pregnant, and to clean out fluid: take some lesser alkanet, mastic, cumin, galingale, wild gourd, red soda, Egyptian salt, and greater alkanet: make all these smooth and strain them through a linen cloth; take honey, heat it over a weak fire, and when it comes to a boil mix in wax and resin. Mix all this together, pour in oil, set it aside, warm it, wrap it in a piece of wool with a loop for a handle, and apply to the patient's uterus until it is cleaned.

(16) An agent to promote pregnancy: if you want a woman to become pregnant, have her employ purgatives in the fasting state, and when she needs to approach her husband, mix ten black bayberries, three pinches of frank-

---

293 Del. Ermerins.   294 χρήσθω Ermerins: χρῆται codd.
295 ἢν δέῃ Ermerins: ἢν δεῖ Θ: δεῖ M.

νωτοῦ τρεῖς δραχμίδας, καὶ κυμίνου ὀλίγον μίξας μέλιτι, ἐς εἰρίον πινῶδες ἐγκυλίων, ἅπαξ τῇ ἡμέρῃ²⁹⁶ τῇ
αὐτῇ [καὶ]²⁹⁷ προστιθέσθω, καὶ ἀφαιρείσθω ἅπαξ ἡμέρας δὲ τέσσερας, καὶ ἔπειτα ἀσιτείτω τὰς ἴσας.

76. Ἀτόκιον· ἢν μὴ δέῃ κυΐσκεσθαι, μίσυ ὅσον
κύαμον διεὶς ὕδατι, πιεῖν δοῦναι, καὶ ἐνιαυτόν, ὡς
ἔπος εἰπεῖν, οὐ κυΐσκεται.

77. (1) Ὠκυτόκια δυστοκούσῃσι· δάφνης ῥίζην²⁹⁸
ξύσας ἢ τοὺς κόρους²⁹⁹ ὅσον ἥμισυ ὀξυβάφου, δὸς
πιεῖν ἐφ᾽ ὕδατι θερμήνας.

(2) Ὠκυτόκιον· δικτάμνου ὅσον δύο ὀβολοὺς τρίψας
ἐν ὕδατι θερμῷ πινέτω.

Ἢ ἀβροτόνου τριόβολον³⁰⁰ καὶ κεδρίδας καὶ ἄννησον τρίψας ἐν γλυκέος οἴνου κυάθῳ, παραχέας ὕδατος
παλαιοῦ κύαθον, δὸς ἐκπιεῖν· τοῦτο καλὸν δίδοται, ἢν
πρὸ τῶν ὠδίνων δοθῇ.

Ἢ δικτάμνου ὀβολόν, σμύρνης ὀβολόν, ἀννήσου
δύο ὀβολούς, λίτρου ὀβολόν· ταῦτα τρίψας λεῖα, γλυκέος οἴνου ἐπιχέας κύαθον καὶ ὕδατος θερμοῦ δύο
κυάθους, δὸς ἐκπιεῖν, εἶτα³⁰¹ λοῦσον θερμῷ.

(3) Ὠκυτόκιον· ῥητίνην τερμινθίνην, μέλι, ἔλαιον
διπλάσιον τούτων, οἶνον εὐώδεα ὡς ἥδιστον, ταῦτα
172 συμμίξας, | χλιήνας, διδόναι ἐκπιεῖν πλεονάκις· κατα-
στήσει δὲ καὶ τὰς μήτρας, ἢν φλεγμαίνωσιν.

²⁹⁶ τῇ ἡμέρῃ Littré: τῆς ἡμέρης codd.
²⁹⁷ Del. I.
²⁹⁸ ῥίζην Froben: ῥίζης Θ: ῥίζας MV.

incense, and a little cumin with honey, roll this up in a piece of greasy wool, and have her apply it as a suppository once daily, and remove it once daily for four days; then have her fast for the same.

76. A contraceptive: if a woman must not become pregnant, give her misy to the amount of a bean dissolved in water to drink, and they say she will not become pregnant for a year.

77. (1) Agents to accelerate birth in women having a difficult delivery: shred root of laurel or its sprouts to the amount of one half oxybaphon, and give in water after heating it.

(2) An agent to accelerate birth: crush two obols of dittany in hot water and have the patient drink it.

Or grind three obols of southernwood together with juniper berries and anise in a cyathos of sweet wine, add a cyathos of standing water, and give to drink: this is beneficial if given before the birth pangs.

Or an obol of dittany, one of myrrh, two of anise, and one of soda: knead these smooth, add one cyathos of sweet wine and two of warm water, and give to drink; then wash the patient in hot water.

(3) An agent to accelerate birth: take turpentine and honey, twice this amount of olive oil, and some very pleasant fragrant wine, mix together, warm, and give many times to drink; this will also settle the uterus if it is inflamed.

---

299 κόρους Θ: κόκκους MV.
300 τριόβολον Θ: δραχμὴν MV.
301 εἶτα om. Θ.

(4) Ὠκυτόκιον· τοῦ σικύου τοῦ ἀγρίου, ὅστις ἂν ᾖ
λευκός, τὸν καρπὸν ἐμπλάσας κηρῷ, εἶτα εἰρίῳ ἐνελίξας φοινικέῳ, περίαψον περὶ τὴν ὀσφῦν.

(5) Ἢν δὲ κύουσα πολὺν χρόνον ἐπέχηται καὶ μὴ
δύνηται τεκεῖν, ἀλλ᾽ ὠδίνη πλείους ἡμέρας, νέη δὲ ᾖ
καὶ ἀκμάζῃ καὶ πολύαιμος, τάμνειν τὰς ἐν τοῖσι σφυ-
ροῖσι φλέβας καὶ ἀφαιρέειν τοῦ αἵματος, πρὸς τὴν
δύναμιν ὁρῶν· καὶ μετὰ ταῦτα λοῦσαι θερμῷ ὡς ὑπὸ
θερμοῦ[302] διαφλύωνται· πιεῖν δὲ διδόναι ἄγνου καρπὸν
καὶ δίκταμνον Κρητικὸν ἴσον ἑκατέρου ἐν οἴνῳ λευκῷ
ἢ ὕδατι· προσθετὸν δὲ ποιήσας πρόσθες χαλβάνην
καὶ δαφνίδας καὶ ῥόδινον ἔλαιον, ἐν εἰρίῳ ἐνελίξας.

(6) Ὠκυτόκιον· τῆς δρυοπτέριδος τὴν ῥίζαν τρίψας
ἐν οἴνῳ, δὸς πιεῖν· ἢ καὶ ἀδίαντον τρίβειν ἐλαίῳ, καὶ
διεῖσα πίνειν[303] ἐν οἴνῳ ἀκρήτῳ.

78. (1) Λεχοῖ τὰ λοχεῖα καθαίρει ἄμεινον· χελώνης
θαλασσίης ἧπαρ χλωρὸν ἔτι ζώου ἐν γάλακτι γυναικείῳ τρίβειν, καὶ ἴρινον, καὶ οἶνον ἀναδεῦσαι καὶ
προστιθέναι· ἢ λινόζωστιν τριπτὴν ἐν εἰρίῳ προστι-
θέναι· ἢ ἀρτεμισίην τρίβων ἐν εἰρίῳ προστιθέναι· καὶ
λινοζώστιος καὶ σικύης ὀλίγον τρίψας, οἴνῳ καὶ μέλιτι δεύσας, προστιθέναι.

(2) Ἐκ τόκου καθαρτήριον λοχείων· πυροὺς τριμη-
174 νιαίους ǀ ἐρείκειν ὅσον ἡμιχοίνικος, ἕψειν δὲ ἐν ὕδατος
κοτύλαις τέσσερσιν· ὅταν δὲ ζέσῃ, δὶς ἢ τρὶς δοῦναι
ῥυφῆσαι.

---

302 ὡς ὑπὸ θερμοῦ om. MV.
303 καὶ δ. πίνειν Θ: διεὶς ἀλείφειν MV.

(4) An agent to accelerate birth: take fruit of the wild cucumber that has already turned white, plaster it with wax, wrap this in a piece of deep red wool, and fasten it around the patient's loin.

(5) If a pregnant woman who is held back for a long time from giving birth and unable to do so, but she has had birth pangs for several days, and she is young, in her prime, and rich in blood, incise the vessels at her ankles and draw off blood, keeping an eye on her strength. Next wash her with hot water so that she will be permeated by the heat. Then give her fruit of the chaste tree and Cretan dittany—an equal amount of each—in white wine or water to drink and apply a suppository made with all-heal juice, bayberries, and rose oil wrapped in a piece of wool.

(6) An agent to accelerate birth: grind root of the black oak-fern in wine and give to drink; or also grind maiden-hair in olive oil, dissolve this in undiluted wine, and give to drink.

78. (1) In a woman who has just given birth the following cleans her lochia very effectively: take the fresh liver of a still living sea turtle, grind it up in woman's milk, suffuse this with iris unguent and wine, and apply as a suppository. Or use ground mercury herb in a piece of wool. Or grind up artemisia and apply it in a piece of wool. Also grind a small amount each of mercury herb and bottle gourd, soak them in wine and honey, and apply.

(2) An agent to clean the lochia after childbirth: bruise a half-choinix of three-month wheat and boil it in four cotyles of water: when it is boiled, give the gruel two or three times.

(3) Λοχεία καθῆραι· τῆς ἀκτῆς τὰ φύλλα ἑψῆσαι ἐν ὕδατι, ἀποχέαι καὶ πίνειν· ἐσθιέτω δὲ κράμβας ἑφθὰς καὶ πράσα καὶ μάραθον καὶ ἄνηθον καὶ πουλύποδας καὶ καράβους· ἢ τοῦ ῥοῦ τὰ φύλλα καὶ ἐρύσιμον[304] ἐν οἴνῳ, ἄλφιτα ἐπαλύνων, δοῦναι πιεῖν· ἢ μίσυ ὅσον δύο ὀβολοὺς τρίψας, οἴνῳ φορύξας, προσθεῖναι.

(4) Καθαρτήριον ἐπιμηνίων καὶ λοχείων μάλιστα, καὶ ὕδωρ ἄγει καὶ τἆλλα· στρουθίου ῥίζαν λεῖον κεκομμένην,[305] ὅσον τοῖσι τρισὶ δακτύλοισι μέλιτι δεύσας, προσθεῖναι· φύεται δὲ οἷον τὸ ἐν Ἄνδρῳ ἐν τοῖσιν αἰγιαλοῖσιν.

(5) Λέχοι πάσῃ· ἐρύσιμον ἕψειν ἐν ὕδατι, καὶ ἐπιχεῖν ἔλαιον, ὅταν ἀναβλύῃ, καὶ ψύχειν, καὶ ὑποθυμιᾶν· ἀγαθὸν δὲ καὶ τὸν χυλόν, καὶ σιτίοισι μαλθακοῖσι χρῆσθαι.

(6) Λοχεία καθαίρει· ἐρύσιμον τρίβειν ἡσύχως, καὶ ἀποφυσᾶν τὸ κέλυφος· ὅταν δὲ καθαρὸν ποιήσῃς, τρίβειν λεῖον, καὶ ὕδωρ παραστάζειν, καὶ ἅλας καὶ ἔλαιον μίσγειν· εἶτα ἐπιπάσας ἄλητα, ἕψε, καὶ[306] ῥυφείτω.

(7) Ἢν δὲ μὴ ᾖ[307] κάθαρσις ἡ λοχείη, λαβὼν σικύης ἐντεριώνης ὅσον τριώβολον, καὶ ἀρτεμισίην ποίην, καὶ λιβανωτοῦ ὀβολόν, τρίψας, ἐν μέλιτι μίξας, ἐς εἴριον ἐνελίξας, πρόσθες πρὸς τὸ στόμα τῆς μήτρης, νυκτὸς καὶ ἡμέρης, εἶτα μέχρι ἐφ' ἡμέρας πέντε τοῦτο ποιεῖν.

176   Ἢ τὸ ἀμπέλιον τρίβων χλωρόν, ἐν | μέλιτι ἐπιχέων, ἐς εἴριον ἑλίξας, προστιθέναι τὸν αὐτὸν τρόπον.

(3) To clean the lochia: boil elder leaves in water, pour the water off, and give to drink; have the woman also eat boiled cabbage, leeks, fennel, anise, octopus, and crayfish. Or take sumac leaves and hedge mustard in wine, sprinkle with barley meal, and give this to drink. Or grind two obols of misy, moisten in wine, and apply as a suppository.

(4) An agent that cleans the menses and the lochia in particular, and also draws water and other things: mix a pinch of finely ground soapwort root with honey, and apply. The plant grows like the one on the island of Andros by the seashore.

(5) For any parturient: boil hedge mustard in water, pour in olive oil, and when this boils cool and use as a fumigation from below; the juice too is good, and also employ emollient foods.

(6) This cleans the lochia: gently mill some hedge mustard and blow away the hulls; when they are clean, grind them fine, sprinkle with water, and add salt and oil; then sprinkle meal over this, boil, and give to the patient to take.

(7) If the lochial flow fails to occur, take three obols of the insides of a bottle gourd, some artemisia herb, and an obol of frankincense, knead, mix into honey, wrap in a piece of wool, and apply against the mouth of the patient's uterus both night and day: continue for five days.

Or crush green grape vine, mix into honey, wrap in a piece of wool, and apply in the same way.

---

304 Add. λίην Θ.   305 λ. κ. Littré after Calvus' *struthive radix tenuiter trita*: λείου -μένου codd.

306 καὶ om. Θ.

307 ᾗ V: ἡ Θ: ἴη M.

Ἢ τῆς κυπαρίσσου τὸν καρπὸν καὶ λιβανωτὸν τρίψας ἐν τῷ αὐτῷ, ῥοδίνῳ μύρῳ διεὶς καὶ μέλιτι, εἰς εἴριον ἐλίξας, πρόσθες.

Ἢ ἀβρότονον ὅσον τριώβολον,[308] καὶ σικύης ἐντεριώνην ὅσον ὀβολὸν τρίψας ἐν μέλιτι, ἐς εἴριον ἐλίξας, προστιθέναι.

Ἢ ἐλατηρίου ὀβολὸν καὶ σμύρνης ὀβολὸν τρίψας ἐν μέλιτι, ἐς εἴριον ἐλίξας, πρόσθες.

Ἢ κυπαρίσσου καρπὸν καὶ σικύης ἐντεριώνην καὶ λιβανωτὸν[309] μέλιτι μίξας, ἐν εἰρίῳ πρόσθες.

(8) Καθαρτήριον, ἢν ἐκ τόκου μὴ καθαρθῇ· τοῦ τριφύλλου ἐν οἴνῳ λευκῷ πίνειν· καὶ τὰ καταμήνια καταρρήγνυσι τωὐτὸ καὶ προσθετὸν καὶ ἔμβρυον ἐκβάλλει.

(9) Μητρέων καθαρτικόν, ὅταν παιδίου ἐναποθανόντος αἷμα ἐμμείνῃ· κολοκύντην ἀγρίην τρίψασα ἐν μέλιτι λειχέτω, ἢ προσθέσθω.

(10) Λέχοι πάσῃ, ἢν μὴ καλῶς ἴῃ· ἐρύσιμον ἕψειν καὶ ἄλφιτον, καὶ ἔλαιον ἐπιχέας, ὅταν ᾖ ἐφθόν, ῥυφείτω, καὶ σιτίοισιν ὡς μαλθακωτάτοισι χρήσθω.

Ἢ σκαμωνίην τρίψας ἐν γάλακτι γυναικείῳ, εἰρίῳ ἀνασπογγίσας, προσθεῖναι· ἢ τῆς γλυκείης ῥίζης μέλιτι δεύσας καὶ ῥοδίνῳ μύρῳ ἢ Αἰγυπτίῳ ἐν εἰρίῳ προσθέσθω· ἄλητον πλυτὸν ὁμοίως προσθέσθαι· πινεῖν δὲ κρήθμου καρπὸν καὶ σεσέλιος, καὶ πηγάνου καρπόν, ἑκάστου ὀβολοὺς δύο τρίβειν ὁμοῦ ἐν ἀκρήτῳ· ἢν μὴ πυρεταίνῃ, διδόναι.

Or knead cypress berries and frankincense together, dissolve in rose unguent and honey, wrap in a piece of wool, and apply.

Or knead three obols of southernwood and an obol of the insides of a bottle gourd into honey, wrap in a piece of wool, and apply.

Or knead an obol of squirting cucumber juice and an obol of myrrh into honey, wrap in a piece of wool, and apply.

Or mix cypress berries, the insides of a bottle gourd, and frankincense in honey, and apply in a piece of wool.

(8) A cleaning agent: if a woman is not cleaned after she has given birth, give her clover in white wine to drink. A suppository of the same also causes the menses to break out, and expels a fetus.

(9) An agent to clean the uterus when, after the death of a fetus, blood remains inside: have the woman knead wild gourd with honey, and take this as an electuary or apply it as a pessary.

(10) For any parturient, if she is not doing well: boil hedge mustard and barley meal, add olive oil, and when this has boiled give it to the patient to take; she should also employ the most emollient foods.

Or knead scammony in woman's milk, soak this up in a piece of wool, and apply. Or dissolve some licorice in honey with rose unguent or Egyptian unguent, and have the patient apply it on a piece of wool. Apply washed barley meal in the same way. Have a woman drink samphire seed, hartwort seed, and rue seed—two obols of each— ground together into undiluted wine, unless she has fever.

---

308 τριώβολον Θ: δραχμὴν MV.
309 Add. τρίψαι ἐν ταὐτῷ Θ.

(11) Χορίον ἐκβάλλει προσθετόν· τῆς ἀκτῆς τὸ φύλλον, προπυριᾶν δὲ καὶ προστιθέναι τὸ σὺν τῇ κανθαρίδι τοῦτο καὶ ὅσα ἠπεδανὰ ἰᾶται· ἢν δὲ ἀμύσ-

178 σηται καὶ δάκνῃ, αὐτίκα ἀφαιρέειν, καὶ ἐς | ῥόδινον εἴριον ἀποβάπτουσα προσθέσθω.

(12) Προσθετὸν χορίον ἐξάγει, καὶ ἐπιμήνια κατα-σπᾷ καὶ ἔμβρυον ἀπόπληκτον· κανθαρίδας πέντε ἀποτίλας τὰ πτερὰ καὶ τὰ σκέλεα καὶ τὴν κεφαλήν, εἶτα τρίβολον παραθαλάσσιον κόψας σὺν τῇ ῥίζῃ καὶ τοῖσι φύλλοισιν ὅσον κόγχην,[310] καὶ τὸ βοάνθεμον[311] χλωρὸν τρίψας τὸ ἔξω σκληρὸν ὅσον κόγχην,[312] καὶ σελίνου σπέρμα τὸ ἴσον, καὶ σηπίης ᾠὰ πεντεκαίδεκα ἐπ᾽ οἴνῳ γλυκεῖ κεκρημένῳ προσθεῖναι, καὶ[313] ἐπὴν ὀδύνη ἔχῃ, ἐν ὕδατι θερμῷ καθήσθω, καὶ μελίκρητον ὑδαρὲς πινέτω, καὶ οἶνον γλυκύν, καὶ τοῦ τετριμμένου ὅσον στατῆρα Αἰγιναῖον ἐν οἴνῳ πίνειν γλυκεῖ· ἐπὴν δὲ ὀδύνη ἔχῃ, λευκοὺς ἐρεβίνθους καὶ ἀσταφίδας ἑψή-σας ἐν ὕδατι ψῦξαι καὶ διδόναι πίνειν, καὶ ὅταν ἡ στραγγουρίη ἔχῃ, ἐν ὕδατι χλιαρῷ ἐγκαθήσθω.

(13) Ἐκβόλιον ὑστέρων· σικύου ἀγρίου τὸν ὀπὸν ὅσον πόσιν ἐς[314] μαζίον ἐμπλάσας προστιθέναι, προ-νηστεύσασα ἐπὶ δύο ἡμέρας, οὐκ ἂν εὕροις τοῦδ᾽ ἄμεινον.

(14) Πειρητήριον· μώλυζαν σκορόδου ἀποζέσας προσθεῖναι.

(15) Πειρητήριον· νέτωπον ὀλίγον εἰρίῳ ἐνελίξας προσθεῖναι, καὶ ὀρὴν ἢν διὰ τοῦ στόματος ὄζῃ.

(11) This suppository expels the placenta: elder leaves. First apply a fomentation, and then apply the leaves together with a blister beetle and whatever gently heals; if, however, the suppository causes laceration and irritation, remove it immediately and have the woman dip a piece of wool in rose (sc. oil) and apply it.

(12) A suppository that expels the placenta and brings down the menses or a paralyzed fetus: remove the wings, legs and head from five blister beetles, then pound a cotyle of water chestnut together with its roots and leaves, and grind a cotyle of the hard outside of green oxeye together with the same amount of celery seed and fifteen cuttlefish eggs: apply in diluted sweet wine. When pain is present, have the patient take a sitz bath in hot water, and drink diluted melicrat and sweet wine; she should also drink an Aeginetan stater of the ground mixture in sweet wine. When pain is present boil white chickpeas and grapes in water, cool, and give to drink; when there is strangury, have the woman take a sitz bath in warm water.

(13) A medication to cause expulsion from the uterus: a draft of juice from the squirting cucumber daubed on a barley scone: apply after having the patient fast for two days. You would never find anything better than this.

(14) Test (sc. for fertility): boil a head of garlic and insert it.

(15) Test: wrap a little oil of bitter almonds in a piece of wool, insert it, and see whether the patient emits the smell through her mouth.

---

310 ὅσον κόγχην om. Θ.   311 βοάν- Θ: εὐάν- MV.
312 χλωρὸν . . . κόγχην om. MV.
313 προσθεῖναι, καὶ om. Θ.
314 πόσιν ἐς Potter: ποσι Θ: ποεῖν ὡς M: ποιεῖν ὡς V.

(16) Προσθετά· σκορπίου θαλασσίου τὴν χολὴν ἐν
εἰρίῳ ξηρήνας ἐν σκιῇ προστίθει.

Ἢ γλήχωνα ξηρήνας, λείην ποιήσας, ἐν μέλιτι
δεύσας, ἐν εἰρίῳ προστιθέναι.

Ἢ ἄνθος χαλκοῦ ἐν μέλιτι ἐς ὀθόνιον ἐνδῆσαι καὶ
προσθεῖναι.

Ἢ σικύου σπέρμα καὶ ὄστρακον κατακαύσας,[315] ἐν
οἴνῳ τε δεύσας, ἐν λαγῴαις θριξὶ καὶ εἰρίῳ προσθεῖναι.

(17) Ἄλλο· στυπτηρίην Αἰγυπτίην ἐν εἰρίῳ κατελί-
180 ξας | προσθεῖναι.

Ἄλλο· ἢ κανθαρίδας τρίψας οἴνῳ δεύσας, ἐν εἰρίῳ[316]
προστίθει.

Ἢ τὴν ἀρτεμισίην ποίην οἴνῳ δεύσας προστίθει.

Ἢ μελάνθιον τρίψας ἐν οἴνῳ δεύσας,[317] ἐν εἰρίῳ
προστιθέναι.

Ἢ βόλβιον τὸ ἐκ τῶν πυρῶν τρίψας ἐν οἴνῳ δεύσας, ἐν εἰρίῳ προσθεῖναι.

Ἢ οἴνου παλαιοῦ λευκοῦ τὴν τρύγα κατακαῦσαι,
καὶ κατασβέσαι οἴνῳ λευκῷ, καὶ τρῖψαι ἐν οἴνῳ, καὶ
πρόσθες ἐν ὀθονίῳ.

Ἢ χαλβάνην καὶ νέτωπον καὶ μίσυ ἐν ῥοδίνῳ
μύρῳ, ἐν ὀθονίῳ προσθεῖναι.

(18) Ἄλλο· ἐλατηρίου δύο πόσιας καὶ κηρίον ἐν
οἴνῳ ἐν ὀθονίῳ προστίθει. ἢ βούτυρον καὶ στυπτηρίην μέλιτι δεύσας ὁμοίως.

Ἢ ὀπὸν σκαμωνίης καὶ στέαρ ἐν μάζῃ συμμίξας,
οἴνῳ δεύσας, ἐν ὀθονίῳ προσάγειν.

(16) Suppositories: dry some sculpin bile in a piece of wool set in the shade, and apply it.

Or dry pennyroyal, grind it fine, knead it into honey, and apply it in a piece of wool.

Or tie some flower of copper in honey in a piece of linen, and apply it.

Or burn cucumber seed and cuttlefish bone, dissolve them in wine, and apply them in hare's hair and a piece of wool.

(17) Another: wrap Egyptian alum in a piece of wool and apply it.

Another: or crush blister beetles, dissolve them in wine, and apply in a piece of wool.

Or dissolve artemisia herb in wine and apply.

Or grind black cumin, dissolve it in wine, and apply it in a piece of wool.

Or crush one of the bulbs that grow in wheat, soak this in wine, and apply in a piece of wool.

Or burn the lees of aged white wine, quench it in white wine, knead it in the wine, and apply it in a piece of linen.

Or all-heal juice, oil of bitter almonds, and misy in rose unguent: apply in a piece of linen.

(18) Another: two drafts of squirting cucumber juice with honeycomb in wine: apply in a piece if linen. Or butter and alum soaked in honey in the same way.

Or mix together scammony juice and fat in a barley cake, dissolve in wine, and apply in a piece of linen.

---

315 -καύσας Θ: -κλύσας MV.

316 ἐν εἰρίῳ om. MV.

317 δεύσας om. MV.

(19) Ἄλλο· κολοκυνθίδος ἀγρίης τρίψας τὸ ἔνδον ἐν γάλακτι γυναικείῳ ἐν ὀθονίῳ προσθετόν.

(20) Χορίον ἐκβάλλει ποτά· ὀπὸν σιλφίου ὅσον κύαμον ἐν οἴνῳ πίνειν. ἐκβόλιον ἄλλο· ὑπὸ τὰς μασχάλας λαβόντα σείειν ἰσχυρῶς.

(21) Ποτὰ δὲ διδόναι ἄγνου πέταλα ἐν οἴνῳ· ἢ δίκταμνον Κρητικὸν ὅσον ὀβολὸν ἐν ὕδατι τρίβειν καὶ διδόναι· ἢ κονύζης τῆς ἡδυόσμου[318] ὅσον χεῖρα πλέην διεῖναι πράσου χυλῷ, καὶ νέτωπον, ὅσον χήμην ἁδρήν· ταῦτα ἐν οἴνῳ δοῦναι πιεῖν τρίψαντα λεῖα. ἢ δᾷδα πιοτάτην σὺν γλυκεῖ ἕψειν, οἴνου ἐπιχέας ὅσον τρεῖς κυάθους, καὶ χαλβάνην ὅσον τριώβολον, καὶ σμύρναν, ἑψήσας, ἐπειδὰν παχὺ | γένηται, δὸς πιεῖν χλιαρόν. ἢ πράσου χυλὸν καὶ σμύρναν καὶ ⟨οἶνον⟩[319] γλυκὺν ὁμοῦ. ἢ ἀνδράχλης καρπὸν λεαίνειν, δίδου δὲ ἐν οἴνῳ λευκῷ παλαιῷ. ἢ αἰγείρου Κρητικῆς κόκκους ἐννέα τρίψας ἐν οἴνῳ πινέτω. ἢ βατραχίου τοῦ φύλλου καὶ τοῦ ἄνθεος τετριμμένου ὅσον δραχμὴν Αἰγιναίαν ἐπ᾽ οἴνῳ πίνειν γλυκεῖ.

(22) Ἢν δὲ ἐνέχηται τὸ χορίον, λεβηρίδος ὅσον ὀβολὸν τρίβειν ἐν οἴνῳ καὶ πῖσαι. ἢ σελίνου ῥίζαν καὶ μύρτα ἑψέτω, πίνειν δὲ ἡμέρας τέσσερας. ἢ κύμινον Αἰθιοπικόν, καὶ τὸ καστόριον ὅσον ὀβολόν, καὶ κανθαρίδος σμικρόν· ταῦτα ἐν οἴνῳ δὸς πίνειν. ἢ μαράθου ῥίζαν ἐν οἴνῳ καὶ ἐλαίῳ καὶ μέλιτι ἀναζέσαντα πιπίσκειν.

(23) Ἢν τὸ χορίον μὴ ὑποχωρέῃ, κόνυζαν τρίψας ἐν εἰρίῳ πρόσθες, πίνειν δὲ ἄμεινον. ἢ κονύζης ὅσον

(19) Another: knead the insides of a wild gourd with woman's milk: make a suppository in a piece of linen.

(20) Potions that expel the placenta: silphium juice to the amount of a bean, give to drink in wine. Another means to expel: take hold of the woman under her axillae and shake her violently.

(21) Give potions made from chaste-tree leaves in wine; or knead an obol of Cretan dittany in water and give it; or dissolve a handful of fragrant fleabane in leek juice, and a full cheme of oil of bitter almonds: knead these fine and give them in wine to drink. Or boil very oily pine wood with three cyathoi of sweet wine and three obols of all-heal juice and some myrrh; when it becomes thick, give it warm to drink. Or leek juice, myrrh and sweet wine together. Or crush purslane fruit and give it in aged white wine. Or crush nine Cretan black poplar berries into wine and have the patient drink this. Or grind an Aeginetan drachma of ranunculus leaf and flower in sweet wine for the patient to drink.

(22) If the placenta is held back inside, grind an obol of serpent skin in wine and give this to drink. Or have the patient boil celery root and myrtle berries, and drink this for four days. Or Ethiopian cumin, an obol of castoreum, and a small blister beetle: give these in wine to drink. Or give fennel root boiled in wine, olive oil and honey to drink.

(23) If the placenta is not expelled, grind fleabane and apply it in a piece of wool, although to give it to drink is

---

318 ἡδυόσμου Θ: δυσόδμου MV.
319 Add. I.

χανδάνει χείρ, πράσου χυλὸν καὶ νέτωπον συμμίξαι
ὅσον χηραμύδα, ταῦτα ἐν οἴνῳ πινέτω.

(24) Χορίον δὲ ἐξάγει, καὶ ἐπιμήνια κατασπᾷ, καὶ
ἔμβρυον ἡμίεργον ἐξάγει· κανθαρίδας πέντε, ἀποτίλας τὰ πτερὰ καὶ τὰ σκέλεα καὶ τὴν κεφαλήν, ἔπειτα
τρίβολον παραθαλάσσιον κόψας σὺν τῇ ῥίζῃ καὶ
τοῖσι φύλλοισιν, ὅσον κόγχην, καὶ τὸ εὐάνθεμον τὸ
χλωρὸν τρῖψον ἴσον πλῆθος, καὶ σελίνου σπέρμα, καὶ
σηπίης ᾠὰ πεντεκαίδεκα ἐν οἴνῳ γλυκεῖ κεκρημένῳ
ταῦτα ὁμοῦ, καὶ ἐπειδὰν ἡ ὀδύνη ἔχῃ, πίνειν· καὶ ἐν
ὕδατι θερμῷ ἐγκαθιζέσθω, πινέτω δὲ μελίκρητον ὑδαρὲς καὶ γλυκὺν οἶνον λευκόν.

(25) Ὕστερον ἐκβάλλει· ὁλοκωνίτιδος[320] τῆς ἡδείης
ῥίζα, ἔστι δὲ ὡς βολβός, σμικρὸν δὲ ὡς ἐλαίη, ταύτην
τρίβειν ἐν οἴνῳ καὶ διδόναι πίνειν· ἢν μὲν ᾖ μικρή,
δύο, ἢν δὲ μείζων, μία ἀρκεῖ· παραμίσγειν δὲ καὶ τῶν
σπερμάτων κύμινον Αἰθιοπικὸν καὶ σέσελι Μασσα-
184 λιωτικόν, ἢ φύλλον | τὸ Λιβυκὸν ξηρὸν ἡμιχοινίκιον
σὺν οἴνῳ κοτύλῃσι τρισὶν, ἕψειν καὶ ἄγειν ἐς τὸ
ἥμισυ, καὶ ἀπὸ τοῦδε διδόναι.

(26) Ἄλλο· λύγου καρπόν, σεσέλιος ἴσον, σμύρνης,
τρίβειν ὁμοῦ, καὶ ἐν ὕδατι διδόναι πίνειν.

(27) Ἐκβόλια· ἄγνου λευκῆς νέης ὅσον ὀξύβαφον,
ἐν οἴνῳ λευκῷ εὐώδει δὸς πιεῖν, τρίψας λεῖον.

(28) Ἕτερον· καστορίου ἢ σαγαπήνου ὀβολόν,
ἀσφάλτου δραχμὴν μίαν, νίτρου δραχμὰς δύο, τρίψας ἐν γλυκεῖ <οἴνῳ>[321] καὶ ἐλαίῳ ὅσον ἡμικοτυλίῳ,
δὸς πιεῖν νήστει ὀβολοὺς δύο, καὶ λοῦε θερμῷ.

better. Or take a handful of fleabane, mix it in a cheramys of leek juice and oil of bitter almonds, and have the patient drink this in wine.

(24) This expels the placenta and brings down the menses or a half-formed fetus: remove the wings, legs and head from five blister beetles, then pound a conche of water chestnut together with its root and leaves, and grind the same amount of green chamomile, celery seed, and fifteen cuttlefish eggs together in diluted sweet wine. When pain is present, give this to drink. Also have the patient take a sitz bath in hot water, and drink dilute melicrat and sweet white wine.

(25) This expels the afterbirth: take root of sweet earth-almond in the shape of a bulb, but small like an olive, grind this into wine and give to drink—if it is small take two, but if larger one will suffice. Mix together seeds of Ethiopian cumin and hartwort from Marseille, or a hemichoinix of dry Libyan leaf (i.e., silphium) with three cotyles of wine, boil, reduce to half, and give some of this (sc. to drink).

(26) Another: fruit of agnus castus, the same amount of hartwort and myrrh, knead together, and give in water to drink.

(27) Expulsives: an oxybaphon of tender white agnus castus: grind fine and give to drink in fragrant white wine.

(28) Another: take an obol of castoreum or ferula, one drachma of asphalt, and two drachmas of soda, grind into a half cotyle of sweet wine and olive oil, and give two obols to the patient to drink in the fasting state; wash her with hot water.

---

320 ὁλοκ- Littré: ὁμοκ- codd.
321 Add. I.

(29) Ἄλλο· ἐχίνους θαλασσίους τρεῖς τρίψας ὅλους ὡς ἔχει λείους ἐν οἴνῳ εὐώδει, δὸς πιεῖν.

(30) Ἄλλο· μίνθης δεσμίδα μικρὴν καὶ πηγάνου καὶ κοριάννου, καὶ κέδρου ἢ κυπαρίσσου πρίσματα,[322] ἐν οἴνῳ εὐώδει δὸς πιεῖν· καὶ τῶν ἐχίνων, ἢν ἔχῃ, ῥοφείτω ἐπὶ τὸ πλεῖστον ὡσαύτως· εἶτα λοῦσον θερμῷ.

(31) Ἄλλο· ἄννησον, κεδρίδας, σελίνου καρπόν, Αἰθιοπικὸν κύμινον, σέσελι, ἑκάστου ἥμισυ ὀξυβάφου δὸς ἐκπιεῖν ἐν οἴνῳ λευκῷ τρίψας λεῖον.

(32) Ἄλλο· δικτάμνου δεσμίδα καὶ δαύκου καρποῦ δραχμὰς δύο, καὶ μελάνθιον ἴσον, ἐν οἴνῳ λευκῷ, τρίψας λεῖον, δὸς ἐκπιεῖν, καὶ λοῦσον θερμῷ πολλῷ· διδόναι δὲ πρὸς τὴν ἰσχὺν τοῦ νοσήματος.

(33) Ἄλλο· χαλβάνην ὅσον ἐλαίην τρίψασα ἐν κεδρίνῳ ἐλαίῳ προσθέσθω· τοῦτο ἐκβάλλει καὶ διαφθείρει, ἢν ᾖ νωχελές.

(34) Ἐκβόλιον ἔγχυτον ὑστερῶν· ὅταν σαπῇ νεκρωθὲν ὑπὸ ψύχεος, ὅταν ἄνεμος ψυχρὸς ᾖ, κρόκον τρίψας λεῖον ὅσον ὁλκήν, ἐν χηνὸς στέατι ἐγχέαι, καὶ ἐᾶν ὡς πλεῖστον χρόνον.

186   (35) Ποτόν· κόνυζαν τὴν | ἡδύοσμον, μέλι καὶ ῥητίνην τρίψας λεῖον ἐν οἴνῳ εὐώδει, ἢ ἐν συρμαίῃ, δοῦναι πιεῖν, καὶ λοῦσαι θερμῷ.

(36) Ἄλλο ποτὸν τοῦ παιδίου καὶ τῶν ἐνόντων κακῶν· ἰὸν χαλκοῦ τρίψας ἐν μέλιτι καὶ συρμαίῃ δὸς πιεῖν.

(37) Ἄλλο προσθετόν· χαλκοῦ ῥινήματα ἐνθεὶς ἐς

(29) Another: grind three whole sea urchins very smooth in fragrant wine and give to drink.

(30) Another: take a small handful of mint, rue, or coriander and sawdust of a cedar tree or cypress: give in fragrant wine to drink. Also sea urchins, if they are available: have the woman take as much soup made from these in the same way as she can. Then bathe her in hot water.

(31) Another: take a half oxybaphon each of anise, juniper berries, celery seed, Ethiopian cumin, and hartwort, grind them fine, and give in white wine to drink.

(32) Another: a handful of dittany, two drachmas of dauke seeds, and the same of black cumin: grind these fine and give them to drink in white wine; also bathe the patient in copious hot water. Make your administration in accordance with the strength of the disease.

(33) Another: have the patient knead all-heal gum the size of an olive into cedar oil and apply it as a suppository; this is expulsive and provokes an abortion if (sc. the fetus) is sluggish.

(34) Expulsive injection for the uterus: when, as the result of a cold wind, a fetus dies from cold and putrefies, grind a drachma of saffron fine, infuse it in goose grease, and leave it inside for a long time.

(35) Potion: knead fragrant fleabane along with honey and resin smooth into fragrant wine or radish juice, and give to drink; also wash the patient in hot water.

(36) Another potion for the fetus, or for internal ills: grind verdigris in honey and radish juice: give to drink.

(37) Another suppository: place copper filings in a

---

322 πρίσματα Θ: σπέρμα MV.

ὀθόνιον μαλθακὸν πρόσθες πρὸς τὸ στόμα τῶν μητρέων.

(38) Ἐκβόλιον προσθετόν, ἢν ἐναποθνήσκῃ· ὄστρακον νέον καὶ στέαρ χήνειον τρίψας, προσθέσθω.

(39) Ἄλλο προσθετόν· λίτρον ἑψήσας σὺν ῥητίνῃ,
ποιήσας βάλανον, βάπτων ἐς ὄρνιθος στέαρ, προστίθει.

(40) Ἄλλο προσθετόν, χάριες οὔνομα ῥίζα· πρὸς
τὸν ὀμφαλὸν πρόσθες μὴ πολὺν χρόνον.

(41) Ἄλλο· ἀγρίην κολοκύντην καὶ μυῶν ἀπόπατον
τρίψας λεῖα προσθέσθω.

(42) Ἄλλο ἐπιδετόν· ῥητίνην καὶ ὄρνιθος στέαρ
τρίψασα ἐπιδησάσθω ἐπὶ τὸν ὀμφαλὸν καὶ τὴν γαστέρα.

(43) Ἄλλο προσθετόν· τοῦ κισσοῦ τοῦ λευκοῦ τὸν
καρπὸν καὶ κέδρου πρίσμα τρίψασα, βαλάνια ποιήσασα, προστιθέσθω.

(44) Ἄλλο προσθετόν· χελώνης θαλασσίης τὸν ἐγ-
κέφαλον καὶ Αἰγύπτιον κρόκον καὶ ἅλας Αἰγύπτιον
τρίψας καὶ συμμίξας ποιέειν βαλάνους, καὶ προσθέσθω.

(45) Ἐκβόλιον θυμίημα, καὶ αἷμα γαστρὸς ἐξελά-
σαι· ἰτέης φύλλα ἐπὶ πῦρ ἐπιτιθεὶς θυμιῆν, καὶ περικαθίσας τὴν γυναῖκα ἑῆν ἄχρι ἂν ὁ καπνὸς ἐς τὴν
μήτρην ἐνδύνῃ.

(46) Ἄλλο ἔγχυτον· ὅταν ἡ γυνὴ ἐκτρώσῃ καὶ τὸ
παιδίον μὴ ἐξίῃ, ἤν τε σαπῇ καὶ οἰδήσῃ, ἢ ἄλλο τι
188 πάθῃ τοῖον, | πράσα καὶ σέλινα ἐκθλίψας τὸν χυλὸν

piece of soft linen and apply against the mouth of the uterus.

(38) An expulsive suppository: if the fetus dies inside, knead fresh cuttlefish bone in goose grease, and have the patient apply it.

(39) Another suppository: boil soda together with resin, form a suppository, dip in chicken fat, and apply.

(40) Another suppository: the root they call "gracious": apply to the navel for a limited time.

(41) Another: knead wild gourd and mouse's excrement smooth and have the woman apply it.

(42) Another application for under a bandage: have the patient knead resin and chicken fat, and bind it over her navel and belly.

(43) Another suppository: have the patient knead white ivy berries and cedar sawdust together, form this into small suppositories, and apply them.

(44) Another suppository: knead together the brain of a sea turtle, Egyptian saffron, and Egyptian salt, form suppositories, and have the woman apply.

(45) An expulsive fumigation, which also drives blood out of the belly: set willow leaves on a fire and fumigate: have the patient sit down over it and stay there until the smoke enters her uterus.

(46) Another infusion for when a woman has suffered a miscarriage but the fetus has not been expelled, and it putrefies and swells up or suffers something else like that: take leeks and celery—squeeze the juice out of these and

διὰ ῥάκεος ἀμφοτέρων, ῥοδίνου ἐλαίου κοτύλην, καὶ
χηνὸς στέαρ ὅσον τεταρτημόριον, καὶ ῥητίνης ὀβολοὺς τρεῖς κατατήξας ἐν ἐλαίῳ, καὶ ποιήσας πρὸς
ποδῶν ὑψηλοτέρην, ἔγχεον ἐς τὰς μήτρας· καὶ ἐχέτω
χρόνον ὅτι πλεῖστον· ἔπειτα κάθισον ἡμέρας τέσσε-
ρας· ἢν ἐξέρχεται· εἰ δὲ μή, λαβὼν ἅλας Αἰγυπτίους
καὶ κολοκύντην ἀγρίην χλωρήν, μέλιτι μίξας, τρίψας,
δοῦναι καταφαγεῖν, καὶ ἐπὴν φάγῃ, κινεέσθω τῇδε
καὶ τῇδε.

(47) Ἐκβόλιον προσθετόν· ἅλας Αἰγυπτίους, καὶ
μυόχοδα, καὶ ἀγρίην κολοκύντην, μέλιτος ὅσον τεταρτημόριον ἐπιχεῖν ἡμίεφθον, καὶ λαβὼν ῥητίνης
δραχμὴν μίαν ἔμβαλε εἰς τὸ μέλι καὶ τὴν κολοκύντην
καὶ τὰ μυόχοδα, ξυντρίψασα πάντα, ποιήσασα βαλάνους, πρὸς τὴν μήτρην προσθέσθω, ὡς ἂν δοκέῃ καιρὸς εἶναι.

(48) Ἐκβόλιον ποτὸν παιδίον ἐκβάλλειν ὁλόκλη-
ρον·[323] ἐλξίνην ἐν οἴνῳ τρίψας πότισον.

(49) Ἐκβόλιον ποτὸν παιδίον ἐκβάλλει πελιδνόν·[324]
τοῦ ἐκτόμου τὰς ῥίζας τρίψας λεπτά, τοῖσι τρισὶ
δακτύλοισι, καὶ σμύρνης ὅσον κύαμον, ἐν οἴνῳ γλυκεῖ πίπισκε.

(50) Ἐκβόλιον· κορίαννον τῇ ῥίζῃ καὶ νίτρον[325] καὶ
νέτωπον προσθεμένη περιπατείτω.

(51) Ἔγχυτον ἔμβρυον, ἢν ἐναποθάνῃ, ὥστε ἐκ-
βάλλειν· κρόκον τρίψας καὶ ἐπιχέας χηνὸς ἔλαιον,
καὶ διηθήσας, ἐγχέαι ἐς τὰς μήτρας, καὶ ἐᾶν ὡς πλεῖστον χρόνον.

pass it through a piece of cloth—a cotyle of rose oil, a quarter portion of goose grease, and three obols of resin dissolved in olive oil: position the patient with her feet higher (sc. than her head), pour the mixture into her uterus, and have her retain it for a good long time. Then have her sit down for four days: if the fetus comes out, (sc. fine): if not, take Egyptian salt and green wild gourd, mix with honey, knead, give to the patient to eat, and, after she has done so, have her move herself back and forth.

(47) An expulsive suppository: take Egyptian salt, mouse's excrement, wild gourd, and pour on a fourth portion of half-boiled honey; add one drachma of resin to the mixture of honey, gourd, and mouse's excrement, have the patient knead this together, form it into suppositories, and apply it against her uterus for as long as seems right.

(48) An expulsive potion to remove the fetus intact: crush bindweed in wine and give this to drink.

(49) An expulsive to remove the fetus when it is livid: grind a pinch of roots of black hellebore fine together with myrrh the size of a bean, and give this in sweet wine to drink.

(50) An expulsive: have the patient apply a suppository made from coriander together with its root, soda, and oil of bitter almonds, and then walk about.

(51) Infusion to expel the fetus if it has died inside: crush saffron, pour on goose oil, filter, and infuse into the uterus; leave in place for a good long time.

---

323 ἐκβάλλειν ὁλόκληρον Θ: βλητὸν ἐκβάλλει M: ἐκβάλλει βλῆτον V.

324 πελιδνόν MV: τέλειον Θ.

325 καὶ νίτρον om. MV.

(52) Ἔμβρυον ἀκίνητον φθεῖραι καὶ ἐκβάλλειν·
στυπτηρίης σχιστῆς δραχμὴν μίαν, σμύρνης ἴσον,
ἐλλεβόρου μέλανος τριώβολον τρίψας λεῖα ἐν οἴνῳ
μέλανι, βαλάνια ποιέειν, καὶ προστιθέναι, ἄχρι κατ᾽
ὀλίγον ἀπολυθῇ.

(53) Κλυσμοὶ καθαρτήριοι μητρέων, ἢν ἐκ τόκου
190 ἑλκωθέωσιν | ἢ φλεγμασίης· ὀλύνθους χειμερινούς,
ὕδωρ ἐπιχέασα καὶ ζέσασα, ἀφεῖναι, καταστῆναι,
εἶτα ἔλαιον ἐπιχέαι χλιαρὸν καὶ μῖξαι, κλύσαι δὲ δύο
κοτύλῃσι τὸ πλεῖστον· πάντα δὲ τὰ κλύσματα μὴ
πλείονι κλύζειν. καὶ σιδίοισι καὶ μάννῃ, ἐν οἴνῳ μέλανι αὐστηρῷ ἡψημένῳ, εἶτα ἀποχέας τὸν οἶνον,
τούτῳ κλύζε.

(54) Ἄλλος κλυσμός· τρύγα οἴνου κατακαύσας τῆξαι ἐν ὕδατι· σίδια, μύρτα, σχοῖνον εὐώδεα, φακοὺς
ἑψήσας ἐν οἴνῳ, ἀποχέας τὸν οἶνον, κλύσαι.

(55) Ἄλλος κλυσμός· βούτυρον, λιβανωτόν, ῥητίνην, μέλι τήξας[326] ἐν τῷ αὐτῷ, οἶνον ἐπιχέας, κλύζε
χλιαρῷ.

Ἡ ἀκτῆς καρπὸν ἑψήσας ἐν ὕδατι, καὶ ἀποχέας τὸ
ὕδωρ, τρίψας ἐν τῷ αὐτῷ σέλινον, σμύρναν, ἄννησον,
λιβανωτόν, ἐπιχέας οἶνον ὡς εὐωδέστατον ἴσον τῷ
ὕδατι, διηθήσας δι᾽ ὀθονίου, χλιήνας, κλύσαι.

(56) Ἄλλος· κράμβην, καὶ λινόζωστιν, καὶ λίνου
σπέρμα, καὶ χλωρὸν τὸ λίνον ἑψήσας ἐν ὕδατι, ἀπηθήσας, κλύσαι τῷ ὕδατι.

Ἡ[327] μυρσίνης τῶν φύλλων ὀξύβαφον, σμύρνης,
ἀννήσου, μέλι, ῥητίνην, μύρον Αἰγύπτιον, τρίψας

(52) This aborts and expels a fetus that fails to move: one drachma of powdered alum, the same of myrrh, and three obols of black hellebore ground fine in dark wine: make into small suppositories and apply them until it is gradually released.

(53) Injections to clean the uterus after childbirth if it becomes ulcerated or inflamed: have the patient take winter wild figs, pour water over them, and boil; decant, let stand, and then pour in warm olive oil and mix together; flush with up to two cotyles of this infusion. Do not inject more than this of any infusion. Also take pomegranate peels and frankincense powder in boiled dark dry wine, pour off the wine, and flush with this.

(54) Another injection: burn wine lees and dissolve them in water; then boil pomegranate peels, myrtle berries, fragrant mastic, and lentils in wine; pour off the wine and flush with the residue.

(55) Another injection: Melt butter, frankincense, resin, and honey together, add wine, and flush with this warm.

Or boil elderberries in water, pour the water off, and knead the rest together with celery, myrrh, anise, and frankincense; add very fragrant wine equal in amount to the water, filter through a piece of linen cloth, warm, and infuse.

(56) Another: take cabbage, mercury herb, linseed, and green flax, boil in water, filter, and flush with the liquid.

Or mix an oxybaphon of myrtle leaves, myrrh and anise together with honey, resin, and Egyptian unguent: knead

---

326 μέλι τήξας Θ: μέλιτι τήξας M: μέλιτι μίξας V.
327 ἢ om. MV.

πάντα ἐν τῷ αὐτῷ, ἐπιχέας οἴνου λευκοῦ ὡς εὐωδεστάτου κοτύλας δύο, διηθήσας, χλιήνας, κλύσον τούτῳ.

Ἢ δάφνης καρπὸν καὶ γλήχωνα ἑψήσας ἐν ὕδατι,
μύρον ῥόδινον ἐπιχέας, τούτῳ κλύζε χλιήνας.

Ἢ χηνὸς στέαρ ῥητίνῃ μίξας ἐς τωὐτό, ἐπιχέας
οἶνον, καὶ χλιήνας κλύσαι.

(57) Ἄλλος· βούτυρον καὶ κέδρινον ἔλαιον ἐν μέλιτι
ὀλίγῳ χλιήνας, συμμίξας, κλύσαι.

Ἢ μέλι, βούτυρον, σχοῖνον, κάλαμον εὐώδεα,
βρύον θαλάσσιον ἕψειν ἐν οἴνῳ, ἀπηθήσας, τούτῳ
κλύζε.

Ἢ σελίνου καρπόν, σέσελι, σμύρναν, ἄννησον,
192 μελάνθιον ἐν | οἴνῳ, ἀπηθήσας τὸν οἶνον, κλύσαι.

Ἢ κέδρον ἑψήσας, κλύσαι τῷ οἴνῳ.

Ἢ κισσὸν ὁμοίως ἐν ὕδατι, κλύσαι τούτῳ.

Ἢ ἐλατήριον ὅσον δύο πόσιας, ἕψειν ἐν ὕδατι ὅσον
δύο κοτύλῃσι, καὶ κλύσαι χλιαρῷ.

Ἢ σικύης ἐντεριώνην ὅσον διδάκτυλον ἑψήσας ἐν
ὕδατος δύο κοτύλῃσι, ἐπὶ τὸ ὕδωρ ἐπιχέας μέλι καὶ
ἔλαιον, τούτῳ κλύσαι.

Ἢ θαψίης ῥίζης ὅσον δύο πόσιας τρίψας λεῖον,
ἐπιχέας μέλι ἔλαιον, διεὶς ὕδατι χλιαρῷ ὅσον δύο
κοτύλῃσι, κλύσαι.

Ἢ ἐλλεβόρου μέλανος, ὅσον δύο πόσιας διεὶς οἴνῳ
γλυκεῖ καὶ ὕδατι, κλύζειν.

Ἢ κόκκους Κνιδίους ὅσον ἑξήκοντα τρίψας λείους,
ἐπιχέας μέλι ἔλαιον ὕδωρ, κλύσαι.

(58) Κλυσμὸς κρατυντήριος, ἢν ἕλκεα ᾖ καθαρά·

all these together, add two cotyles of the most fragrant white wine, filter, warm, and flush with this.

Or boil bayberry and pennyroyal in water, add rose unguent, warm, and flush with this.

Or mix goose grease together with resin, pour on wine, warm, and inject.

(57) Another: warm butter and cedar oil in a little honey, mix together, and inject.

Or boil honey, butter, rushes, fragrant reeds, and seaweed in wine, filter it off, and flush with this.

Or mix celery seed, hartwort, myrrh, anise and black cumin in wine, filter off the wine, and inject.

Or boil cedar wood and flush with the wine.

Or prepare ivy in water the same way and flush with this.

Or boil two drafts of squirting cucumber juice in two cotyles of water, and inject this warm.

Or boil a two-inch piece of the insides of a bottle gourd in two cotyles of water, pour honey and olive oil into the water, and flush with this.

Or crush two drafts of thapsia root smooth, pour in honey and olive oil, dilute this in two cotyles of warm water, and inject.

Or dilute two drafts of black hellebore in sweet wine and water, and infuse.

Or crush sixty Cnidian berries smooth, add honey, oil and water, and infuse.

(58) A strengthening injection: if the patient's ulcers

ὀλύνθους χειμερινοὺς τρίβειν, καὶ ἐπιχέαι ὕδωρ, βρέ-
ξον δ' ἡμέρην, καὶ ἔλαιον ἐπιχέαι, καὶ κλύσαι.

Ἢ σιδίοισι καὶ λωτοῦ πρίσμασιν, ἐν οἴνῳ δὲ μέλανι ἕψειν.

(59) Ὅταν δὲ ἀκάθαρτα φέρηται, τρύγα καίειν οἴνου καὶ τῷ ὕδατι κλύζειν.

Ἢ σιδίῳ, ῥόῳ βυρσοδεψικῇ, μυρσίνης φύλλοισι
<καὶ>³²⁸ βάτου, ἐν οἴνῳ μέλανι ἑψεῖν, καὶ κλύζειν.

(60) Πρὸς τὰ παλαιά· χυλῷ κράμβης ἡψημένης
κλύζειν· καὶ λινόζωστιν ὁμοίως· καὶ λίτρον παρά-
μισγε ὀλίγον ἐρυθρόν.

Σμύρνης ὀξύβαφον, λιβανωτόν, σέσελι, ἄννησον,
σελίνου σπέρμα, νέτωπον, ῥητίνην, μέλι, χήνειον
στέαρ, ὄξος τὸ λευκόν, μύρον τὸ λευκὸν Αἰγύπτιον, ἐν
τωὐτῷ τρίβειν ἴσον ἑκάστου λεῖα, εἶτα οἴνῳ διεὶς
λευκῷ κοτύλῃσι δυσί, χλιηρῷ³²⁹ κλύζειν.

194 Ἢ λινόζωστιν | ἐν ὕδατι ἕψειν καὶ ἀπηθεῖν.

Ἢ σμύρνης ὀξύβαφον, λιβανωτόν, νέτωπον, ἴσον
ἑκάστου ἐν ταυτῷ, χλιαρῷ κλύζειν.

Ἢ ἐλελίσφακον καὶ ὑπερικόν, ἐν ὕδατι ἑψήσας, τῷ
ὕδατι κλύζειν.

Ἢ ἀκτῆς καρπόν, δαφνίδας, ἑκατέρου τὸ ἴσον, ἐν
οἴνῳ ἕψε, εἶτα τούτῳ οἴνῳ κλύζε.

Ἢ γλήχωνος τῷ ὕδατι κλύσον.

Ἢ χηνὸς ἔλαιον, ῥητίνην συντήξας, ὁμοίως κέδρινον ἔλαιον παραμίξας ὀλίγον καὶ μέλι διατῆξαι
ταῦτα, κλύζειν χλιερῷ.

are clean, crush winter figs, pour on water, soak for a day, add olive oil, and infuse.

Or infuse with pomegranate peels and sawdust of the nettle tree boiled in dark wine.

(59) But when ulcers remain unclean, burn wine lees, and flush with the fluid.

Or take pomegranate peel, the sumac used in tanning, myrtle leaves, and bramble leaves, boil these in dark wine, and inject.

(60) Against longstanding (sc. ulcers): inject with the juice of boiled cabbage; or that of mercury herb in the same way. Also mix in a little red soda.

Take an oxybaphon of myrrh together with frankincense, hartwort, anise, celery seed, oil of bitter almonds, resin, honey, goose grease, white vinegar, and white Egyptian unguent: knead an equal amount of each of these together until the mixture is smooth, then dissolve it in two cotyles of white wine and inject this warm.

Or boil some mercury herb in water and filter it off.

Or take an oxybaphon of myrrh and an equal amount each of frankincense and oil of bitter almonds together: inject this warm.

Or salvia and hypericum boiled in water: inject the water.

Or elderberries and bayberries: boil the same amount of each in wine, and then flush with this.

Or flush with water extracted from pennyroyal.

Or take goose oil, melt resin with it, and likewise mix in a little cedar oil and honey, melt these together, and inject this warm.

---

328 Add. I. 329 κοτύλῃσι δ. χλιηρῷ om. Θ.

Ἢ ἀργύρου ἄνθος ἐν οἴνῳ μέλανι ἢ μέλι καὶ τυρὸν[330] τηκτόν, ἢ κύπαιρον ἢ σχοῖνον[331] καὶ κάλαμον,
ἃ δὴ ἐς μύρον μίσγεται, ἶριν, βρύον ἐν οἴνῳ ἕψων,
κλύζειν.

Ἢ σελίνου καρπόν, ἄννησον, σέσελι, σμύρναν,
μελάνθιον ἐν οἴνῳ ἑψῆσαι, ἢ κέδρον Κρητικὴν ἐν οἴνῳ
ἕψειν, καὶ κλύζειν· ἢ κισσὸν Κρητικὸν ἐν ὕδατι, ταὐτὸ
δρᾷ.

Ἢ ἐχέτρωσιν καὶ σμύρναν διεὶς ἐν ὕδατι κλύσαι.

Ἢ ἐλατήριον ὅσον δύο πόσιας ὕδατι κλύσαι.

Ἢ κολοκυντίδας ἀγρίας δύο ἐν οἴνῳ ἢ ἐν γάλακτι
ἑφθῷ ἀποβρέξας ὅσον τέσσερας κοτύλας, καὶ ἀπη-
θεῖν, καὶ κλύζειν.

Ἢ σικύης ἐντεριώνην ὅσον παλαιστὴν ἑψήσας ἐν
ὕδατι κοτύλαις τέσσερσι, μέλι ἔλαιον ὑποχέαντα
ἐνεργεῖν.

Ἢ τῆς θαψίης ῥίζης ὅσον δύο πόσιας καὶ μέλι καὶ
ἔλαιον διεὶς ὕδατος κοτύλῃσι δυσί, κλύσαι χλιαρῷ.

Ἢ ἑλλεβόρου ὅσον[332] δύο πόσιας ἐν οἴνῳ διεὶς
γλυκεῖ ὡς δύο κοτύλῃσιν, ἢ θλάσπιος ὅσον ὀξύβαφον
μέλι παραχέας, ὕδατι διεὶς ὅσον δύο κοτύλῃσι, χλιερῷ.

Ἢ σικύης ὅσον παλαιστήν, ⟨ἢ⟩ κνεώρου[333] ὅσον
πόσιν ἑψήσας ἐν ὕδατος κοτύλῃσι πέντε, μέλι ἔλαιον
παραχέας κλύσαι. |

196   Ἢ κόκκους Κνιδίους ὅσον ἑξήκοντα, μέλι, ἔλαιον
ὕδατι διείς, κλύσαι.

(61) Ἢν ἐκ τόκου διαρροίη ληφθῇ, πινέτω ἀστα-

Or take flower of silver in dark wine, or melted honey and cheese, or galingale, or rushes and reeds, mix into an unguent by boiling with iris and tree moss in wine: inject.

Or boil celery seed, anise, hartwort, myrrh, and black cumin in wine, or boil Cretan cedar wood in wine, and inject. Or Cretan ivy in water does the same.

Dissolve bryony and myrrh in water, and inject.

Or inject two drafts of squirting cucumber juice together with water.

Or soak two wild gourds in four cotyles of wine or boiled milk, filter it off, and inject.

Or boil the insides of a bottle gourd as wide as the palm of your hand in four cotyles of water, add honey and oil, and apply.

Or take two drafts of thapsia root together with honey and olive oil, dissolve in two cotyles of water, and infuse this warm.

Or dissolve two drafts of hellebore in about two cotyles of sweet wine, or an oxybaphon of shepherd's purse to which you have added honey in two cotyles of water: (sc. apply) warm.

Or boil a bottle gourd as wide as the palm of your hand or a draft of spurge flax in five cotyles of water, add honey and oil, and inject.

Or dissolve sixty Cnidian berries with honey and oil in water, and inject.

(61) If after childbirth a woman suffers from diarrhea,

330 τυρὸν Θ: κηρὸν MV.
331 σχοῖνον Froben after Calvus' *iuncum*: σχῖνον codd.
332 ὅσον om. Θ.
333 ἢ add. Potter: κνεώρου Θ: ἐκνεώρου MV.

φίδα μέλαιναν, καὶ σίδια γλυκείης ῥοιῆς τὸ ἔνδον,
καὶ πυτίην ἐρίφου, ταῦτα διεὶς οἴνῳ μέλανι, καὶ τυρὸν
αἴγειον ἐπιχύσας[334] καὶ ἄλφιτα πύρινα ἐπιπάσσειν·
καὶ δίδου πιεῖν· τοὺς πυροὺς δὲ ἐπ᾽ ὀλίγον φῶξαι.

(62) Ἢν ἐκ τόκου αἷμα ἐμέῃ, ταύτῃ ἡ σύριγξ τοῦ
ἥπατος τέτρωται· πινέτω γάλα ὄνου, ἔπειτα βοός, εἰ
εὐμαρές, τεσσεράκοντα ἡμέρας, καὶ σήσαμον τριπτόν, ἄχρι ἂν εὖ[335] ἔχῃ· πινέτω δὲ τὸ γάλα νῆστις.

(63) Ἢν ἐκ τόκου τὴν ἕδρην ἀλγέῃ, ἀρκεύθου καρπὸν καὶ λίνου ῥίζας ἕψειν, καὶ πίνειν ἡμέρας τέσσε-
ρας· καὶ θρίδακος σπέρμα τρίψασα ἐν χηνείῳ ἀλείφατι ἐσθίειν.

(64) Ἢν ἐκ τόκου αἱ μῆτραι φλεγμήνωσιν, στρύ-
χνου χυλὸν ἐγχέειν ἢ σεύτλου ἢ ῥάμνου.

(65) Ἢν δ᾽ ἐκ τόκου τὸ σκέλος ἀπὸ ὑστερέων χω-
λωθῇ, ἀνίστασθαι δὲ μὴ[336] δύνηται· πίνειν ὑοσκυάμου
καρποῦ ὅσον χηραμύδα ἐν οἴνῳ μέλανι τρεῖς ἡμέρας·
παραφέρεται δὲ ὁ πίνων· λύσις, γάλακτος ὀνείου
πῖσαι ὅσον κύλικα, ἔπειτα τοῦ φαρμάκου, ὑφ᾽ οὗ
φλέγμα καθαίρεται· σανδαράκῃ δὲ καὶ κηρωτῇ καὶ
λαγωοῦ θριξὶ θυμιάσθω τρεῖς ἡμέρας.

(66) Ἢν ἐκ τόκου φλεγμήνωσιν αἱ ὑστέραι, στρύ-
χνου χυλὸν ἐγχέαι ἐς τὰ αἰδοῖα, ἢ σελίνου, ἢ ῥάμνου,
ἢ σεύτλου, ἢ κολοκύντης χυλὸν ἐκπιέσας ἐγχέαι· καὶ
αὐτῆς τὸ μέσον ἁπαλώτατον ἐὸν περιξέσας μακρὸν
ἔνθες. ἢ ἀψίνθιον ἐν ὕδατι τρίβειν, εἰρίῳ δὲ ἀνασπογ-

---

[334] ἐπιχύσας om. MV. [335] εὖ om. Θ. [336] μὴ om. Θ.

have her drink a potion of dark grapes, the insides of sweet pomegranate peel, and kid's rennet: dissolve these in dark wine, grate goat's cheese over it, sprinkle on wheat meal, and then give it to drink—toast the wheat a little.

(62) If after childbirth a woman vomits blood, the pipe on her liver is injured. Have her drink ass's milk—and then cow's milk if it is available—for forty days, and take ground sesame until she is doing well. While she is drinking the milk she should fast.

(63) If after childbirth a woman suffers pain in her seat, boil Phoenician juniper berries and linen roots, and give this to drink for four days; also grind lettuce seeds into goose grease and have her eat this.

(64) If after childbirth a woman's uterus swells up with phlegm, apply an infusion made from nightshade juice, beets, or buckthorn.

(65) If after childbirth a woman's leg is made lame by her uterus, and she cannot stand up, have her drink a cheramys of henbane seed in dark wine for three days. Anyone who drinks this becomes deranged; solution: have her first drink a cylix of ass's milk and then some of the medication that cleans phlegm. She should also fumigate herself with red arsenic, a cerate, and hare's hair for three days.

(66) If after childbirth a woman's uterus swells up with phlegm, infuse nightshade juice into her genitalia, or juice from celery, buckthorn, or beets; squeeze the juice out of a gourd, and inject this. Also scrape out the softest material in the middle of a gourd and introduce a long piece of this. Or crush wormwood in water, soak it up with a piece

198 γίζειν, ἢν δὲ φρίξῃ, | ἀφαιρέειν. ἢ κοτυληδόνος φύλλα
καὶ πράσα ἕψειν ἐν πυρῶν κρίμνοισιν, ἔλαιον ἐπιχέας, δίδου.

79. Χολὴν καθαίρει ἐκ μήτρης· σικύης τὴν ἐντεριώνην λείην τρίψας, μέλιτι φυρήσας, βάλανον ποιέων,
προστίθει· φάρμακον δὲ χρὴ διδόναι καὶ ἄνω καὶ
κάτω καθαίρειν, καὶ λούειν τῷ θερμῷ, προστιθέναι δὲ
ἄνηθον[337] ἢ μελάνθιον.

Ἢ κολοκυνθίδος ἀγρίης τὸ ἔνδον λεῖον ποιέειν, καὶ
μέλιτι φυρᾶν, καὶ προστιθέναι.

Ἢ ἐλατηρίου ὅσον πόσιας τέσσερας, συμμίξας
στέαρ μήλειον[338] ἢ αἴγειον, βάλανον εὐμηκεστέρην
προστιθέναι.

Ἢ λίτρον καὶ κύμινον καὶ σκόροδον καὶ σῦκον,
λεῖα ποιήσας καὶ μέλιτι δεύειν, καὶ προστίθεσθαι·
θερμῷ δὲ λούσθω, καὶ τοῦ θερμοῦ πινέτω.

Ἢ θλάσπιν λείην ποιέων, μέλιτι φυρῶν, προστιθεῖ.

Ἢ σύκου Φιβαλίου[339] τὸ πῖον ξύσας, μίσγειν πόσιας ἐλατηρίου δύο, καὶ λίτρον ὅσον τὸ ἐλατήριον,
μέλιτι δεύσας, προστίθει.

Ἢ πευκεδάνου ὁπόσον τρεῖς κυάθους δοῦναι πιεῖν.

Ἢ ἄννησον καὶ μελάνθιον διεὶς οἴνῳ, πιεῖν δίδου.

Ἐλατηρίου πόσιας τέσσερας μῖξαι στέατι μηλείῳ,
ἀφελομένη δὲ διανιζέσθω ὕδατι εὐώδει,[340] ἠρέμα στύ-
φοντι.

337 ἄνηθον Θ: ἄννησον Μ: ἄνισσον V.
338 μὴ λίον Θ: χήνειον MV.

of wool, (sc. and apply); but if the woman has a chill, remove this. Or boil leaves of the navelwort and of leeks with wheat crumbs, pour on oil, and give.

79. The following cleans bile from the uterus: knead the insides of a bottle gourd smooth, mix with honey, form into a suppository, and apply. You must give a medication to clean both upward and downward, wash with hot water, and apply a suppository of anise or black cumin.

Or knead the insides of a wild gourd smooth, mix with honey, and apply.

Or take four drafts of squirting cucumber juice, mix it with goose grease or goat's fat, and apply as a suppository of a good length.

Or take soda, cumin, garlic, and fig, knead them smooth, mix with honey, and apply. Have the patient bathe herself with hot water and take a hot potion.

Or knead shepherd's purse smooth, mix it with honey, and apply.

Or scrape out the fat of a Phibalian fig, add two drafts of squirting cucumber juice and the same amount of soda, combine with honey, and insert.

Or give three cyathoi of sulphurwort to drink.

Or dissolve anise and black cumin in wine, and give this to drink.

Have the patient mix four drafts of squirting cucumber juice with sheep's fat, (sc. apply) and then remove this suppository and wash herself thoroughly with fragrant, slightly astringent water.

---

339 Φιβαλίου (cf. Bourbon, p. 40, n. 1): φιαλίου Θ: παλαιοῦ MV.

340 εὐώδει om. Θ.

Ἢ ἐλατηρίου πόσιας τρεῖς, σὺν μηλείῳ στέατι, βάλανον περίπτερον ποιέειν· ἢν δ' ἀφέληται, διανιζέσθω ὕδατι εὖ ἱκανῷ.[341]

Ἢ θαλάσσιον[342] ὅσον πόσιν σὺν μέλιτι δίδου.

80. (1) Κλυσμός, ἢν χολώδης ᾖ· ἐλατηρίου δύο πόσιας ὕδατι διείς, ἐπιχέαι ἔλαιον ναρκίσσινον, καὶ κλύζειν χλιερῷ.

Ἢ κολοκυνθίδας | δύο ἀγρίας ἀποβρέξας ἐν ὀνείῳ[343] γάλακτι ἐφθῷ ὅσον τέσσερας κοτύλας, μίαν ἀπηθεῖν, καὶ συμμίσγειν ἔλαιον ναρκίσσινον, καὶ κλύζειν.

(2) Ἄλλος χολώδης καὶ φλεγματώδης κλυσμός· σικύης ἐντεριώνης ὅσον παλαιστὴν ἑψήσας ἐν ὕδατι ποτῷ τέσσερσι κοτύλῃσι, μέλι μίξας καὶ ἔλαιον ἄνθινον, κλύζειν.

(3) Φλέγμα καὶ χολὴν καθαίρει· κόκκους Κνιδίους ἑξήκοντα, μέλι ἔλαιον ἄνθινον μίξας, κλύζειν ἐν ὕδατι.

Ἢ κνῆστρον ἑψήσας ἐν ὕδατι ποτῷ ἐν πέντε κοτύλῃσιν, καὶ ἀποχέας δύο κοτύλας, συμμῖξαι μέλι καὶ ἔλαιον ἄνθινον σὺν ναρκίσσῳ, καὶ κλύσαι.

(4) Κλυσμοὶ καθαρτήριοι· ὄλυνθοι χειμερινοὶ καέντες, καὶ τεγγόμενοι ἐν ὕδατι· ἀποχέαι δὲ τὸ ὕδωρ, καὶ ἔλαιον συμμῖξαι, καὶ κλύζειν, μετακλύζειν δὲ σιδίοισι, κηκίδι, λωτοῦ πρίσμασιν, ἐν οἴνῳ δὲ χρὴ μέλανι ἕψειν.

Ἢ τρύγα καίων τῷ ὕδατι, μετακλύζειν δὲ τοῖσι μυρσίνης φύλλοισι, ῥόῳ τῇ βυρσοδεψικῇ, ἕψειν δὲ

Or take three drafts of squirting cucumber juice together with sheep's fat, make a suppository around a feather, (sc. and apply); if this is removed, have the patient wash herself thoroughly with ample water.

Or give a draft of brine with honey.

80. (1) Injection, if a woman is suffering from bile: take two drafts of squirting cucumber juice diluted with water, add narcissus oil, and flush with this warm.

Or take two wild gourds well steeped in four cotyles of boiled ass's milk, filter off one cotyle, mix this with narcissus oil, and inject.

(2) Another injection to treat bile and phlegm: boil a hand's breadth of the insides of a bottle gourd in four cotyles of drinking water, mix into this honey and lily oil, and inject.

(3) (sc. This injection) cleans phlegm and bile: take sixty Cnidian berries, mix into this honey and lily oil, and inject with water.

Or boil cnestron in five cotyles of drinking water, pour off two cotyles of this and add honey and lily oil mixed with narcissus; inject.

(4) Cleaning injections: burn winter figs and soak them in water; pour off the water, add oil, and inject. Then inject again with pomegranate peel, oak gall, and sawdust of nettle tree wood you have boiled in dark wine.

Or (sc. inject) burned wine lees in water, and then inject again myrtle leaves and the sumac used in tanning

---

341 εὖ ἱκανῷ I: ευκαν Θ: εὖ ἱκανόν M: ἐν ἱκανῷ V.

342 θαλάσσιον ΘM: θλάσσιος V. See *Index Hipp.* s.v. θαλάσσιος [ὕδωρ].   343 ὀνείῳ L. Servin in Foes' *Variae Lectiones*: ὀνίῳ Θ: οἴνῳ MV.

οἴνῳ μέλανι εὐώδει· μετακλύζειν δὲ καὶ σχίνου φύλλα
καὶ[344] ὑπερικὸν καὶ ἐλελίσφακον, ἑψῆσαι ἐν οἴνῳ μέλανι εὐώδει· ἢ κράμβης ὕδατι, κἂν τῷδε ἕψειν λινόζωστιν, λίτρου ἐρυθροῦ ὀλίγον, καὶ κλύζειν.

Ἐλατηρίου ὅσον πόσις, σὺν ναρκισσίνῳ ἐλαίῳ, ἢ
ἀνθίνῳ, καὶ κλύσαι χλιαρῷ.

(5) Ἢν δὲ χολώδης ᾖ, κολοκυντίδας δύο ἀποβρέξαι
ἐν γάλακτι ὀνείῳ ἐφθῷ ὅσον τέσσερσι κοτύλῃσιν,
ἀπηθήσαντα κλύσαι, συμμῖξαι δὲ ἔλαιον ναρκίσσινον ἢ ἄνθινον.

Ἢ τὸ διὰ τῆς σικύης· τῆς ἐντεριώνης ὅσον παλα-
στὴν ἕψειν ἐν ὕδατι ποτῷ τέσσερσι κοτύλησι, καὶ
μέλι παραχέαι, καὶ ἔλαιον ἄνθινον· οὗτος φλεγματώ-
δει καὶ χολώδει ξυμφέρει. |

202   Φλέγμα ἄγει· κόκκον[345] τοῦ μανδραγόρου ξὺν
ὕδατι.

81. (1) Κάθαρσις πολλή τε καὶ παντοίη ὑπὸ τούτου
γίνεται· σκορόδου μώλυζαν, λίτρον, σύκου τὸ ἔνδον
τὸ πῖον, ἴσον τρίψας λεῖα, ὅσον κηκίδα ποιέειν τὸ
μέγεθος, καὶ προστιθέναι.

Ἢ κυμίνου ἐν οἴνῳ φύλλα τρίψας, ἐν εἰρίῳ πρόσθες.

Ἢ γῆς λευκῆς ὅσον πόσιν.

Ἢ λευκὴν ῥίζαν τρίβειν λείην, μέλι ἐπιχέας, ἀναζέσας, βάλανον[346] προστιθέναι.

Ἢ ὀπὸν σιλφίου σύκῳ μαλάσσειν, καὶ βάλανον
ποιέειν· ἀγαθὸν δὲ καὶ σικύης σπέρμα τρίβειν ὁμοίως.

Ἢ χολὴν ταύρου καὶ λίτρον ἐρυθρόν, νέτωπον καὶ

214

boiled in fragrant dark wine. Also inject again with mastic leaves, hypericum, and salvia; boil these in fragrant dark wine; or inject with cabbage juice in which you have boiled mercury herb and a little red soda.

Prepare a draft of squirting cucumber juice with narcissus or lily oil, and inject this warm.

(5) If a woman is suffering from bile, soak two gourds in four cotyles of boiled ass's milk, filter it off, and inject, also adding narcissus or lily oil.

Or an injection made with a gourd: boil a hand's breadth of its insides in four cotyles of drinking water, and add honey and lily oil. This is beneficial against phlegm and bile.

An injection to attract phlegm: mandrake berries with water.

81. (1) To bring about a major, wide-ranging cleaning: take a head of garlic, some soda, and the fatty insides of a fig, knead an equal amount of each of these smooth, form it into a suppository the size of an oak gall, and insert.

Or apply cumin leaves crushed in wine in a piece of wool.

Or white earth to the amount of a draft.

Or grind white root fine, add honey, boil, and apply as a suppository.

Or soften a fig with silphium juice and form it into a suppository; it is also good to grind bottle gourd seeds in the same way.

Or have the woman take bull's gall, red soda, oil of bit-

---

344 καὶ om. Θ.
345 Add. τρίβειν καὶ MV.
346 Add. ποιέοντα MV.

κυκλάμινον, τούτων ὅσον κηκίδα, τῆς δὲ κυκλαμίνου
πλέονα μοῖραν, μέλιτι συμμίξασα, προστιθεῖναι.

(2) Προσθετὸν κυκλαμίνου ὅ[347] τὴν κεφαλὴν καθαί-
ρει· ὕδατι τρῖψαι, καὶ ἐς ἄχνην ἀναφορύξαντα προστίθεσθαι.

Ἢ σμύρναν, ἅλας, κύμινον, χολὴν ταύρου σὺν
μέλιτι ὁμοίως.

Ἢ κόκκους ἐκλέψαντα ὅσον τρεῖς Μηδικοῦ φαρμάκου τοῦ τῶν ὀφθαλμῶν, ὃ καλέεται πέπερι, καὶ τοῦ
στρογγύλου τρία ταῦτα λεῖα τρίβειν, ἐν οἴνῳ παλαιῷ[348] διείς, βάλανον περὶ πτερὸν ὄρνιθος περιτιθέ-
ναι, καὶ ὧδε προσάγειν.

Ἢ τιθυμάλλου ὀπὸν μέλιτι ἀναδεῦσαι, ἢ σκίλλης
ῥίζαν ὅσον ἑξαδάκτυλον ἑλίξας δύο δακτύλους εἰρίῳ,
πρόσθες.

Ἢ αὐτὴν τὴν σκίλλαν[349] ἄνευ ῥίζης τρίβειν, καὶ
ὡσαύτως ἑλίσσειν εἰρίῳ, ἔπειτα προστιθέναι.

82. Κλυσμός, ἢν φλεγματώδης ᾖ· ἐλλεβόρου δύο
204 πόσιας | ἐν οἴνῳ διεῖναι γλυκεῖ, ὅσον δύο κοτύλησι,
καὶ κλύζειν μίσγοντα οἶνον.

Ἢν δὲ καθάρσιος δέῃ, πράσα ἕψειν, καὶ ἀκτῆς
καρπόν, ἄννησον, λιβανωτόν, σμύρναν, οἶνον, ταῦτα
πάντα τρίβειν, τῷ χυλῷ τούτων κλύζε.

Ἢ κράμβην ἑψήσας ἐν ὕδατι, ἐν τῷ χυλῷ ταύτης
ἔψε τὴν λινόζωστιν, καὶ σμικρὸν ἀποχέας κλύζε.

Ἢ κνήστρου πόσιν ἐν μέλιτι διεὶς ἐνιέναι.

83. Σκέπτεσθαι δὲ χρὴ τὰ ἐπιμήνια, ἤν τε χολώδεα
ἤν τε φλεγματώδεα ᾖ· ψάμμον ἐς τὸν ἥλιον ὑποβάλ-

ter almonds, and cyclamen—each to the amount of an oak gall, except for the cyclamen which should have an extra portion—mix these with honey, and apply.

(2) Suppository: the cyclamen that cleans the head: crush this in water, mix it up into a foam, and apply.

Or prepare myrrh, salt, cumin, and bull's gall in honey the same way.

Or shell three berries of the Median eye medication called pepper, grind the three corns of these finely into old wine, dilute, wind the suppository around a bird's feather, and apply in this state.

Or take spurge juice mixed into a paste with honey; or a six-finger length of squill root: wrap two fingers of this with wool and apply it.

Or crush the squill alone without its roots, wrap it with wool in the same way, and apply.

82. Injection if a woman is suffering from phlegm: dissolve two drafts of hellebore in two cotyles of sweet wine, mix, and inject.

If a cleaning is called for, boil leeks and elderberries, and add anise, frankincense, myrrh and wine, knead all this together, and inject with this fluid.

Or boil a cabbage in water, add mercury herb, boil, pour off a little of this, and inject it.

Or dissolve a draft of cnestron in honey, and inject with this.

83. You must examine to see whether the menses are bilious or phlegmatic: spread out some fine dry sand in the

---

347 ὁ Θ: om. MV.
348 Add. χλιηρῷ MV.
349 σκίλλαν MV: σχῖνον Θ.

λειν λεπτήν, ξηρήν, καὶ ὅταν τὰ καταμήνια ἴῃ, ἐπιχέαι τοῦ αἵματος, καὶ ἐὰν ξηρανθῆναι· καὶ ἢν μὲν
χολώδης[350] ᾖ, ἐπὶ τῇ ψάμμῳ ξηραινόμενον τὸ αἷμα
χλωρὸν γίνεται· ἢν δὲ φλεγματώδης, οἷον μύξαι· ἢν
δὲ φύσει[351] φλεγματώδης ᾖ, κνήστρου ὅσον πόσιν διεῖναι σὺν μελικρήτου κοτύλῃ, καὶ κλύσαι.

84. (1) Καθαρτικὸν μαλθακὸν ὕδωρ ἄγει καὶ δέρματα[352] καὶ ἰχῶρα ὕφαιμον, καὶ καταμήνια κατασπᾷ,[353] ἢν μὴ πολυχρόνια ᾖ, καὶ στόμα μαλθάσσει·
μύρον ναρκίσσινον καὶ κύμινον ὃ ἐσθίεται, σμύρναν,
λιβανωτόν, ἀψίνθιον, Κύπριον ἅλας, ῥόδινον
ἄλειφα,[354] τούτων ἴσον ἑκάστου, τοῦ δὲ ναρκισσίνου
τέσσερας μοίρας, ἐπικτένιον ὠμόλινον συμμίξας,
πάντα τρῖψαι, ποιῆσαι δὲ βάλανον, περὶ πτερῷ ῥάκος
λεπτὸν περιθείς, καταδῆσαι, καὶ ἐμβάψαι ἐς ἄλειφα
λευκὸν Αἰγύπτιον, καὶ προσθεῖναι, καὶ ἐὰν τὴν ἡμέ-
ρην· λουσαμένη δὲ καὶ ἀφαιρευμένη διανιζέσθω τῷ |
ὕδατι τῷ εὐώδει.

(2) Καθαρτικόν, ὕδωρ ἄγει καὶ δέρματα καὶ μύξας
καὶ ἰχῶρα ὕφαιμον· σμύρναν, ἅλας, κύμινον, χολὴν
ταύρου, ταῦτα ξυμμῖξαι, μέλιτι δὲ ἀναφυρῆσαι καὶ ἐς
ῥάκος ἐνθεῖναι, ἡμέρην δὲ ἐντίθεσθαι, λουσαμένη δὲ
καὶ ἀφαιρεομένη διανιζέσθω ὕδατι τῷ εὐώδει.

Ἢ ἅλας, κύμινον, χολὴν ταύρου μέλιτι ἀναποι-

---

350 -ώδης Θ: -ώδεα MV.
351 δὲ φύσει Θ: οὖν MV.

sun, and when the menses are passing pour some of the blood on this and let it dry. If the woman is bilious, when the blood on the sand dries it will be green; but if she is phlegmatic it will look like mucus. If the woman is phlegmatic by nature, dissolve a draft of cnestron in a cotyle of melicrat, and infuse.

84. (1) An emollient cleaning agent that expels water, pieces of skin, and bloody serum, brings down the menses—unless they are chronically absent—and softens the mouth of the uterus: take narcissus unguent, edible cumin, myrrh, frankincense, wormwood, Cyprian salt and rose unguent—an equal amount of each of these except for four portions of the narcissus—add some tow of raw flax left in the comb, knead all these together, and form into a suppository; wind a piece of fine cloth around a feather, attach the suppository, dip it in white Egyptian unguent, apply, and leave in place for one day. Then have the woman bathe herself, remove the suppository, and wash herself out thoroughly with fragrant water.

(2) A cleaning agent that expels water, pieces of skin, mucus and bloody serum: take myrrh, salt, cumin, and bull's gall, mix these together, add honey, place in a piece of cloth, and insert for a day; then have the woman bathe herself, remove the suppository, and wash herself out thoroughly with fragrant water.

Or take salt, cumin, and bull's gall, make this into a

---

352 δέρματα ΘΜ: φλέγματα V.

353 κατασπᾷ om. Θ.

354 αλιφα Θ: ἄλειφα ἄλφιτα Μ: ἄλφιτα V.

έειν,[355] καὶ προστίθεσθαι, λουσαμένην, ἀφελομένην δὲ[356] τῷ εὐώδει ὕδατι διανίζεσθαι.

Ἢ[357] σίλφιον σύκῳ μῖξαι καὶ προστίθεσθαι, εἶτα διανίψαι μύρῳ ῥοδίνῳ.

Ἢ κόκκους ἐκλέψαντα ἕψειν,[358] καὶ βάλανον ποιέειν, ἐπὴν δ᾽ ἀφέληται, προστιθέσθω ῥόδινον.[359]

Ἢ σκόροδον, λίτρον ἐρυθρόν, σῦκον, τούτων ἑκάστου ἴσον μίξας μέλιτι, δοῦναι προστίθεσθαι τὴν ἡμέρην, καὶ ἐπειδὰν ἀφέληται, ἐλάφου στέαρ προστίθεσθαι, τήξας ἐν οἴνῳ.

Ἢ πεπέριος κόκκους πέντε σὺν ἐλατηρίῳ ὀλίγῳ μῖξαι,[360] παραστάζειν δὲ γάλα γυναικὸς ἐν εἰρίῳ· ἐς μύρον ἀποβάψασα, προστίθεσθαι δὲ ἀφελομένην ὡσαύτως.

Ἢ σύκου τὸ πιότατον σὺν ἐλατηρίου πόσει καὶ λίτρου ἐρυθροῦ τὸ ἴσον καὶ μέλιτος ἴσον, ὡσαύτως.

Ἢ χολὴν ταύρου καὶ λίτρον ἐρυθρόν, καὶ νέτωπον, κυκλαμίνου ὅσον κηκίδα ἐν μέλιτι.

Ἢ χολὴν ταύρου, ἐς Αἰγύπτιον ἔλαιον βάψας προστίθεσθαι, καὶ ἀφελομένη, ῥοδίνῳ.

Ἢ σικύης τῆς μακρῆς ἐντεριώνην τὸ σπέρμα ἐξελὼν σὺν γάλακτι γυναικὸς κουροτρόφου, καὶ σμύρναν ἄκρητον καὶ μέλι ὀλίγον καὶ μύρον Αἰγύπτιον, τρίβειν, καὶ προστιθέναι.

355 ἀναποιέειν Θ: ἀναφυρέειν MV.
356 ἀφελομένην δὲ Θ: om. M: ἀφαιρεομένην V.
357 λουσαμένην . . . ἢ om. M: ἢ om. V.

suppository and apply it; then have the woman bathe herself, remove the suppository, and wash herself out thoroughly with fragrant water.

Or mix silphium with a fig and insert it, and then wash thoroughly with rose unguent.

Or remove the hulls of (sc. Cnidian) berries, boil the berries, make them into a suppository (sc. and apply); when the patient removes this, have her employ rose (sc. oil).

Or take garlic, red soda, and a fig: mix an equal amount of each of these with honey, give this suppository to the patient to insert during the day, and when she removes it, have her apply deer's fat dissolved in wine.

Or mix five peppercorns with a little squirting cucumber juice, and dribble some woman's milk on to a piece of wool; then have the patient dip this in unguent, and apply, removing it in the same way.

Or use the greasiest part of a fig together with a portion of squirting cucumber juice, and equal amounts of red soda and honey in the same way.

Or bull's gall, red soda, oil of bitter almonds, and cyclamen to the amount of an oak gall, in honey.

Or dip bull's gall into Egyptian oil and apply; on removing this, have the patient employ rose (sc. oil).

Or remove the seeds from the insides of a long bottle gourd and knead this together with the milk of a woman who is nursing a male child, some unmixed myrrh, a little honey and also Egyptian unguent: apply.

---

358 ἕψειν Θ: ποσὶν τρίψαντα M: ποσὶν τρίψαντα ἕψειν V.
359 ἐπὴν . . . ῥόδινον om. MV.
360 μῖξαι om. MV.

*Ἡ τὴν ἐντεριώνην τῆς σικύης αὐῆναι, κόψαι δὲ ἄνευ τοῦ σπέρματος καὶ μέλιτι ἀναζέσαι, καὶ βάλανον ποιέειν εὐμήκεα, καὶ βάπτειν ἐς ἔλαιον λευκόν. ἢ 208 τὴν κολοκύντην | τὴν ἀγρίην ὡσαύτως.

Ἐλατηρίου τρεῖς πόσιας ἐν στέατι[361] τρίβειν, καὶ βάλανον ποιέειν.

(3) Καθαρτικόν· ὕδωρ ἄγει καὶ μύξας καὶ δέρματα· κόκκους ἐκλέψας, τρῦψον[362] ὅσον πόσιν, ἑψῆσαι, καὶ μέλι ἐπιχέαι, καὶ προσθέσθω, ἢ ῥόδινον ἄλειφα καὶ προσθέσθω.[363]

Ὅσαι δὲ βάλανον προστίθενται· ἐμπήξαντα χρὴ τὸ πτερὸν[364] ἐς τὴν βάλανον, ἔπειτα ῥάκος περιθεῖναι λεπτὸν ἐν εἰρίῳ,[365] βάψας ἐς ἄλειφα Αἰγύπτιον προστίθεσθαι· ἄγει δὲ μάλιστα δέρματα τὰ ἀφαιρεόμενα.

(4) Προσθετὰ καθαρτικὰ ἰσχυρά, ὕδωρ ἄγει καὶ δέρματα καὶ μύξας καὶ αἱματώδεα[366] μᾶλλον τῶν πρόσθεν· τοῦ πεπέριος κόκκους τέσσερας τῶν μεγίστων, ἢ δέκα τῶν μικρῶν, ἐλατηρίου πόσιν συμμίξας, τρῖβε λεῖα, παραστάζων γυναικὸς γάλα, καὶ μέλι ὀλίγον, ἀναφυρήσας τοῦτο ἐς εἴριον μαλθακὸν καθαρόν, περὶ πτερὸν περιελίξας, ῥάκει[367] καταλαβὼν προστιθέναι, ἐς λευκὸν ἄλειφα Αἰγύπτιον βάπτων· προσκείσθω δὲ ἡμέρην, καὶ ἐπειδὰν ἀφέληται, προστιθέσθω τὸ στέαρ τοῦ ἐλάφου.[368]

---

361 στέατι ΘV: σταιτὶ Μ. 362 τρῦψον Θ: τρεῖς MV.
363 καὶ πρ. om. MV.
364 τὸ πτερὸν Θ: πρότερον MV.

Or dry the insides of a bottle gourd, knead it without its seeds, boil together with honey, form this into a long suppository, and dip it in white oil. Or wild gourd in the same way.

Knead fat in three drafts of squirting cucumber juice and make a suppository.

(3) A cleaning agent that expels water, mucus and pieces of skin: hull (sc. Cnidian) berries, crush a draft of these, boil them, add honey, and have the patient insert this suppository; or use rose oil and have her insert this.

Any women that are applying a suppository: she must fix a feather in the suppository, then wrap a piece of fine cloth in wool around it, dip this in Egyptian unguent, and insert. This suppository is very effective in carrying off pieces of skin that have lifted off.

(4) Powerful cleaning suppositories that expel water, pieces of skin, mucus, and bloody serum better than the ones above: take four peppercorns of the largest size or ten small ones, mix these with a draft of squirting cucumber juice, grind fine, add some drops of woman's milk and a little honey; soak this in a piece of soft clean wool, wrap it around a feather, cover with a rag, and, after dipping it in white Egyptian unguent, apply it. Leave the suppository in place during the day, and when the woman removes it, she should apply deer's fat.

---

365 ἔπειτα ῥάκος . . . εἰρίῳ ΘMV: Ermerins comments *non recte habere videtur* and deletes the whole passage from ὅσαι δὲ βάλανον to ἀφαιρεομένα as a reader's note. 366 καὶ αἱμα-τώδεα om. M. 367 ῥάκει Ermerins followed by Fuchs: ῥάκεα καὶ codd. 368 M transposes 84.6 Προσθετὰ . . . προστιθέναι to this position in the text.

(5) Προσθετὸν καθαρτικόν, χορία[369] ἐκβάλλει, καὶ
καταμήνια κατασπᾷ, καὶ ἔμβρυον ἐξάγει ἀπόπλη-
κτον· κανθαρίδας πέντε ἀποτίλας καὶ τὰ πτερὰ καὶ τὰ
σκέλεα καὶ τὴν κεφαλήν, εἶτα τρίβολον παραθαλάσσιον κόψας σὺν τῇ ῥίζῃ καὶ τοῖσι φύλλοισι καὶ τὸ
χλωρὸν τὸ ἔξω, τρίψας ἴσον ὄγκον καὶ σελίνου σπέρματος καὶ σηπίης ᾠὰ πεντεκαίδεκα ἐν οἴνῳ γλυκεῖ
κεκρημένῳ, καὶ ὕδατι θερμῷ καθίννυσθαι, καὶ μελίκρητον ὑδαρὲς πινέτω καὶ οἶνον γλυκύν, καὶ τοῦ τε-
210 τριμμένου | ὅσον στατῆρα Αἰγιναῖον ἐν οἴνῳ πίνειν
γλυκεῖ· ἐπὴν δὲ ὀδύνη ἔχῃ, λευκοὺς ἐρεβίνθους καὶ
ἀσταφίδα ἑψήσας ἐν ὕδατι, ψύξας, διδόναι πιεῖν· καὶ
ὅταν στραγγουρίη ἔχῃ, ἐν ὕδατι χλιερῷ καθήσθω,
καὶ οἶνον γλυκὺν πινέτω.

(6) Προσθετὰ καθαρτικά· ἢν μὴ τὰ ποτὰ καθαίρῃ·
λινόζωστιν τρίψας καὶ σμύρναν, προστιθέναι.[370]

(7) Πάσης νούσου καὶ ἀναστομῶσαι καὶ καθῆραι·
σμύρναν ὀπτὴν[371] καὶ ἐλελίσφακον καὶ ἄννησον τρίψας. . . .

85. Καθαρτικὸν ἄτοκον καθῆραι, ἢν τὸ στόμα ὀρθῶς ἔχῃ· πυριὴν βόλβιτον ξηρὸν κόψας καὶ διασήσας ὅσον τέσσερας χοίνικας, ὄξους δὲ κοτύλας δέκα,
καὶ οὔρου βοείου ἕτερον τοσοῦτον, καὶ θαλάσσης[372]
κοτύλας εἴκοσι, πυριὴν δὲ βληχρῇσι πολὺν χρόνον,
εἶτα λουσαμένη φάκιον πιοῦσα καὶ μέλι καὶ ὄξος
συμμίξασα ἐμεσάτω, καὶ ῥυφεῖν ἄλευρον δοῦναι, καὶ
ἐπιπεῖν οἶνον εὐώδεα παλαιὸν ἐξηθριασμένον, σίτου
δὲ μὴ γεύεσθαι τὴν ἡμέρην ταύτην, τῇ δὲ δευτέρῃ

(5) A cleaning suppository that expels the placenta, brings down the menses, and removes a paralyzed fetus: remove the wings, legs and head from five blister beetles, then pound a water chestnut plant together with its root, leaves and green outside part, and grind the same amount of hartwort seed and fifteen cuttlefish eggs in diluted sweet wine; give the patient a hip bath in hot water, and have her drink watery melicrat with sweet wine: she should take an Aeginetan stater of the ground things in the sweet wine. When pain is present, boil white chickpeas and grapes in water, cool, and give this to drink. When there is strangury, have the patient sit down in warm water and drink sweet wine.

(6) Cleaning suppositories for use if the drinks fail to clean: knead mercury herb and myrrh: apply.

(7) To open the mouth (sc. of the uterus) and clean it in every disease: knead burned myrrh, salvia and anise. . . .

85. An agent to clean a woman who has not given birth, if the mouth (sc. of her uterus) is as it should be: Foment with four choinixes of dried cow's excrement pounded and sifted, ten cotyles of vinegar, an equal amount of cow's urine, and twenty cotyles of brine; employ mild fomentations for a long time, and then have the patient wash herself, drink a lentil decoction mixed with honey and vinegar, and vomit. Then give her wheat meal to eat, and after that fragrant aged wine that has been exposed to the open air to drink. On this day she must not taste solid food, but on

---

369 χορία MV: λοχεῖα Θ.

370 See note 368 above.

371 ὀπτὴν Θ: ὀλίγην MV.

372 θαλ. Θ: θαλάσσης ἄλμης M: θαλασσίης ἄλμης V.

κόκκον δοῦναι κατάποτον, τῇ δὲ τρίτῃ διουρητικόν,
ἀσταφίδα καὶ ἐρεβίνθους λευκούς, δύο χοίνικας τῶν
ἐρεβίνθων, χοίνικα δὲ τῆς ἀσταφίδος, ἐπιχέαι δὲ τοῦ
ὕδατος τρία ἡμίχοα, καὶ ἔπειτα ἀποχέας, πρὸς τὴν
αἰθρίην θεῖναι, καὶ τῇ ὑστεραίῃ πίνειν, καὶ τοῖσι
προσθέτοισι χρῆσθαι.

86. Καθαρτικόν·[373] τὸ βόλβιτον πλάσαι οἷον
σκαφίδα, φυρᾶν δὲ συμμίσγων τῆς κυπαρίσσου τὰ
πρίσματα, ξηραίνειν δ᾽ ἐν τῷ ἡλίῳ, καὶ ἐς τοῦτο θυ-
μιήματα ἐμβάλλειν. |

87. Ἔγχυτον καθαρτικόν, ἢν μὴ[374] ἴῃ τὰ καταμήνια·
ἀκάνθης λευκῆς φύλλα τρίψας καὶ ἐξηθήσας καὶ ἐπι-
χλιήνας ἐγχεῖν.

88. (1) Κατάχριστον καθαρτικὸν ὥστε μήτρας ἐκ-
καθαίρειν· πράσου σπέρμα καὶ καρδάμου[375] τρίψας, ἐν
οἴνῳ διεὶς καὶ γάλακτι ἑφθῷ, διαχρίειν τὴν νείαιραν
γαστέρα.

(2) Μαλθακτήρια, ὕδωρ ἄγει καὶ μύξας καὶ δέρματα καὶ λοχεῖα, καὶ οὐχ ἑλκοῖ· σμύρνης ὡς ἀρίστης
ἥμισυ, καὶ ἁλὸς χόνδρον ὁμοίως, πίσσαν ἡδυντὴν
φλάσας, μὴ θλίψας λεῖα, ἔστω δὲ τῆς σμύρνης τὸ
ἥμισυ τοῦ ἁλὸς καὶ τῆς πίσσης, ἐμβαλεῖν δ᾽ ἐς ῥάκος
τῆς πεφλασμένης πίσσης μέγεθος ὅσον κηκίδα μεγά-
λην· δύο δ᾽ εἶναι,[376] τὸ μὲν ἡμέρης ἔχειν, τὸ δὲ εὐφρό-
νης, ἔστ᾽ ἂν κατατακῇ· λουέσθω δὲ θερμῷ, καὶ ἀφαιρέουσα διανιζέσθω ὕδατι εὐώδει.

the second day give her some (sc. Cnidian) berry to eat, and on the third day grapes and white chickpeas as a diuretic: take two choinixes of the chickpeas and one choinix of the grapes, pour on three hemichoes of water, pour off the fluid, set it in the open air, and on the next day give this to the patient to drink, and employ suppositories.

86. A cleaning agent: shape cow's excrement like a little bowl, knead it, and mix in sawdust of cypress wood; dry this in the sun, and add fumigants.

87. A cleaning infusion for use if the menses fail to pass: crush some leaves of the fish thistle, filter, warm and inject.

88. (1) An ointment to clean the uterus: grind some seeds of leek and of cress, dissolve this in wine and boiled milk, and anoint this over the lower abdomen.

(2) Nonulcerating emollients that expel water, mucus, pieces of skin and the lochia: knead together a half portion of the best myrrh, a lump of salt, and seasoned pitch, without rubbing them smooth—let the amount of the myrrh be half that of the salt and the pitch—and place the kneaded pitch to the amount of a large oak gall in a rag. There are to be two of these, one for the day and the other for the night, and they are to remain in place until they melt. The woman should bathe herself in hot water, and on removing the suppository wash herself thoroughly with fragrant water.

---

373 Καθαρτικόν Θ: ἔγχυτον καθ. καταμηνίων MV.
374 μὴ om. MV.
375 κ. Θ: καρδαμώμου MV.
376 δ᾽ εἶναι Θ: δὲ ἔστω MV.

89. (1) *Ην[377] ἄτοκον θεραπεύῃς, βόλβιτον αὖον κόψας[378] καὶ διασήσας ὅσον τέσσερας χοίνικας Ἀττικάς, ὄξους δὲ κοτύλας δέκα καὶ ὀροβίου χοίνικα καὶ θαλάσσης κοτύλας εἴκοσι, πυριῆσαι ἀλέην πολὺν χρόνον· ἔπειτα φακίον ποιῆσαι, καὶ μέλι καὶ ὄξος μῖξαι, καὶ ἐμείτω, καὶ ῥυφεῖν ἄλητον, καὶ ἐπιπίνειν οἶνον εὐώδεα· σίτου δὲ μὴ ἅπτεσθαι ταύτῃ τῇ ἡμέρῃ· τῇ δ' ὑστεραίῃ κόκκον δοῦναι κατάποτον· τῇ δὲ τρίτῃ διουρητικόν·[379] σταφίδα καὶ ἐρεβίνθους | λευκοὺς δύο χοίνικας, καὶ ἐπιχέας ὕδατος[380] τρία ἡμίχοα, ἀφελεῖν δὲ τὸ ἥμισυ, εἶτα ἐς τὴν αἰθρίην θεῖναι, καὶ τῇ ὑστεραίῃ πίνειν ἐκ τούτου κατ' ὀλίγον, καὶ τοῖσι προσθέτοισι χρήσθω.

(2) *Ην βούλῃ γυναῖκα κυῆσαι, καθῆραι αὐτὴν καὶ τὰς μήτρας, ἔπειτα δίδου ἄνηθον[381] ἐσθίειν νήστει, καὶ οἶνον ἐπιπίνειν ἄκρητον, καὶ προστιθέναι λίτρον ἐρυθρὸν καὶ κύμινον καὶ ῥητίνην μέλιτι δεύσας, ἐν ὀθονίῳ προσθέσθω· καὶ ὅταν τὸ ὕδωρ ἀπορρυῇ, τοὺς μέλανας πεσσοὺς προστιθέσθω μαλθακτήριον, καὶ τῷ ἀνδρὶ συνέστω.

(3) *Ην δὲ μεμύκῃ, προστιθέσθω καὶ ὀπὸν συκῆς, μέχρι ἀναστομωθῇ· καὶ ὕδατι αὐτίκα περινίζεσθαι· ἴρηκος δὲ ἄφοδον τρίβειν ἐπ' οἴνῳ γλυκεῖ καὶ πίνειν νῆστιν, καὶ αὐτίκα συγγίνεσθαι τῷ ἀνδρί.

---

[377] Before *Ην add. Καθαρτικὸν ἄτοκον καθῆραι ἢν τὸ στόμα ὀρθῶς ἔχῃ M: add. Καθαρτικὸν ἄτοκον V.

[378] κόψας Index Hipp. s.v. ἥλιος C. III: λιου Θ: ἡλίου MV.

89. (1) If you are treating a woman who has failed to become pregnant, pound and sift four Attic choinixes of dried cow's excrement, and add ten cotyles of vinegar, a choinix of vetches, and twenty cotyles of brine: foment thoroughly with this over an extended time. Then make a lentil decoction, add honey and vinegar, and give this emetic; then the patient should take a meal soup and drink fragrant wine. Food is not to be touched that day, but on the next day give a (sc. Cnidian) berry potion. On the third day, prescribe a diuretic: e.g., if you prefer, take some white raisins and two choinixes of white chickpeas, pour on these three hemichoes of water, dispose of half, and set the rest in the open air; the next day have the patient drink some of this a little at a time, and also employ suppositories.

(2) If you want a woman to become pregnant, first clean her and her uterus; then give her dill to consume in the fasting state, have her drink undiluted wine, and apply a suppository of red soda, cumin, and resin soaked in honey and wrapped in a piece of linen. When fluid runs out (sc. from her vagina), the woman should employ black pessaries as an emollient, and have intercourse with her husband.

(3) If it (i.e., the mouth of a woman's uterus) is closed, she should inject fig juice until the mouth opens, and then immediately conduct a thorough cleaning: crumble some hawk's excrement into sweet wine, have her drink this in the fasting state, and then immediately have intercourse with her husband.

---

379 Add. ἐὰν δὲ λευκὸν βούλῃ MV.

380 Add. λευκοῦ MV.   381 ἄνηθον Θ: ἄλητον MV.

Ἢ ὅταν τὰ καταμήνια παύηται, χηναλώπεκος ἄφοδον ἐν ῥοδίνῳ μύρῳ τρίβειν, καὶ τὸ αἰδοῖον χρίεσθαι, καὶ συγκοιμᾶσθαι.

90. (1) Πρὸς ἕλκεα κλυσμός,[382] ἢν ἕλκεα ᾖ δριμέα·[383] κλυσμὸς ἢν φλεγμήνῃ· χηνὸς ἔλαιον καὶ ῥητίνην μίσγειν, καὶ διεὶς ὕδατι χλιαρῷ κλύσαι. ἢ μέλι, βούτυρον τῆξαι, καὶ κλύσαι.

Ἢ ἐχετρώσιος ξῦσαι ὅσον σκαφίδα μικρὴν καὶ σμύρνης καὶ μέλιτος ὁμοίως, ταῦτα διεῖναι ἐν οἴνῳ μέλανι εὐώδει, καὶ κλύζειν χλιαρῷ.

(2) Ἢν ἑλκωθῇ ἢ φλυκταινῶν ἀνάπλεα ᾖ ἐν τῇ καθάρσει ἄκρα τὰ χείλεα, σάρκα βοός, ἢ πικέριον, ἢ χήνειον ἄλειφα, καὶ ἄννησον, ἢ κρόκον, ἢ σποδὸν κυπρίην, ταῦτα τρίβειν[384] πάντα, καὶ περιαλεῖψαι τὴν σάρκα, καὶ προστίθεσθαι.

(3) Ἢν ἕλκεα γένηται καὶ ἀδαξᾶται, βοὸς σάρκα, στέαρ ἐπαλείφειν, καὶ τὴν | σάρκα προστιθέναι, καὶ ἐκκλύζειν. ἢν ῥυπαρὰ ᾖ, καὶ πυριὴν συκῆς ἀπὸ ῥίζης, καὶ μετέπειτα ἄπιον[385] ἐν γλυκεῖ πινέτω.

(4) Ἢν ἐν τοῖσιν αἰδοίοισιν ἕλκεα ἐγγένηται, βόειον στέαρ ἐπαλείφειν, καὶ μυρσίνην ἐν οἴνῳ ἀφέψων[386] διακλῦσαι τῷ οἴνῳ, ἢ ἐλαίης φύλλα καὶ βάτου καὶ ῥοιῆς· ταῦτα δὲ ποιέει καὶ περσέης φύλλα καὶ οἶνος Πράμνιος, τὰ[387] φύλλα τρίψαι λεῖα, καὶ πρὸς τὰ αἰδοῖα προστιθέναι.

---

382 Πρὸς . . . κλυσμός om. MV.   383 Add. πρὸς ἕλκεα MV.
384 ταῦτα τρίβειν Θ: τρίψαι ταῦτα Μ: ταῦτα τρύψαι V.

Or when a woman's menses are stopping, knead the excrement of an Egyptian goose into rose unguent, and have her anoint this on her genitalia, and sleep with her husband.

90. (1) An injection for ulcers, if they are irritating; an injection if they become inflamed: mix goose oil and resin, dissolve this in warm water, and inject; or melt honey and butter, and inject this.

Or shred a small bowl of bryony with the same amount of myrrh and honey, dissolve this in fragrant dark wine, and inject warm.

(2) If the extremities of the lips (sc. of the uterus) become ulcerated or covered with blisters in the (sc. menstrual) cleaning, take some beef, knead together either butter or goose grease with anise, saffron or copper scoria, anoint this on to the meat, and insert it.

(3) If ulcers develop which cause irritation: anoint with beef and fat, apply the beef as a suppository, and flush out (sc. the vagina). If the ulcers are unclean, also foment with fig tree root, and afterward have the patient drink a pear potion in sweet (sc. wine).

(4) If ulcers develop on the genitalia, anoint with cow's fat and flush thoroughly with wine in which you have boiled myrtle, or olive leaves, brambles and pomegranates. Persea leaves and Pramnian wine have the same effect: grind the leaves fine and apply them against the genitalia.

---

385 ἄπιον Littré: ἄπιους Θ: ἀπιοῦσα M: καὶ πίους V.

386 ἀφεψῶν Θ: καθεψῶν M: καθεψεῖν V.

387 ταὐτὰ δὲ . . . τὰ Θ: καὶ οἶνος Πράμνιος· ταῦτα δὲ ποιέειν· περσαίης MV.

Ἢ ἀνήθου καρπὸν καὶ σελίνου τρίψας ἐπίχριε.

(5) Ἢν δὲ ἀφθήσῃ τὰ αἰδοῖα, ὧδε ἰᾶσθαι· σάρκα βοὸς ὡς δύο παλαιστῶν μῆκος, πάχος δὲ ὡς τηλίας,[388] προστιθέσθω μέχρι ἑσπέρης, τὴν δὲ νύκτα οὐ χρή· τῇ δ' ὑστεραίῃ πάλιν προστίθεσθαι μέχρι μεσημβρίης, καὶ πινέτω οἶνον γλυκύν, μέλι ξυμμίσγουσα.[389]

(6) Κλυσμὸς ἢν ἑλκωμέναι ἔωσιν αἱ μῆτραι καὶ στραγγουρίη ἐπιλάβῃ· πράσα, ἀκτῆς καρπόν, σέσελι, ἄννησον, λιβανωτόν, σμύρναν, οἶνον ἴσον τῷ χυλῷ, μίξας ταῦτα καὶ ἀναζέσας, ψύχειν, καὶ μετρίως κλύζειν. ἢ μέλι, βούτυρον, μυελόν, κηρὸν ἐνιέναι.

Ἅσσα ἐν αἰδοίῳ ἕλκεα ἔνι καὶ φύεται, φύλλα ἐλαίης καὶ βάτου καὶ κισσοῦ καὶ ῥοιῆς γλυκείης τρίβειν λεῖα, διεὶς δὲ οἴνῳ παλαιῷ, εἰρίῳ προστίθεσθαι τὴν νυκτὰ πρὸς τὰ αἰδοῖα, καὶ καταπλάσσειν τούτοις· ὅταν δὲ ἡμέρη ᾖ, εἰρύσαι, ἀφέψειν δὲ μύρτα ἐν οἴνῳ καὶ διακλύζεσθαι. ἢ χηνὸς ἔλαιον καὶ ῥητίνην τῆξαι, καὶ κλύσαι. ἢ βούτυρον καὶ κέδρινον ἔλαιον, μέλι παραμίσγειν σμικρόν, καὶ κλύσαι.

218   (7) Ἕλκεα | ἐν στόματι ἀποξηραίνει· ἀργύρου ἄνθος ἐν οἴνῳ τρῖψαι καὶ κλύσαι. ἢ βούτυρον μετὰ μέλιτος κλύσαι.[390] ἢ ἐχέτρωσιν καὶ σμύρναν καὶ μέλι οἴνῳ διεῖναι οἰνώδει μέλανι χλιερῷ, καὶ κλύζειν τῇ ὑστεραίῃ· σχίνου δὲ φύλλα ἑψήσας ἐν ὕδατι, τούτῳ μετακλύζειν.

(8) Ἢν ἑλκωθῇ τὸ στόμα πάμπαν[391] τῶν μητρέων·

Or grind dill and celery seeds, and anoint with this.

(5) If aphthae develop on the genitalia, treat them as follows: take a piece of beef two handbreaths in length and as thick as a board, have the patient apply this in the evening, but at night it is not necessary. On the next day apply this again until noon, and have the patient drink sweet wine mixed with honey.

(6) An injection for use if the uterus becomes ulcerated and strangury follows: take leeks, elderberries, hartwort, anise, frankincense, myrrh and wine equal in amount to the fluid, mix, boil, cool, and inject a moderate amount. Or inject honey, butter, marrow and wax.

For any ulcers that form in the vagina and increase there: crush together leaves of olive, bramble, ivy and sweet pomegranate, dissolve in aged wine, and apply wrapped in a piece of wool against the genitalia over night, and also plaster with the same mixture. At daybreak remove the suppository, and then boil some myrtle berries in wine and flush thoroughly with this. Or melt goose oil together with resin and inject. Or take butter and cedar oil, add a little honey, and inject.

(7) To dry ulcers in the mouth (sc. of the uterus): grind flower of silver into wine, and inject. Or inject butter with honey. Or dissolve bryony, myrrh and honey in strong, dark wine that has been warmed, and inject this on the following day; after that boil mastic leaves in water and inject again.

(8) If the mouth of the uterus becomes completely

βούτυρον, λιβανωτός, σμύρνα, ῥητίνη, μυελὸς ἐλά-
φειος, τούτοισι κλύζε. ἢ φακῆν ἕψειν ἐν ὕδατι, καὶ
ἀποχέαι, τούτῳ κλύζε.

Ὅταν δὲ ὕδωρ ἐκ τῶν ὑστερέων ῥέῃ καὶ ἕλκεα ᾖ[392]
καὶ δάκνηται, χηνὸς στέαρ καὶ ᾠὸν διαχρίεσθαι· ἢ
ὄϊος ἢ σύειον,[393] καὶ φακὸν ἕψειν ἐν οἴνῳ κεκρημένῳ
ἴσον τῷ ὕδατι, τούτῳ κλύζε.

(9) Τὰ δ᾽ ἐν αἰδοίῳ ἕλκεα οἴνῳ[394] καταιονᾶν· ἐπιπάσσειν δὲ μάνναν, βάτον, πίτυος φλοιόν, καὶ τῷ
ὕδατι τούτων νιψάσθω.

91. (1) Διαφθόρια, ἢν ἐναποθάνῃ·[395] χαλβάνης ὅσον
ἐλαίην ἐνελίξας ἐς ὀθόνιον, ἐς κέδριον ἐμβάψας,
προστιθέσθω πρὸς τὸ στόμα τῆς μήτρης.

(2) Ἕτερον· κάλαμον τὸν εὐώδεα καὶ σικύης ἐντεριώνην τρῖψαι ἐν χηνὸς στέατι· ἐπίδησον δὲ τὸν ὀμφαλὸν καὶ τὸ ἦτρον· καὶ μικρὸν ἀπ᾽ αὐτοῦ ἐνστάξας
ἐς εἴριον, προστιθέσθω πρὸς τὸ στόμα τῆς μήτρης·
ἐκ τούτου[396] κατ᾽ ὀλίγον ἔρχεται.

(3) Ἄλλο· ἐρευθεδανὸν κόψας καὶ κέδρου πρίσματα,
ὕδωρ ἐπιχέας, θὲς ἐς τὴν αἰθρίην, εἶτα πρωῒ δὸς πρὸς
τὰς ἀλγηδόνας.

(4) Ἄλλο· σιλφίου ὁκόσον δραχμὴν μίαν, καὶ πρά-
220 σου χυλὸν ὅσον ὀξύβαφον, παραμίξας | κέδρινον
ἔλαιον ἥμισυ κυάθου σμικροῦ, δὸς ἐκπιεῖν.

---

[392] ᾖ om. Θ.
[393] συιον Θ: συός MV.
[394] οἴνῳ Θ: τούτῳ MV.

ulcerated, flush with butter, frankincense, myrrh, resin, and deer's marrow. Or boil lentils in water, pour the fluid off, and inject it.

When fluid runs out of the uterus and there are ulcers which cause irritation, anoint generously with goose fat and an egg. Or boil sheep's (sc. fat) or lard with lentils in wine mixed with an equal amount of water, and flush with this.

(9) On ulcers in the genitalia pour wine: then sprinkle on frankincense powder, brambles, and pine bark, and have the patient wash herself with a fluid made from these.

91. (1) Abortive agents, if a child dies in the uterus: wrap all-heal juice to the amount of an olive in a piece of linen, dip this in cedar oil, and have the patient apply it against the mouth of her uterus.

(2) Another: knead some fragrant reeds and the insides of a bottle gourd into goose grease, and bind this against the patient's navel and lower abdomen. Also pour a few drops of this on to a piece of wool, and have her apply it against the mouth of her uterus: this makes the fetus gradually pass out.

(3) Another: pound together some madder and sawdust of cedar wood, add water, set in the open air, and at dawn apply it against pains.

(4) Another: mix one drachma of silphium and an oxybaphon of leek juice in a small hemicyathos of cedar oil, and give to the patient to drink.

---

395 Διαφ. ἢν ἐναποθάνῃ Θ: ἢν ἀποθάνῃ MV, διαφθόριον add. M in marg. sup.

396 τούτου Aldina: τοῦ codd.

(5) Ἄλλο· ταύρου χολῆς ὅσον ὀβολόν, ἢ ἡμιωβέλιον, τρίβων ἐν οἴνῳ δός· ἢ σταιτὶ περιπλάσσων πάλιν καταπιεῖν.

(6) Ἄλλο· καρκίνους ποταμίους πέντε καὶ λαπάθου ῥίζαν καὶ πηγάνου ῥίζαν,[397] καὶ αἴθαλον ἀπὸ τοῦ ἱπνοῦ τρίψας ὁμοῦ πάντα καὶ ἐν μελικρήτῳ, ὑπαίθριον θεῖσα, πινέτω νῆστις τρίς.

(7) Ἄλλο· σικύης ἐντεριώνην τρίψας λείην ἐν κεδρίνῃ πίσσῃ ἐς εἴριον ἐνελίξασα, προσδησάτω πρὸς τὸ πτερὸν λίνῳ, προσθέσθω ἔσω· τοῦ δὲ πτεροῦ τὸ σκληρὸν[398] προεχέτω μικρὸν ἔξω ἐκ τοῦ εἰρίου· ὅταν δὲ αἷμα φανῇ, ἀφελέσθω.

(8) Ἄλλο· ἐλλεβόρου μέλανος λαβὼν ῥαβδίον ὅσον ἑξαδάκτυλον περιέλιξον εἰρίῳ, τὸ δὲ ἄκρον ἔα ψιλὸν εἶναι, εἶτα προσθέσθω ἔσω ὅτι μάλιστα· ὅταν δὲ αἱμαχθῇ τὸ ἄκρον, ἀφελέσθω.

(9) Ἄλλο· ἐλλέβορον μέλανα καὶ κανθαρίδας καὶ κονίαν[399] τρίψας ἐν ὕδατι, ποιήσας βαλάνιον μαλθακὸν ὅσον ἑξαδάκτυλον, ἐπειδὰν σκληρὸν γένηται, εἰρίῳ περιελίξασα προσθέσθω, τὸ ἄκρον κεδρίᾳ χρισάτω, καὶ ἔστω ψιλόν· ὅταν δὲ αἷμα φανῇ, ἀφελέτω.

(10) Ἐκβόλιον· ἢν ἔμβρυον τεθνεὸς ἢ ἀπόπληκτον ᾖ, βατράχιον καὶ ἐλατηρίου μικρὸν μῖξαι ἐν ὄξει εὐκρήτῳ, καὶ δοῦναι πῖσαι. ἢ κράμβης καυλὸς ἁπαλός, ἄκρον χριόμενος νετώπῳ.[400] |

---

397 ῥίζαν καὶ πηγ. ῥίζαν Θ: καὶ πηγ. ῥίζαν Μ: ῥίζαν V.
398 τὸ σκλ. om. Θ.

(5) Another: knead a whole or half obol of bull's gall in wine and give to drink; or plaster the patient with spelt flour dough, and have her take the drink again.

(6) Another: take five river crabs, the roots of monk's rhubarb and rue, and oven soot, and crush all these together in melicrat: have the patient set this out under the open air and drink from it three times in the fasting state.

(7) Another: knead the insides of a bottle gourd smooth in cedar pitch: have the patient wrap this in a piece of wool, attach it to a feather with a linen thread, and apply it inside—the hard stalk of the feather should project a little out of the wool. When blood appears, have her remove the suppository.

(8) Another: take a small six inch shoot of black hellebore, wrap it in a piece of wool leaving the tip exposed, and then have the patient insert it as far in as she can: when the tip becomes bloody, have her remove the suppository.

(9) Another: grind black hellebore, blister beetles and ashes together in water, make a soft suppository six inches long, and when this becomes hard have the patient wrap it in a piece of wool and insert it; she should leave the tip exposed, anoint it with cedar oil, and when blood appears remove the suppository.

(10) Expulsive agent, if a fetus has died or become paralyzed: mix some ranunculus and a little squirting cucumber juice together in diluted vinegar, and have the mother drink it. Or take a tender stalk of cabbage and anoint the tip with oil of bitter almonds.

---

399 κονίαν Θ: κονύζη M: κονύζην V.
400 The treatise ends in M at this point.

222   92. (1) Βηχὸς[401] παιδίοισι· θαψίην ἐπ᾽ ἀλφίτοισι[402] ψωμίζειν.

(2) Ἕτερον· ᾠὸν ὀπτήσαντα, τὴν λέκιθον ἐξελόντα, τρίψαι· καὶ σήσαμον λευκὸν πεφρυγμένον καὶ ἅλες, ἐν μέλιτι ἐλλείχειν.

(3) Τὴν κοιλίην λῦσαι παιδίου· εἴριον ἄπλυτον εἰς μέλι βάψας ἐντιθέναι· ἢν δὲ γεραίτερον ᾖ, κρομμύου τὰ ἔσωθεν τρίψας ἐντιθέναι· ἢν δὲ μή, κλύσαι γάλακτι αἰγός, συμμίξας μέλι· ἐὰν δὲ γάλα μὴ ᾖ, σητάνιον ἄλητον[403] ἐκπλύνας, μέλι καὶ ἔλαιον μίξας, χλιερῷ κλύσαι.

(4) Ἄσθμα τοῖς παιδίοις· λιβανωτὸν ἐν οἴνῳ γλυκεῖ, ἀλουσίη. . . .

(5) Καθαρτήριον· βαλανίδας ποιέειν, κοτύλην μέλιτος, ἀννήσου ὀξύβαφον, ἀσφάλτου δύο δραχμάς, χολὴν βοός, σμύρνης τρεῖς δραχμάς, πόσιν ἐλατηρίου· ἕψειν ἐν χαλκίῳ, μίσγειν δ᾽ ἔλαιον χηνός, καὶ ὅταν μέλλῃς χρῆσθαι, ἀλείφειν τὰς βαλάνους τῷ χηνείῳ μαλθακῷ· εἰρίῳ δὲ χρῆσθαι ἔξω νετώπῳ,[404] ἐλαίῳ σχινίνῳ, τούτῳ μίσγειν κιννάβαρι.

93. Ἔμετον λύει· ὠκίμου χυλὸς ἐν οἴνῳ λευκῷ.

Ἕτερον· ἀλήτου σητανίου κεχυλισμένου τὸ ὕδωρ, ἢ ῥοιῆς γλυκείης καὶ ὀξείης ἀποχυλώσας, εἶτα μέλιτι μίξας.

---

[401] Before Βηχὸς ms. Θ adds Ἢν ἄτοκον θεραπεύῃς· κόψας βόλιτον αὖον καὶ διασήσας ὅσον τέσσερας (cf. the beginning of ch. 89 above).

238

92. (1) For use against a cough in children: give pieces of thapsia spread on barley meal.

(2) Another: bake an egg, and remove and crumble its yolk: mix this with toasted white sesame and some salt, and give as an electuary in honey.

(3) To move the cavity of a child: dip a piece of unwashed wool in honey, and insert this. If the child is older, knead the insides of some onions, and apply as a suppository. If it is not older, flush with goat's milk to which you have added honey. If no milk is available, wash meal of spring wheat, add honey and olive oil, and flush with this warm.

(4) For use against shortness of breath in children: frankincense in sweet wine, without bathing. . . .

(5) Cleaning agent. To make suppositories: boil a cotyle of honey, an oxybaphon of anise, two drachmas of asphalt, some bull's gall, three drachmas of myrrh and a draft of squirting cucumber juice in a copper vessel, add goose oil, and when you are about to employ them anoint the pessaries with soft goose (sc. grease). Apply on a piece of wool covered with oil of bitter almonds and mastic oil, and mix in some cinnabar.

93. The following relieves vomiting: basil juice in white wine.

Another: take some fluid boiled out of spring wheat meal or juice squeezed from sweet and sour pomegranates and mix this with honey.

---

402 ἀλφ. Θ: ἀλείφιτοις V.
403 ἄλητον Θ: ἄλευρον V.
404 ἔξω νετ. Θ: τῷ οἴῳ V.

94. Τὸ σηπτικὸν ὧδε ποιέεται· ἐλλέβορος μέλας, σανδαράκη, λεπὶς χαλκοῦ, ἴσον ἑκάστου τρίβειν χωρίς· ὅταν δὲ λεῖον ᾖ, παραμῖξαι μιᾶς μερίδος διπλάσιον τίτανον, καὶ δεύσας κεδρίνῳ, χρῶ.

95. Τὸ ὀπτὸν φάρμακον ὧδε ποιέεται· ἄνθος κεκαυμένον καθαρῶς, ἄχρι οὗ φοινικοῦν γένηται, τρίψας λεῖον, τούτῳ χρῶ.

96. Τὸ μέλαν φάρμακον· λεπίς, ἄνθος, χωρὶς τρίβειν ἑκάτερον· ὅταν δὲ λεῖον τρίψῃς οὕτω, μίσγειν·
224 ποιέειν δύο ἢ τρία εἴδη τοῦ φαρμάκου, | τὸ μὲν ἰσχυρότατον τὸ ἄνθος τριτημόριον τῆς λεπίδος, τὸ δὲ δεύτερον, τεταρτημόριον, τὸ δὲ τρίτον, πεμπτημόριον· τοῦτο τὸ φάρμακον ἐπίπαν ἁρμόζει.

97. Διαλειπτὸν πρὸς τοὺς συνάγχους· κάχρυς, ἀσταφὶς ἀγρία, ἀψίνθιον, ἐλατήριον, μέλι.

98. Ἐν τοῖσι ποδαγρικοῖσιν ὀδυνήμασι τὰ ἀφιστάμενα ἁλὶ καταπλάσσειν[405] ὕδατι φυρῶντα λεῖον, καὶ μὴ λύειν τριῶν ἡμερέων· ὅταν δὲ λύσῃς, αὖθις λίτρον ὠμόν, τοῦ ἐρυθροῦ τρίψαντα καὶ μέλι ὀλίγον, τούτῳ ὥσπερ ἁλὶ χρῆσθαι τὸν ἴσον χρόνον· ἐς χύτρην ἅλας ἐμβαλὼν λείους, εἶτα στυπτηρίην ὀλίγην ἐπιπάσαι, εἶτα τιθέναι τὰ χαλκία, καὶ αὖθις ἐπιπάσσειν τοὺς ἅλας καὶ τὴν στυπτηρίην, εἶτα καταλείψας ὑποκάειν νύκτα καὶ ἡμέρην.

---

[405] -πλάσσειν V: -πάσσειν Θ.

94. An agent to promote sepsis can be made as follows: take black hellebore, red sulfide of arsenic, and scales of copper, grind an equal amount of each separately, and when they are ground fine mix with one portion of these a double amount of gypsum; dissolve this in cedar oil and employ it.

95. The burned (sc. copper) medication is made as follows: flower (sc. of copper) is burned pure until it becomes red, ground fine, and employed thus.

96. The black medication: grind scales and flower (sc. of copper) each separately, and when they are ground fine mix them together. Make two or three varieties of the medication: a) the strongest, with one third as much flower as scales; b) the second, with one quarter as much; c) the third with one fifth as much. These medications are applicable in every case.

97. A liniment for use against anginas: frankincense plant, stavesacre, wormwood, squirting cucumber juice, and honey.

98. In the case of gout pains apply to the swollen parts a plaster of salt crushed fine with water, and do not undo it for three days. After you remove it, grind some raw soda of the red variety and a little honey and apply this in the same way as the salt, and for the same length of time. Pour the finely crushed salt into an earthen pot, sprinkle a little alum over it, and then place copper vessels (sc. over a fire); sprinkle in salt and alum again, and then leave a fire burning under it for a night and a day.[18]

[18] This seems to be a kind of double boiler consisting of an earthen pot set in a copper pan of boiling water.

99. Τὴν ἕδρην ἐμβάλλει· ἀσταφίδι λείῃ, τετριμμένῃ, ξηρῇ, ἐπαλείφειν τὴν ἕδρην.

100. Τὰ πεπωρωμένα διαχεῖ σανδαράχην ἐν σταιτί.

101. Θρίδακος τῆς ἐρυθρῆς ὀπὸς ὀδύνην λύει ἅπασαν ἐν ὕδατι, σταθμὸν ἡμιωβέλιον Ἀττικόν.

102. Ὀφθαλμικά· χαλκὸς κεκαυμένος, ἰός, σμύρνα χολῇ αἰγὸς λύεται· ταῦτα πάντα ὁμοῦ τρίψας λεῖα, οἴνῳ διεῖναι λευκῷ· εἶτα ξηρᾶναι πρὸς τὸν ἥλιον ἐν χαλκίῳ· ἔπειτα ἐς κάλαμον ἐμβαλών, ξηρῷ χρῆσθαι.

103. (1) Ἔμπλαστρον·[406] μίσυ κατακαύσας, τρίβειν ἐν ἴγδῃ· συμμίσγειν | δ᾽ αὐτῷ σποδὸν χρυσῖτιν πεπλυμένην· εἶναι δὲ τῆς σποδοῦ τρία μέρη, τοῦ μίσυος ἕν· τὸ δὲ μίσυ κατακαίειν ἐς μᾶζαν, φυλασσόμενος ὅπως μὴ ἐκρυῇ· ὀπτώμενον γὰρ ἐξυγραίνεται· ὅταν δὲ καλῶς ὀπτὸν ᾖ, φοινίκεον γίνεται.

(2) Ἔμπλαστρον·[407] ψιμύθιον τὸν αὐτὸν τρόπον μισγόμενον τῷ μίσυϊ ὀπτωμένῳ, ὅπερ ἐν τῇ χρυσίτιδι σποδῷ τὸ μίσυ γίνεται.

(3) Ἕτερον ἔμπλαστρον[408] ἰσχυρότερον τούτων· σποδὸς κυπρίη ἐκ τοῦ ἀσβόλου πεπλυμένη, καὶ ψιμύθιον, καὶ μίσυ ὀπτόν· εἶναι δὲ δύο μοίρια τῆς σποδοῦ καὶ τοῦ ψιμυθίου, ἓν τοῦ μίσυος.

104. (1) Ὑγρὸν ἀνεμώνης, τὰ φύλλα κόψαντα, ἐκπιέσαι, καὶ ἐς τὸν ἥλιον θεῖναι ἐν χαλκῷ ἐρυθρῷ κατακαλύψαντα, ὅπως μηδὲν ἐμπεσῆται· ὅταν δὲ παχὺ ᾖ, διαπλάσσειν φθόεις, εἶτα ξηραίνειν· ὅταν δὲ ξηρανθῇ, κατακαίειν ὡς δυνατὸν μάλιστα· εἶτ᾽ ἐπει-

99. An agent to reduce the seat:[19] take finely ground raisins, dry, and anoint the seat with this.

100. An agent to dissolve calluses: realgar in fat.

101. Juice of red lettuce in water resolves any pain: the dose is an Attic half-obol.

102. Agents for the eyes: dissolve burned copper, verdigris and myrrh in goat's gall, grind them all together fine, and dissolve this in white wine; then place this in a copper vessel and expose it to the sun, and after that put it into a reed and employ it dry.

103. (1) A plaster: grind burned misy in a mortar and mix with this washed gold scoria (i.e., litharge)—let there be three portions of the scoria and one of the misy. Then bake the misy into a barley cake, taking care that it does not run out, since as the cake bakes it becomes moist: when the cake is baked through it will turn red.

(2) A plaster: white lead mixed the same way with baked misy as misy is prepared in gold scoria.

(3) Another plaster, stronger than the preceding ones: use washed copper scoria from (sc. oven) soot, white lead, and baked misy: there should be two portions of the scoria and the white lead, and one of the misy.

104. (1) Anemone juice: pound the leaves, squeeze the (sc. fluid) out of them, and set this in the sun in a covered red copper vessel so that nothing can fall in. When this has become thick, mold pastilles from it and then dry them. Once they are dry, heat them from below as strongly as

19 I.e., anal prolapse.

---

406 Ἔμπλαστρον V: Ἔμπαστον Θ.   407 Ἔμπλαστρον V: Ἔμπαστον Θ.   408 Ἔμπλαστρον V: Ἔμπαστον Θ.

δὰν ψυχθῇ, τρίβειν λεῖα, καὶ μίσγειν σποδὸν πεπλυ-
μένην τὴν ἐκ τοῦ ἀσβόλου ἴσην πρὸς ἴσον, εἶτα
παραστάζων νέτωπον σμικρὸν τρίβειν, εἶτα μέλιτι
διεῖναι· εἶτα ξηρήνας, ἐς χαλκῆν κιστίδα τούτῳ χρῶ.

(2) Ξηρὸν μαλθακόν· σποδὸς κυπρίη, χαλκῖτις
ἄπλυτος, λεῖα τετριμμένη, καὶ ἀφρὸς[409] χαλκοῦ, ταῦτα
ἴσα μίσγειν καὶ τρίβειν λεῖα.

(3) Ἕτερον ξηρόν· σποδὸς κυπρίη, χαλκῖτις λεῖα
τετριμμένη, καὶ σποδὸς χρυσῖτις ἄπλυτος, ἐν ᾗ ἀφέ-
ψεται τὸ χρυσίον, ἴσα ἀλλήλοις λεῖα τρίβειν.

(4) Ἕτερον ξηρόν· σποδὸς πεπλυμένη χρυσῖτις καὶ
ἀφρὸς χαλκοῦ ἴσα λεῖα.

(5) Ἕτερον· ὀμφακὸς χυλός, σποδὸς κυπρίη· τὴν
ὄμφακα ἀκμάζουσαν χρὴ ἐκπιέσαι τὸν χυλὸν δι᾽ ὀθονίου ἐς χαλκὸν ἐρυθρόν, καὶ μῖξαι ὄξεος τρίτον μέρος
λευκοῦ ὡς ὀξυτάτου, καὶ οὕτω καθέψειν ἐν τῷ ἡλίῳ,
καὶ ἀναταράσσειν πεντάκις τῆς ἡμέρης· ὅταν δὲ πα-
228 χὺς γένηται ὁ χυλός, σποδὸν τῆς Κυπρίης τῆς χαλ-
κίτιδος λείην ἐμβάλλειν καὶ ἀναμῖξαι· ἐμβάλλειν δὲ
τὴν σποδόν, ὅταν ἑκταῖος ἢ ἑβδομαῖος ὁ χυλὸς ἐν τῷ
ἡλίῳ κείμενος ᾖ, ἐς κοτύλην Ἀττικὴν τοῦ χυλοῦ τῆς
σποδοῦ δραχμὰς ὀκτώ· ἐὰν δὲ βούλῃ δριμύτερον εἶναι, ἐλάσσω τὴν σποδόν· ἐὰν δὲ μαλθακώτερον,
πλείω· μετὰ δὲ ταῦτα ξηραίνεις, ἄχρις οὗ δυνατὸν
διαπλάσαι φθόεις· εἶτ᾽ ἐγξηραίνειν, κρεμάσαι δὲ[410]
ὑπὲρ καπνοῦ, καὶ οὕτω ξηραίνειν μέχρι οὗ ὀστρακῶ-
δες γένηται, ὥστε τριβόμενον μὴ ξυστρέφεσθαι, εἶθ᾽
οὕτως χρῶ· κείσθω δὲ ὅπου ἰκμάδα μὴ ἔξει.

possible. Then, when they have cooled, grind them fine and add washed scoria from soot—an equal amount of each—dribble a little oil of bitter almonds over this, grind, and then dissolve in honey. Then, after you have dried this, put it into a copper basket, and make use of it.

(2) A dry emollient: take copper scoria, unwashed copper ore finely ground, and copper scum, mix an equal amount of each of these, and grind this fine.

(3) Another dry agent: take copper scoria, finely ground copper ore, and unwashed gold scoria from which the gold has been boiled off—an equal amount of each: grind fine.

(4) Another dry agent: washed gold scoria and copper scum—equal amounts: (sc. grind) fine.

(5) Another with juice of wild grapes and copper scoria: squeeze the juice of ripe grapes through a linen cloth into a red copper vessel, and mix with this one third as much very acidic white vinegar; boil this down in the sun, stirring it five times a day. When the juice has become thick, introduce fine scoria of Cyprian copper ore and mix this together: introduce the scoria when the juice has been sitting in the sun for six or seven days, (sc. and let there be) eight drachmas of scoria added to an Attic cotyle of the juice. If you want the agent to be sharper, use less scoria, if milder, use more. After that dry the mixture until you can form it into pastilles; then dry them by suspending them over smoke, and continue this drying procedure until they become like shards that on being ground do not crumble: then apply in this state. Store them where they will not be exposed to moisture.

---

409 ἀφρὸς Θ: ἄνθος V.
410 δὲ Θ: ἄνω V.

(6) Ἕτερον ξηρόν· σποδός χαλκῖτις ὄξει πεφυρη-
μένη λευκῷ, εἶτα φθόεις ποιήσας ξηρῆναι· ὅταν δὲ
ξηρήνῃς, λεῖον τρίβειν.

105. (1) Ὑπαλείφειν ὀφθαλμῶν· μέλι ὡς κάλλιστον
καὶ οἶνον παλαιὸν γλυκὺν ἑψεῖν ὁμοῦ.

(2) Ἐς ἄργεμον· αἰγείρου δάκρυον, γάλα γυναι-
κεῖον μίξας χρῶ.

(3) Ἐὰν ὀφθαλμὸς δακρύῃ καὶ ὀδύνη ἔχῃ· ῥοιῆς
γλυκείης τὸν χυλὸν ἐκπιέσας, ἐν χαλκίῳ ἕψειν ἐν πυρὶ
μαλθακῷ, μέχρι οὗ παχὺ γένηται καὶ μέλαν ὥσπερ
πίσσα· ἐὰν δὲ θέρος ᾖ, ἐς τὸν ἥλιον τιθέναι· εἶτα
ὑγρῷ ὑπαλείφειν.

(4) Ἐὰν δὲ δακρύῃ καὶ γλαμυρὸς[411] ᾖ ὁ ὀφθαλμός,
ὅταν ἡ σταφυλὴ ἡ λευκὴ πέπειρας ἰσχυρῶς ᾖ καὶ
ἰσχνὴ ἐπὶ τῇ ἀμπέλῳ, ἐπιδρέψας ἐξηθῆσαι, εἶτα ξηραίνειν ἐν τῷ ἡλίῳ· ὅταν δὲ ξηρὸν ᾖ, ἀποξῦσαι, μῖξαι
δὲ ἴου ἡμιωβέλιον Ἀττικῷ σταθμῷ· εἶτα τούτῳ ὑπαλείφειν.

(5) Παράπαστον· μόλιβδος κεκαυμένος,[412] σποδοῦ
ἴσον, σμύρνης δέκατον μέρος, ὀποῦ μήκωνος σμικρόν, οἶνος παλαιός· ξηρήνας τρίψας χρῶ. Σκίλλα
κεκαυμένη,[413] σποδοῦ τρίτον μέρος, βυβλίον κεκαυμέ-
νον,[414] ψιμυθίου τρίτον μέρος,[415] σμύρνης δέκατον. |

106. Εἰ βούλει ἐκ τοῦ σώματος τρίχας ἀπελάσαι·
δακρύῳ ἀμπέλου ἀλείφειν ἐλαίῳ· ἢν δὲ καὶ τὸν ὀφθαλ-
μὸν βούλῃ, ἀποδρέψας ἀλείφειν. ἀλκυόνιον κατακαύ-

---

411 γλαμυρὸς Froben: πλαμυζος Θ: γλυκύμυρος V.

(6) Another dry agent: take scoria of copper ore mixed in white vinegar, make this into pastilles, and dry them; after that grind them fine.

105. (1) An ointment for the eyes: boil together the finest honey and sweet, aged wine.

(2) An agent against albugo: mix tears from the black poplar tree with woman's milk, and apply.

(3) To use if an eye sheds tears and is painful: squeeze juice out of a sweet pomegranate and boil it in a copper vessel on a low fire until it becomes thick and dark like pitch. If it is summer, set this in the sun and then anoint with it moist.

(4) If an eye sheds tears and becomes bleary, take some white grapes that are very overripe and dried up on the vine, pluck them and press out their juice, and then dry this in the sun. When it is dry, scrape it out, mix it with a half obol by Attic measure of verdigris, and anoint with this.

(5) A sprinkling powder: take burned lead, an equal amount of scoria, a tenth portion of myrrh, a little poppy juice, and some aged wine: dry, grind, and employ. Burned squill, a third as much scoria and burned papyrus, a third as much white lead, and a tenth as much myrrh.

106. If you wish to remove hairs from the body, anoint with tears of the vine . . . olive oil; if you also wish to remove the eye brows, tear them out and anoint. Burn bastard sponge, grind it fine, dissolve it in wine, and anoint

---

[412] κεκαυμένος om. V.

[413] κεκαυμένη Θ: καὶ V.

[414] βυβλ. κεκ. Θ: καὶ V.

[415] Add. χάρτου κεκαυμένου μέρος V.

σας, ἔπειτα τρίψας λεῖον, οἴνῳ διείς, ἐπαλείφειν· καὶ
ἄπεισι σὺν λεπτῷ δέρματι, καὶ ἔσται ἐρυθρὸν καὶ
εὔχροον.

107. Λειεντερίης· φακούς, πυροὺς σητανίους ὅσον
δύο χοίνικας βρέξας, ἐπειδὰν μαλθακοὶ ὦσι διατρω-
γόμενοι, ἰσχυρῶς ποιῆσαι λείους ἐν ὅλμῳ ἢ ἐν θυίῃ·
ἔπειτα ἐπιχέαι ὕδατος κοτύλας ἕξ, καὶ ἀνακινῆσαι
ἰσχυρῶς· ὅ τι δ᾽ ἂν ἀπέλθῃ, ἐγχέας ἐς χύτρην, ἕψειν,
μέλι ὀλίγον παραχέας· ἐπειδὰν δὲ ἐφθὸν ἰσχυρῶς γένηται, φρύξας διδόναι ἐσθίειν τούτου· ἢν δὲ διψᾷ,
οἶνον πινέτω ὡς παλαιότατον· τούτῳ χρήσθω, ἄχρι
ὅτου[416] ὑγιὴς γένηται.

108. Ἢν δὲ κόρυζαν ἔχῃ, σμύρναν τρίψας λεῖα, καὶ
μέλι μίξας, ὀθόνιον ἀναποιήσας, τὰς ῥῖνας τρίβειν.

109. (1) Κλυσμὸς φλέγμα ἄγει· θαψίης πόσιν, ἢ
ἀσταφίδος ὅσον τεσσεράκοντα κόκκους, ἢ Κνιδίου
πόσιν, ἢ κνῆστρον· μίσγειν δὲ μέλιτος ἡμικοτύλιον,
ἐλαίου ἴσον, διεῖναι θαλάσσῃ, ἢ πίτυρα ἐναφεψήσας
ἢ πτισάνην, μέχρι οὗ λιπαρὰ γένηται, ἢ στέατος, ἢ
τεύτλων χυλῷ μούνῳ, ἢ γάλακτι ἐφθῷ, ἢ ἀκτῆς χυλῷ,
ἢ λινοζώστιος χυλῷ· παραμίσγειν δὲ λίτρου ὅσον
δέκα δραχμάς, ἢ ἁλὸς τρυβλίον, πλὴν[417] ἐς θάλασ-
σαν.

(2) Ἢν δὲ βούλῃ[418] χολὴν ἄγειν· ὀποῦ πόσιν, ἐλατηρίου ὁλκὴν καὶ ἡμίσειαν, κολοκυνθίδας τρῖψαι
δραχμὴν σταθμόν· διεῖναι δὲ τοῖς αὐτοῖς οἷς καὶ τὸ
πρότερον.

with this. Hair is also removed with a fine pellicle, and the skin will remain red and of a good color.

107. An agent for use in lientery: soak lentils and spring wheat, two choinixes, and when this is completely soft, grind it forcefully in a mortar or kneading trough until it is smooth; then add six cotyles of water and stir forcefully. Take what precipitates from this and pour it into an earthen pot; add a little honey, and boil. When this is thoroughly boiled, bake it and give some to the patient to eat. If the patient is thirsty, have them drink very well-aged wine. Employ this until the patient has recovered.

108. If a person has a coryza, knead myrrh smooth, add honey, prepare this in a piece of linen, and rub it against the nostrils.

109. (1) An injection that expels phlegm: take a draft of thapsia, or forty stavesacre berries, or a draft of (sc. Cnidian) berries or of cnestron: mix with this a half cotyle of honey and the same amount of olive oil, and dissolve it in brine in which you have boiled bran or peeled barley until it feels greasy, or in water from flour dough or in just the fluid from beets, or in boiled milk, or in elder juice, or in the water of mercury herb. With this mix ten drachmas of soda, and a tryblion of salt (except in the case of the brine).

(2) If you want the patient to expel bile: take a draft of (sc. silphium) juice, a drachma and a half of squirting cucumber juice, and a drachma of ground gourd: dissolve in the same solvents as in the preceding preparation.

---

416 ὅτου Θ: ἂν V.
417 πλὴν Θ: πλεῖον V.
418 βούλη Θ: θέλης V.

232   (3) Ἢν θέλῃς ἄγειν, σικύης ἐντεριώνην, | τέσσερας δραχμὰς[419] ἀποβρέξας ἐν ὕδατος ἡμικοτυλίῳ, τούτῳ κλύσαι, καὶ ἐὰν ἐξελθὼν δάκνῃ, μετακλύσαι πτισάνης χυλῷ.

(4) Ἕτερον· ἐν γάλακτι ἐφθῷ ὀνείῳ, ἢ[420] ἐν τεύτλου χυλῷ τρισὶ κοτύλαις ἀποβρέχειν τὴν ἐντεριώνην, παραμίσγειν δὲ ἅλας καὶ μέλι καὶ ἔλαιον, μετακλύζειν δὲ πτισάνης χυλῷ.

(5) Ἕτερον· κολοκυνθίδος δραχμὴν τρίψας, ἀποβρέξας[421] ἐν γάλακτι ὀνείῳ, μίσγειν ταῦτα.

(6) Ἕτερον· ἐντεριώνης δραχμήν, ἐλατηρίου πόσιν, †σολόμης†[422] ὅσον τοῖσι τρισὶ δακτύλοις, μέλι, ἔλαιον, διεῖναι θαλάσσῃ.

(7) Ἢν δὲ κόπριον θέλῃς ἄγειν, μηδὲν πίνειν φάρμακον,[423] τοῖσι δὲ ἄλλοισι χρῶ.

(8) Δυσεντερίης κλυσμός· ἐν οἴνῳ σίδια ῥοιῆς γλυκείης ἑψῆσαι ὡς πλεῖστα, ἕψειν δὲ ὡς τὸ ἥμισυ λειφθῇ, μίσγειν δὲ μέλι, ἔλαιον, τεταρτημόριον κοτύλης ἑκατέρου.

(9) Τεινεσμοῦ· λιβανωτοῦ τέσσερας δραχμάς,[424] μύρου ῥοδίνου ἡμικοτύλιον, πτισάνης χυλός, θάλασσα ἐφθή. ἐλλεβόρου δύο πόσιας τρίψας, ὕδατος διεὶς ἡμικοτυλίῳ, ἐλαίου ἴσον κλύζειν.

Χυλὸν[425] ἐγχέας ἐς σκαφίδα, μῆλα κυδώνια κατατέμνειν, καὶ ἐὰν βρέχεσθαι· ἐπειδὰν δὲ τὸ ὕδωρ τὴν ὀσμὴν ἔχῃ, διδόναι πίνειν.

---

[419] τέσσερας δραχμὰς Θ: τέταρτον δραχμῆς V.
[420] ἢ om. Θ.   [421] ἀποβρ- Θ: προβρ- V.

(3) If you want to draw, soak four drachmas of the insides of a bottle gourd in a half-cotyle of water, and flush with this; if this burns as it comes out, irrigate again with barley water.

(4) Another agent: soak the insides (sc. of a bottle gourd) in three cotyles of ass's milk or beet juice, add salt, honey and olive oil, (sc. employ), and then make a second injection with barley gruel.

(5) Another agent: mash a drachma of gourd that has been soaked in ass's milk, and mix the same ingredients as above.

(6) Another agent: take a drachma of the insides (sc. of a bottle gourd), a draft of squirting cucumber juice, a pinch of † . . . †, honey, and olive oil: dissolve in brine.

(7) If you wish to move feces, do not give any purgative to drink, but use other means.

(8) An enema for use in dysentery: boil as much sweet pomegranate peel as you can in wine, continuing until it is reduced to half, and mix this with honey and olive oil—a quarter cotyle of each.

(9) For use in tenesmus: four drachmas of frankincense, a half-cotyle of rose unguent, barley juice, boiled brine. Grind two drafts of hellebore, dissolve this in a half-cotyle of water and the same amount of olive oil, and inject.

Pour barley water into a bowl, cut up quinces, and leave them to macerate; when the water has taken on their smell, give this to the patient to drink.

---

[422] σολόμης V: γολομης Θ (cf. Hesychius s.v. γολομένη· βοτάνη).   [423] πιν. φαρ. V: μίσγειν Θ.

[424] τέσσερας δραχμάς Θ: τέταρτον δραχμῆς V.

[425] χυλὸν Littré: χυλῷ ΘV.

Πτισάνης λεκίσκιον[426] ἐμβαλὼν ἐς χοέα ὕδατος, ἑψεῖν μέχρι λιπαρὸς γένηται, ψύξας δέ, τὰ μῆλα ἐγκατατάμνειν, κηρίον καταβρέξαι δὲ ἐν ὕδατι, καὶ ἀνατρίβειν, ἔστ᾽ ἂν ὑπόγλυκυ ᾖ, καὶ διηθήσας, ἐμβάλλειν σελίνου φύλλα.

(10) Ἕτερον· ἀσταφίδα λευκὴν ἐς ὕδωρ ἐμβάλλων, καλαμίνθην ἢ κορίον ἀνατρίβειν ἐς ὕδωρ ὑπόγλυκυ.

---

[426] λεκίσκιον Potter: λέκισκον Θ: λέκινθον V.

Pour a lekiskion[20] of barley gruel into a chous of water, boil it until it feels greasy, and cool; cut up quinces, moisten honeycomb in water, and knead these together until it is sweetish; then filter and add celery leaves.

(10) Another agent: put some white grapes into water, and grind calamint or coriander into the water until it becomes sweetish.

[20] LSJ: "a small measure or weight."

DISEASES OF WOMEN II

INTRODUCTION

In principle, each chapter[1] of *Diseases of Women II* presents a specific disorder of women and prescribes its therapy. In their fullest form, these accounts have three parts—identification of the disease by name, pathognomonic symptom or cause; description of its clinical manifestation, course, and prognosis; recommendation of one or more applicable treatments—but they vary greatly in their internal arrangement, emphasis, and comprehensiveness. Thus, for example, chapters 1, 10, 24–25, 53, and 92 furnish detailed accounts of their condition's etiology, semeiology, and therapy; chapters 21, 43, 50–51, 74–75, 81, and 95 contain only a succinct account of their condition's course and one prescription; chapters 33, 38–39, 52, 55, 64, 71, 80, 82, and 89 only name the condition and recommend one or more treatments; and chapters 83–88, 97, and 100 take the form of collections of particular treatment modalities (e.g., suppositories, douches, potions, poultices, fumigations, fomentations) applicable for specific conditions (e.g., red flux, aqueous flux, sclerosis of the uterus, pains caused by excessive purgation).

[1] Exceptions are chapters 2 and 29 on general causes, and the first sentence of chapter 1 on the frequency of different fluxes in women of different ages.

The treatise is arranged according to various aspects of the disorders being considered, such as their etiology, clinical presentation, the part of the body they involve, or their mode of therapy; the chapters are arranged in five main groups, and within these in various subgroups.

A. Vaginal discharges

B. Movements of the uterus within the body

E. Conditions not in the uterus

Diseases of Women II is included in all the collected editions and translations of the Hippocratic Collection but has never been the subject of a special study, although N. Countouris provides an edition and translation of many chapters of the work in his *Hippokratische Gynäkologie*, and part of chapter 36 (145 L.) is edited and translated by H. Grensemann.[2]

No editor or translator before Littré and few after him have treated *Diseases of Women II* as a continuation of *Diseases of Women I* in the numbering of its chapters, and Ermerins argues for different authors.[3] I have followed Ermerins and Fuchs in preserving the traditional division into separate works, but I have printed Littré's chapter numbers of the Greek text in parentheses.

[2] Countouris, pp. 1–46; Grensemann, p. 102f.
[3] Ermerins, vol. 2, xciii.

ΠΕΡΙ ΓΥΝΑΙΚΕΙΩΝ Β

1. (110 L.) Ῥόος λευκὸς ἐν τῇσι γεραιτέρῃσι τῶν γυναικῶν μᾶλλον γίνεται ἢ ἐν τῇσι νεωτέρῃσι· ῥόος πυρρὸς ἐν ἀμφοτέρῃσι· ῥόος ἐρυθρὸς ἐν τῇσι νεωτέρῃσιν.

Ῥόος ἐρυθρὸς γίνεται ἐκ πυρετῶν, μᾶλλον δ᾽ ἐκ τρωσμοῦ· γίνεται δὲ καὶ ἐξ ἀπολήψιος ἐπιμηνίων, ὁπόταν ἀποκλεισθέντα ἐξαπίνης καταρραγῇ· γίνεται δὲ καὶ ἐκ τόκων.[1] αἷμα ῥεῖ πάμπολυ, καὶ θρόμβοι ἐκπίπτουσι, καὶ ὀδύνη ἐγγίνεται τῶν κληΐδων καὶ τῶν τενόντων, καὶ νάρκη σώματος[2] ἀπόψυξίς τε τῶν σκελέων· ἐνίοτε δὲ καὶ ἤρεισε τοὺς ὀδόντας. ἢν πλέον ᾖ τὸ αἷμα τὸ ἀπιόν, καὶ ἄναυδοι γίνονται,[3] ἱδρώς τε καταχεῖται πολύς. πρὸς δὲ τούτων καρδιωγμοί τε γίνονται, καὶ περιψύξιες, καὶ πυρετοὶ ἀκρητόχολοι καὶ ἀλυσμώδεες· καὶ τῆς αὐτῆς ἡμέρης πολλάκις ῥιγεῦσι, καὶ αὖτις ἱδρώς,[4] καὶ ἄλλοτε ἀπὸ τῶν ἄνω χωρίων σπασμοὶ σφίσι γίνονται, καὶ ἄλλοτε ἀπὸ τῶν κάτω, καὶ ἐς τοὺς βουβῶνας ὀδύναι σφίσιν ἐμπίπτουσιν ὀξεῖαί τε καὶ ἰσχυραί, φοιτῶσαι ὥσπερ ὠδῖνες· ἐνίοτε

---

[1] καὶ ἐκ πυρετῶν add. MV.   [2] καὶ ἀρτηρίης add. Θ.
[3] καὶ ἄναυδοι γίνονται om. MV.   [4] καὶ αὖτις ἱδρώς om. Θ.

262

DISEASES OF WOMEN II[1]

1. A white flux is more likely to occur in older women than in younger ones, a flame-colored flux occurs in both, and a red flux is commoner in younger women.

A red flux arises from fevers—even more from an abortion—but it can also arise after a stoppage of the menses, when after being held back they suddenly break out downward; it also comes after childbirth. Blood flows copiously, clots are expelled, and pain is present in the collarbones and the tendons there; there is numbness of the body with coldness of the legs, and sometimes the patient also grinds her teeth. If the amount of blood passing is quite great, speech is lost and much sweat is secreted. From these (sc. conditions) there are also heartburn, chills, fevers accompanied by bilious vomiting, and restlessness; on the same day patients often have chills followed by sweating, and at one time convulsions arising from their upper parts, but at another time from their lower parts; pains which are acute and violent befall their groins, intermitting like birth pangs. Sometimes there is also strangury. The mouth is

[1] In *Disease of Women 1*, "*Diseases*" is deduced from the beginning of the subsequent text "Concerning diseases of women": here no such qualification of Γυναικείων is present.

δὲ καὶ στραγγουρίη· καὶ τὸ στόμα ξηρόν, καὶ
δίψα ἔχει, καὶ γλῶσσα τρηχείη, καὶ οἱ δάκτυλοι ξυν-
έλκονται τῶν ποδῶν οἱ μεγάλοι, καὶ τὰς γαστροκνη-
μίας αἰεὶ τῷ μηρῷ ξυντιταίνεται, καὶ τῆς ὀσφύος πε-
236 ριωδυνίαι, καὶ τῶν χειρῶν ἀκρασίη. ὅταν | δὲ τοιαῦτα
γίνηται,[5] τέτανοι φιλέουσι γίνεσθαι ἀπὸ τῶν κληΐδων
κατὰ τὰς σφαγὰς ἐς τὰς γνάθους τε καὶ τὴν γλῶσσαν, ἐκ δὲ τῶν τοιούτων ὀλίγῳ ὕστερον ὄπισθεν ἀπὸ
τῶν τενόντων κατὰ τὴν ῥάχιν ἐς ὀσφῦν,[6] καὶ ὧδε
ἀπόλλυνται κατὰ βίην.[7]

Προλέγειν οὖν δεῖ ἀρχομένων τῶν ῥόων, διαιτᾶν δὲ
τῷδε τῷ τρόπῳ· πρωῒ μὲν διδόναι πρὸς τοὺς ῥόους
φάρμακον πίνειν, ὧν ἂν ἐγὼ γράψω, οὗ ἂν δοκέῃ δεῖσθαι μάλιστα, διδόναι δὲ καὶ δὶς[8] καὶ τετράκις· κἢν
πολὺ ἀπίῃ τὸ αἷμα, διαιτᾶν δέ, ἢν μὲν ἄπυροι ἔωσι,
σιτίοισιν· ἢν δὲ πυρεταίνωσι, ῥοφήμασιν. ἔστι δὲ τῶν
ῥοφημάτων τὰ ἐπιτηδειότατα· ἔλυμος, φακῆ, ἄλητον
ἐφθὸν σητάνιον, χόνδρος κάθεφθος ῥοφητός, ζειὰ
κάθεφθος· τῶν δὲ πωμάτων, πάλη ἀλφίτου ἐφ' ὕδατι,
καὶ τὰ κνήσματα[9] τὰ ἀπὸ τῶν ἄρτων κεκομμένα λεῖα,
καὶ ἄλφιτα προκώνια[10] λεπτὰ βεβρεγμένα ἐν ὕδατι,
ἐλλείχειν ἄναλτα· τῶν δὲ σιτίων, ἄρτος ἔξοπτος ἐν
σποδιῇ· ὄψον δὲ λαγωοῦ κρέας, πελειάδος, φάσσης,
καὶ ἐφθὰ καὶ ὀπτά, ἐρίφου κρέας ὀπτόν, μηδενὶ πέπερι[11] πεποιημένον, ἐς ὄξος ἐμβάπτων, ἧπαρ ἐν σπο-

[5] τότε καὶ add. MV.   [6] ἐς ὀσφῦν om. Θ.
[7] Add. γε δίκην Θ.   [8] δὶς Θ: τρὶς MV.

dry, there is thirst, the tongue is rough, the large toes are retracted, and the patient continually draws her calves against her thighs. These patients have severe pains in the lower back, and their hands are powerless. When such things happen, then spasms are likely to occur, moving from the collarbones along the jugulars toward the mandibles and the tongue, and a little later similar (sc. spasms) spread backward from the tendons and down along the spine as far as the sacrum; in this way patients succumb from the (sc. disease's) violence.

You must make a prognostication when the fluxes are just beginning, and employ a regimen of the following kind. At dawn give whichever potion for the fluxes you think is most needed, from among the ones I will describe below,[2] offering it two and four times. If a great amount of blood passes off, prescribe a regimen of cereals if the patients are without fever, or of gruels if they are febrile. The following kinds of gruels are most suitable: millet, lentil soup, boiled spring wheat meal, boiled spelt-groat gruel, boiled rice-wheat. Among drinks, give one made with the finest barley meal in water, or with finely scraped bread crumbs or fine meal of untoasted barley soaked in water, to be taken as an electuary without salt. Of cereals, give bread well baked in ashes; as a prepared dish, meat of hare, pigeon, or ringdove both boiled and baked, or baked kid's meat made by immersing it in vinegar without any pepper, or liver of goat or beef baked in ashes, or yolks

---

[2] See ch. 83 below.

---

[9] κνήσματα Θ: κλύσματα MV.   [10] προκώνια Θ: προ-
κρήνια MV.   [11] πέπερι om. Θ.

διῇ ὀπτὸν αἰγὸς ἢ βοός, ᾠῶν ὀπτῶν λέκιθοι,[12] τυρὸς
ὀπτὸς[13] ἄναλτος· λαχάνων δὲ μὴ γεύεσθαι, μήτε
ἑφθῶν μήτ᾽ ὠμῶν, λουτρῶν τε ἀπέχεσθαι, καὶ ἀποδεῖν
τὰς χεῖρας εἰρίῳ ῥερυπωμένῳ στρέψαντα καὶ παχετὸν
ποιήσαντα ὑπὲρ τῶν ἀγκώνων καὶ ἰγνύων ὑπὲρ τῶν
γουνάτων, καὶ σικύας ἀείρειν ἐπάρας μαζοὺς ὑπ᾽
αὐτούς, ἄλλοτε ἐς τὰ δεξιά, ἄλλοτε ἐς τὰ ἀριστερά·
ἢν δὲ δύσπνοια γίνηται πρὸς τὴν προσβολὴν τῆς
σικύης, ἀφαιρέειν τὴν | σικύην· αἷμα δὲ μὴ ἀπάγειν·
προσθέτοισι δὲ χρῆσθαι, οἷσιν ἂν ἐγὼ γράψω τῶν
στασίμων τοῦ αἵματος προσθετὰ ποιέοντα τῇ τοιαύτῃ.
καὶ ἢν μὲν περιγένηται ἐκ τῶν ῥόων, ἀπιόντος πολλοῦ
αἵματος, τό τε χρῶμα ἀφυῶδες, καὶ τὸ πρόσωπον μετάρσιον, καὶ τὰ ὑποφθάλμια[14] οἰδήματα, καὶ τὰ σκέλεα ἐπηρμένα, καὶ ὑστέρη ὑγρή, καὶ[15] ἀνεστόμωται
παρὰ λόγον, καὶ τὰ ἀπιόντα ὑδαρέα, οἷον ἀπὸ κρεῶν
ὠμῶν χυμός.

Τῇ τοιαύτῃ χρή, ὅταν ἰσχύῃ, καὶ ἐμέτους ποιέειν
νήστιας, καὶ μετὰ τοὺς ἐμέτους ἄριστον διδόναι.
ταύτῃ ξυμφέρει ὀλιγοποσίη, οἶνος μέλας ἀκρητότε-
ρος, ἀλουσίη, ψυχρολουσίη, περίπατοι, μονοσιτίη,
πᾶσα ξηρασίη. ἢν δὲ πρὸς ταῦτα μὴ καθιστέωνται
μηδὲ συμπίπτωσιν αἱ ὑστέραι, μηδὲ τῶν ῥευμάτων
ἀπαλλάσσωνται, ἀποσκεπτόμενον χρὴ ἐς τὴν δύναμιν τοῦ σώματος, ἢν ᾖ δυνατή, ἐλλεβορίζειν· ἢν δὲ
μὴ ἐνακούσῃ, τὴν κεφαλὴν καθαίρειν· καθήραντα δὲ

---

[12] ᾠὸν λεπτὸν ἢ λέκιθος MV.   [13] ὀπτὸς om. MV.

of baked eggs, or baked cheese without salt. Do not allow the patient to taste vegetables, either boiled or raw, forbid her to bathe, and bind her arms back with uncleaned wool twisted and made thick over her elbows and her hams above the knees; lift up her breasts and apply bloodletting cups directly under them, at one time on the right side and at another time on the left side. If the patient's breathing becomes difficult from the application of the cup, remove it. Do not draw blood. Apply suppositories made such as I will describe as hemostatic for such a patient,[3] and if she survives the fluxes after passing much blood, her color will be whitish, her face swollen, swellings (sc. will be present) under the eyes, her legs will be swollen up, her uterus will be moist and more dilated than it should be, and her discharges will be watery like the juice that comes out of raw meats.

In such a case you must—when the patient is strong—provoke vomiting in the fasting state, and after the emesis give a breakfast. It benefits such a woman to drink a little dark, somewhat diluted wine, to avoid bathing except for a cold wash, to walk, and to eat only once daily and then every kind of dry food. If in response to these measures the uterus neither loses its swelling nor contracts, and the fluxes fail to subside, you must assess the strength of the patient's body and, if she is strong enough, administer hellebore. If she does not respond to this, clean her head, and

---

[3] See ch. 87 below.

---

[14] τ. ὑ. Θ: ἐν τοῖσιν ὑποφθαλμοῖσιν MV.
[15] Add. αἰεὶ MV.

τὰ λοιπὰ διαίτῃ θεραπεύειν τὸν αὐτὸν τρόπον, ὅνπερ
τὰς ἀτέκνους.

2. (111 L.) Σκέπτεσθαι δὲ χρὴ τὰς φύσιας τῶν
γυναικῶν καὶ τὰς χροιὰς καὶ τὰς ἡλικίας καὶ τὰς
ὥρας καὶ τοὺς τόπους καὶ τὰ πνεύματα. αἱ μὲν γὰρ
ψυχραὶ [αἱ δὲ][16] ὑγραὶ καὶ ῥοώδεις, αἱ δὲ θερμαὶ ξηρότεραί τε καὶ στάσιμοί εἰσιν·[17] αἱ μὲν γὰρ ὑπέρλευ-
κοι ὑγρότεραί τε καὶ ῥοωδέστεραι, αἱ δὲ μέλαιναι
ξηρότεραί τε καὶ στριφνότεραι· αἱ δὲ οἰνωποὶ μεσηγύ
τι ἀμφοῖν ἔχουσι. καὶ ἀμφὶ τῶν ἡλικιῶν ὡσαύτως
συμβαίνει· αἱ μὲν γὰρ νέαι ὑγρότεραί τε καὶ πολύαι-
μοι ὡς ἐπὶ τὸ πολύ· αἱ δὲ πρεσβύτεραι | ξηρότεραί τε
καὶ ὀλίγαιμοι· αἱ δὲ μέσαι μέσον τι ἀμφοῖν ἔχουσιν,
ἰσενύουσαι.[18] δεῖ δὲ τὸν ὀρθῶς ταῦτα διαχειριζόμενον
διαγινώσκειν ἑκάστοτε[19] τὰς φύσιας τῶν γυναικῶν[20]
καὶ τὰς ἡλικίας καὶ τὰς ὥρας καὶ τοὺς τόπους καὶ τὰ
πνεύματα.[21]

3. (112 L.) Ἢν ῥόος ἐγγένηται ἐν τῇσι μήτρῃσιν,
αἷμα οἱ ῥεῖ πολλόν, καὶ θρόμβοι πεπηγότες ἐκπίπτουσι, καὶ ὀδύνη ἔχει ἐς τὰς ἰξύας καὶ τοὺς κενεῶνας
καὶ νειαίρην γαστέρα· καὶ σκληρή ἐστι, καὶ θιγγανο-
μένη ἀλγέει· καὶ ῥῖγος καὶ πυρετὸς ὀξὺς λαμβάνει,

240

16 Del. Littré.   17 αἱ μὲν γὰρ ψυχραὶ . . . στάσιμοί
εἰσιν om. Θ· cf. note 21 below.   18 ἰσ. Linden: ἴσαι νῦν
ἐοῦσαι codd.   19 ἑκάστοτε om. Θ.   20 Add. καὶ τούς
καιροὺς MV.   21 Add. αἱ μὲν γὰρ ψυχραὶ (αἱ δὲ Μ) ὑγραὶ
καὶ ῥοώδεες, αἱ δὲ θερμαὶ ξηρότεραί τε καὶ στάσιμοί εἰσιν
ΘΜ; cf. text and note 17 above.

after cleaning the rest of the body employ the same regimen as you would for women without children.

2. You must also consider women's natures, their complexions and their ages, as well as the seasons, the places, and the winds. For cold women are moist and subject to fluxes, whereas warm ones are drier and more subject to stasis;[4] fair women are moister and more subject to fluxes, while dark ones are drier and more constricted; wine-colored women have something of both. The ages of life have the following significance: young women are generally moister and richer in blood, while older women are drier and have less blood; those between the two have something of both, since they are of an intermediate age. A person who manages these matters correctly must distinguish on each occasion women's natures, their ages of life, the seasons, the places, and the winds.

3. If a flux starts from a woman's uterus, her blood flows copiously, congealed clots are expelled, and she has pain in the loins, flanks, and lower belly, which is constipated and on being touched feels pain.[5] Chills and acute fever set in, and weakness develops; everywhere except the

---

[4] The clause "For cold women are moist and subject to fluxes, whereas warm ones are drier and more subject to stasis" is lacking at this point in the Θ text but included at the end of chapter 2 after "winds"; in M it is present in both locations, while in V it is transmitted only here.

[5] So Fuchs: Littré refers the feminine adjective and participle not to the belly, but to the patient: "la malade a le corps rigide; elle souffre si on la touche."

καὶ ἀσθενείη ἐπιγίνεται, καὶ πάντα²² πλὴν ὤμων καὶ
ὠμοπλάτων ἀλγέει, καὶ θέρμη <ἔχει>,²³ καὶ ἐρευθιᾷ,
καὶ τὰ φλέβια σκληρὰ ἀντιτυπεόμενα. ἡ δὲ νοῦσος
γίνεται μάλιστα ἐκ τρωσμοῦ· γίνεται δὲ καὶ ὁκόταν
τὰ ἐπιμήνια μὴ γινόμενα πολλοῦ χρόνου ἐξαπίνης
ῥαγῇ.

Ταύτῃ, ὁπόταν ὧδε ἔχῃ, ὄλονθον ξηρὴν κόψαι καὶ
διασήσας ἐς ὀθόνιον ἐνδῆσαι καὶ προσθέσθαι· καὶ
ἐπὶ τὴν νειαίρην γαστέρα ψύγματα ἐπιτιθέναι, φυλασσόμενος μὴ φρίξῃ. ἐπειδὰν δὲ τὸ ῥεῦμα στῇ,
ἀγριελαίης φύλλα ἑψήσας ἐν ὄξει ὡς ὀξυτάτῳ, διανι-
ψάσθω τὰ αἰδοῖα· πινέτω δὲ γίγαρτα καὶ ῥόον τὴν
242 ἐρυθρὴν ἑψήσασα ἐν ὕδατι | ὅσον ἡμικοτύλιον· ἢ
μόρα τὰ ἀπὸ τοῦ βάτου τὰ ἐρυθρὰ ξηρήνας, τρίψας
λεῖα μετὰ ἀλήτου σητανίου μίσγων ἴσον ἴσῳ, νήστει
διδόναι πίνειν· ἢν θέλῃς ἰσχυρότερον, τιτάνου συμμίξας δύο μοίρας, ἀλήτου ἐπιβαλών, πῖσαι· καὶ μὴ λου-
έσθω· σιτίοισι δὲ χρήσθω ξηροῖσι, καὶ οἴνῳ μέλανι
οἰνώδει. ἢν δὲ οἰδέῃ πεπαυμένου ἤδη τοῦ ῥόου, φάρμακον πιεῖν κάτω·²⁴ μετὰ δὲ τὸ φάρμακον τὰς ὑστέρας
κλύσαι τῷ ἀπὸ τῶν ὀλόνθων, καὶ μετακλύζειν στρυφ-
νοῖσιν· ἢν ταῦτα παθοῦσα ὑγιαίνῃ, θυμιήσθω ἕως ἂν
ἀποξηραίνῃ.

4. (113 L.) Ῥόος ἐρυθρός· ῥεῖ τοιόνδε οἷον αἷμα
νεοσφαγέος, καὶ θρομβία διαλιπόντα,²⁵ ἄλλοτε δὲ καὶ

²² κ. π. om. Θ.   ²³ Add. I.
²⁴ κάτω ΘΜ: ἄνω V.

shoulders and the shoulder blades is painful, there are
warmth and redness, and the small vessels are hard when
pressed. This condition arises most often after abortions,
but it may also follow when the menses fail to pass for a
long time and then suddenly break out.

For such a woman, when the case is such: crush a dry
wild fig, sieve it, attach it inside a piece of linen, and apply
it as a suppository. Apply cold compresses to the lower
belly, but take care not to cause a chill. When the flux
settles down, boil wild olive leaves in very acidic vinegar
and have the patient wash her genitalia with this. She
should take a half-cotyle potion of grape stones and red
sumac she has boiled in water. Or dry some red berries
from brambles, crush them smooth, and mixing them with
spring wheat meal equally diluted, give this to the patient
in the fasting state to drink. If you want something stronger, mix two portions of gypsum with this, sprinkle on
meal, and have the patient drink it. Otherwise the patient
should avoid bathing, employ dry foods, and drink strong
dark wine. If after the flux has ceased she swells, have her
take a laxative potion, and after that flush her uterus with
the juice of wild figs; after that flush again with astringents.
If on having these things done she recovers, have her fumigate herself until she becomes completely dry.

4. Red flux: this flows like the blood of a newly slaugh-
tered animal, together with intermittent[6] clots, although

[6] "Translucent clots," with the reading of Linden.

---

[25] -λιπόντα ΘV: -λείποντα M: -λάμποντα Linden after the
interpretations of Calvus (*perlucentes*) and Cornarius (*pellu-
centes*); cf. Foes note 15.

ῥόον ἐρυθρὸν ἐκβράσσει, καὶ ἡ γαστὴρ ἡ νειαίρη
ἐπαίρεται, καὶ²⁶ λεπτὴ γίνεται, καὶ νηπελεῖ,²⁷ καὶ
σκληρύνεται, καὶ ἀλγέει ψαυομένη ὡς ἕλκεος, καὶ πῦρ
ἔχει καὶ βρυγμός· ὀδύνη τε εἰς αὐτὰ τὰ αἰδοῖα καὶ τὸ
ἐπίσειον καὶ ἐς τὸν κενεῶνα καὶ τὰς ἰξύας καὶ τένοντα
καὶ κοιλίην καὶ στῆθος, καὶ τὰς ὠμοπλάτας καὶ²⁸
πάντα ἀλγέει, καὶ ἀδυναμίη καὶ λιποψυχίη ἔχει, καὶ
ὁ χρὼς τρέπεται. καταρχὰς τῆς νούσου ταῦτα ἐπι-
λαμβάνει· ἢν δὲ μηκύνηται, ταῦτα πάντα ἐπὶ μᾶλλον
ἀνθέει, καὶ δῆλος ἡ νοῦσος, καὶ τὰ κοῖλα ἐπανίσταται, καὶ οἱ πόδες οἰδέουσιν. ἡ δὲ νοῦσος λάζυται μάλιστα ἐκ τόκου, ἤν τι ἐν αὐτῇ διακναισθὲν μὴ ἴῃ, ἀλλ᾽
ἐνσηπῇ τε καὶ τρυχωθῇ τὸ ἔμβρυον.

244 Ταύτῃ κατ᾽ ἀρχὰς ἢν | ἐπιτυγχάνῃς, τῶν σπόγγων
κατατέγγοντα προσθιθέναι ὁπόταν ἡ ὀδύνη ᾖ, καὶ
ὀθόνιον λεῖον μαλθακὸν καθέψων ὕδατι τέγγοντα ψυχρῷ ἐπὶ τὴν γαστέρα ἐπιβάλλειν, καὶ ὕδατι ψυχρῷ
καταχεῖν, καὶ τὴν κλίνην ἀπὸ τῶν ποδῶν ὑψηλοτέρην
εἶναι, καὶ στορέσαι ὧδε· καὶ τῶν γυναικείων πειρώμενος, ὅ τι ἂν μάλιστα προσίηται πιπίσκειν· τοῦ σελίνου τὸν καρπὸν κόψαι καὶ φῶσαι σήσαντα, καὶ ἐρύσιμον ὡσαύτως καὶ μήκωνος καρπὸν σὺν ἀλφίτοισι
σήσας, καὶ κνίδης καρπὸν ὡσαύτως· καὶ τῆς ψώρας
τῆς ἀπὸ ἐλαίης,²⁹ καὶ κηκίδα, καὶ πήγανον, καὶ ὀρίγανον, καὶ γλήχωνα ἐν ἀλφίτοισι σῆσαι καὶ φορύξαι,

sometimes (sc. the woman) discharges a red flux. The patient's lower belly is raised, it becomes narrow, powerless, and hard, and she feels pain on being touched as she would from an ulcer. Fever and chattering of the teeth set in, pain affects the genitalia themselves, the pubes, the flank, the groins, the tendon, the cavity, and the chest, and the patient aches in her shoulder blades and all through her body. There is a loss of strength and consciousness, and the color of the skin changes. These things take place at the beginning of the disease, and as it lengthens they all increase, the disease is revealed, and besides the hollow parts swell and the legs suffer edema. The disease usually attacks after childbirth, if something inside a woman is scraped off but fails to pass out, and the fetus decomposes inside and withers up.

If you chance upon such a woman at the beginning (sc. of the condition), moisten some sponges and apply them as a suppository when pain is present; also boil a piece of soft, fine linen, soak it in cold water, and bind it against her belly; pour on cold water, and after setting her bed with the foot end higher make it up like this. Test the female medications, and have the woman take the most fitting. Pound celery seed, sift and toast it, and do the same with hedge mustard; sift poppy seed together with barley meal and stinging nettle seeds in the same way. Take scab of olive, oak gall, rue, marjoram, and pennyroyal: sift and mix in barley meal, and toast ripe barley meal; grind whole

---

26 καὶ om. Θ.   27 Littré from Galen's gloss (vol. 19, 124), following Cornarius' translation *et impotens fit*: νηνεμεῖ codd.

28 τὰς ὠμ. καὶ MV: ἐκ τῶν ὤμων καὶ πλατας Θ.

29 τῆς ἀπὸ ἐλαίης Θ: ἀγριελαίης MV.

καὶ κρίμνα ἀπ᾽ ἀλφίτων ἁδρὰ φῶξαι, καὶ πύανα[30]
καταλέσας, καὶ τυρὸν αἴγειον περιξύσας τὸ αἶσχος,
τῶν μὲν ἄλλων ἴσον ἑκάστου μίσγε, ὀριγάνου δὲ καὶ
πηγάνου καὶ ψώρας καὶ κηκίδος ἥμισυ, ταῦτα πίνειν
πρωὶ νῆστιν πρὸ τῆς κινήσιος· κιρνάναι δὲ χρὴ καὶ
ἐς χρῆσιν· ἢν δριμέα ἴοι, καὶ κυκεῶνα διδόναι, ἐν μὲν
τοῦ φαρμάκου μέτρον ἔστω, ἐν δὲ τοῦ τυροῦ, ἐν δὲ
τῶν ἀλφίτων· ἐς ἑσπέρην δὲ τοῦ μέλιτος συμμίσ-
γοντα πιπίσκειν. καὶ ἄχρι μὲν ἐν ἀρχῇ ἔχηται τῇ
νούσῳ, καὶ τὸ αἷμα συχνὸν ἴῃ, καὶ διαλίπῃ ὀλίγον
χρόνον, καὶ ὀδύναι ὀξεῖαι ἴσχωσι, ταῦτα χρὴ ποιέειν.
ἢν δὲ τὸ αἷμα ἔλασσον ῥυῇ καὶ δι᾽ ἐλάσσονος χρόνου, πιπίσκειν ἃ δὴ κάτω ὑποχωρέει ἢ ἄνω, καὶ πυριῆν τὰ αἰδοῖα βληχροῖσι πυριήμασιν, ὡς ἂν δοκέῃ
ἑκάστοτε καιρὸς εἶναι, καὶ ζειὰς ἐρίξαντα σὺν τοῖσι
246 κελύφεσι, καὶ ὀλόνθους ἠρινοὺς αὐήναντα | κόψαι, καὶ
σῆσαι, καὶ ἐλαίης φύλλα ὁμοίως, ἴσον ἑκάστου, καὶ
καταπλάσσειν, καὶ γάλα πιπίσκειν καθεφθὸν βόειον,
ἢ ὠμόν, πρὸς τὸ ὀρθῶς ἔχον ὁρῶν, καὶ ὡς ἂν καιρὸς
δοκέῃ εἶναι. ἡ δὲ νοῦσος δοκέει βληχρὴ εἶναι καὶ θα-
νατώδης· παῦραι δὲ διαφεύγουσιν.

5. (114 L.) Ὅταν γυναικὶ αἷμα ῥέῃ ἐκ τῶν ἄρθρων
ὑπὸ τόκου διεφθορυίῃ ἢ ὑπὸ νούσου,[31] οἴονται δ᾽ ἔνιοι
τοῦτο ῥόον εἶναι, τὸ δ᾽ ἐστὶν ἑτεροῖον· τὸ μὲν γὰρ ἐκ
τῶν ἄρθρων κἀκ τῆς ὀσφύος καὶ ἰσχίου κολλῶδες
ὁμοῦ τῷ αἵματι· κεῖνο δὲ ἀπὸ ὑστερέων καὶ κοίλων
φλεβῶν, καθαρὸν αἷμα.

wheat and goat's cheese from which you have scraped off
the crust: mix together the same amount of each of the
others, but half as much of the marjoram, rue, scab, and
oak gall, and have the patient take this potion at dawn in
the fasting state before she goes about—you must mix
these before their usage. If a sharp flux passes, give a cyceon as well; let this have one measure of a medication,
one of cheese, and one of barley meal; toward evening give
a drink mixed with honey. As long as the disease is at its
beginning and blood is passing off in a great amount and
with little intermission, and acute pains are present, this
is what you should do. But if the blood begins to flow less
in amount and for a shorter time, give a potion that is sure
to be passed downward or upward, and foment the genitalia with mild fomentations that seem appropriate at each
stage. Pound coarse rice-wheat with its husks, dry and
crush green wild figs, sieve, do the same to olive leaves—
an equal amount of each—and apply as a poultice. Give
well-boiled cow's milk to drink, or raw milk, keeping an
eye to what is correct and what seems to be appropriate.
The disease manifests itself as slow and deadly: few escape
from it.

5. When in a woman who is injured as the result of
giving birth or having a disease, blood flows out of her
joints, some (sc. physicians) think that this is a flux, but in
fact it is something quite different: this condition arises
from the joints, and is the movement of a gluey material
together with blood from the lower back and hip, whereas
the other condition would be the movement of pure blood
from the uterus and the hollow vessel (*vena cava*).

---

[30] πύανα V: πύρινα ΘΜ. [31] ὑ. ν. Θ: ἀπὸ πόνου MV.

Ταύτην χρὴ ὑποθυμιῆν, ζειὰς κατερίξαντα ὅσον ἡμίεκτον, ὄξει δὲ φυρῆσαι,[32] ὅπως μὴ κατὰ πᾶν ὑγρήνῃς τὰς μήτρας· καὶ τοῦ θείου ὅσον ἡμιωβόλιον μίξας πρὸς τὰς ζειὰς τετριμμένας καὶ τῷ ὄξει φυρήσας, τὴν νύκτα τίθει· πρωὶ δὲ πῦρ πάμπολλον[33] καύσας, καὶ ἐπὶ πῦρ[34] ἐπιτιθέναι· καὶ φλόμου[35] βύσματα ἀπὸ ἐλαιηρῶν κεραμίων, καὶ ἀπὸ τοῦ κνάφου τῶν κναφέων σύμμισγε καθάρσεων,[36] καὶ τοῦ καρποῦ τοῦ ὄφιος· ἀφαιρέων δὲ τὸ πολλὸν τοῦ πυρός, καίειν, καπνίσαις γὰρ ἂν μάλιστα. δίφρον δὲ χρὴ ὀπήεντα εἶναι καὶ ἀμφιέζεσθαι[37] τὴν γυναῖκα περιστειλαμένην εἵμασιν, ὡς μὴ παραπνέῃ· ἐπὶ δὲ τὸ πῦρ ἐπιπάσσει σὺν τῷ ὄξει, καὶ τοῦ καρποῦ τοῦ ὄφιος. καὶ σμύρνα δὲ μισγομένη ἐνεργόν, καὶ παύει τὰ | αἰδοῖα τὸ θυμιητὸν αἱμάσσεσθαι. ἢν δὲ ἅλις ἔχῃ, ἐρυσίμου καρπὸν πεφωγμένον τρῖψαι καὶ ἐν οἴνῳ διδόναι.

6. (115 L.) Ῥόος πυρρὸς ῥεῖ οἷον ἐξ ᾠοῦ εἰδεχθέος πολύ τε καὶ κάκοδμον, καὶ φλεγμαίνουσιν αἱ ὑστέραι, καὶ ὀδύναι ἐκ τῆς ὀσφύος καὶ τῶν βουβώνων, καὶ τὰ ἐπερχόμενα πολλά, καὶ ἢν μὴ ἀπαλλάσσηται, ταχὺ ῥέοντα· ἀλλ᾽ ἢν καὶ[38] χρόνος ἐγγίνηται, τὰ ἀπιόντα σήπει ὡς οἷόν τε μάλιστα· ῥεῖ γὰρ οἷον ἀπὸ κρεῶν ὀπτῶν χυμός· ἅμα δὲ τούτοισι πυρετοὶ ἰσχυροὶ καὶ ῥίγεα· ἐκ δὲ τῶν τοιῶνδε ῥόων αἱ μὲν πολλαὶ ἀπόλλυνται, ὀλίγαι δὲ διαφεύγουσιν.

[32] Add. ὀλίγῳ MV.   [33] πάμπ. Θ: πολλῷ M: πολὸν V.
[34] καὶ ἐπὶ πῦρ om. MV.   [35] κ. φλόμου om. MV.

You must foment such a patient by grinding rice-wheat to the amount of a half sextary and adding an amount of vinegar that does not completely inundate the uterus. Mix a half-obol of sulfur with ground rice-wheat, add the vinegar, and then set this aside for the night. In the morning kindle a full fire and set the fomentation over it: take mullein plugs from olive-oil vessels, and mix them with cleanings from the carding combs of fullers and flower buds of caper. Burn this after removing most of the fire, since this will produce the most smoke. Use a stool with an opening (sc. in its seat), and clothe the woman[7] by wrapping her in garments such that no air will be admitted. Sprinkle the vinegar and flower buds of caper over the fire. Myrrh mixed with this is also effective, and burned as a fumigant prevents the genitalia from becoming bloody. If the fumigation is enough, grind toasted hedge-mustard seed, and give it in wine to drink.

6. A flame-colored discharge resembles what comes out of a rotten egg—copious and ill-smelling—the uterus becomes swollen, pains radiate from the lower back and the groins, much fluid is produced, and unless there is a change for the better the flux flows rapidly. With the passage of time the discharge becomes very putrid, and flows like the juice that comes out of roasted meat; these fluxes are accompanied by violent fevers and rigors. Many women perish from such fluxes, but a few do escape.

---

[7] With the reading of M: "set the woman upon it."

---

36 καθάρσεων M: θαρσέων Θ: θαρσὸν V.
37 ἀμφιίζεσθαι M.
38 ἀλλ᾽ ἢν καὶ Littré: ἀλλὰ Θ: ἀλλὰ καὶ MV.

Ἢν οὖν ἐν ἀρχῇ παραλάβῃς, θεραπεύειν ὧδε χρή·
ἢν μὲν ἀπύρετοι ἔωσι καὶ ἰσχύωσιν, ἐλλεβορίζειν·
ὅταν δὲ ἐλλεβορίζῃς,[39] διαλιπόντα ἡμέρας τρεῖς ἢ
τέσσερας κάτω πεῖσαι φάρμακον. εἶτα μετὰ δὲ τὴν
κάθαρσιν διαιτῆν ὧδε, ὅπως τὰ ῥεύματα ὑδαρέα ἔσται
καὶ λιπαρά· διδόναι δὲ πρωὶ νήστει τῶν φαρμάκων τι
πίνειν ἐπ᾽ οἶνον ἐπιπάσσων ἃ ἐγὼ γράψω πρὸς ῥόον.
μετὰ δὲ τὸ φάρμακον τῇ ἄλλῃ διαίτῃ θεραπεύειν δὲ
καὶ τὰς ὑστέρας ὧδε· ἢν μὲν φλεγμήνωσι καὶ ξυμ-
μεμύκωσι, πυριῆν χρὴ μαλθακῇσι τῇσι πυρίῃσι, μέχρις[40] τὸ στόμα εὔλυτον καὶ ὁ στόμαχος μαλθακὸς
γένηται· μετὰ δὲ τὰς πυρίας, κλύζειν τῶν κλυσμάτων
ὁκοίοισιν ἂν δοκέῃ δεῖσθαι, ἤν τε καθαρτικωτέροισιν
ἤν τε μαλθακωτέροισι· μετὰ δὲ τοὺς κλυσμοὺς μαλθακτήρια προστιθέναι· ἢν δὲ τὸ στόμα μὴ εὔλυτον
γίνηται, πυριῆν καὶ μαλθάσσειν προσθέτοισιν ὧν ἂν
ἐγὼ γράψω, μέχρι ἂν ἀναστομωθῇ. ἢν δὲ πρὸς ταῦτα
250 μὴ παύηται τὰ ῥεύματα, καθαίρειν | τὴν κεφαλήν, καὶ
οὕτω διαιτῆν· ἢν μὲν δυσουρέῃ,[41] ὄνου γάλα πίνειν
λαχάνοισιν ἐφθοῖσι καὶ ἡμέροισι καὶ ἀγρίοισι, πλὴν
σκορόδων καὶ πράσων καὶ κράμβης καὶ ῥεφάνου τῆς
μακρῆς· θαλασσίων βάτῳ τῷ λείῳ, σκορπίῳ, γόγ-
γρῳ, νάρκῃ, ἐγχέλυϊ, ψήσσῃ, κωβιῷ, ἕψειν δὲ χρὴ ἐν
κρομμύοισι καὶ κοριάννοις, ἐν ἅλμῃ γλυκείῃ καὶ λι-
παρῇ δίεφθα· κρεῶν δὲ μάλιστα μὲν συός, δεύτερον
δὲ ἀρνός, ἢ ὄϊος, ἐφθοῖσι μᾶλλον ἢ ὀπτοῖσι, καὶ ζω-

If you take them on at the onset, treat as follows: If they are without fever and in a strong state, give hellebore. After this, leave off for three or four days and then give a purgative to drink. After the cleaning, employ a regimen that will make all the fluxes watery and fatty. At dawn give some of the medications I will describe for fluxes[8] to the patient in the fasting state to drink, after sprinkling them over wine. Then after the medication treat with the rest of this regimen, and for the uterus do the following: if it is swollen and closed together, foment with gentle fomentations until its mouth relaxes and the orifice becomes soft. After giving the fomentations, flush with the agents you think are required—either ones cleaning more intensely or more emollient ones. After the injections, apply emollient suppositories. If the (sc. uterine) mouth does not become fluent, foment and soften it until it dilates, using the suppositories I will describe.[9] If the fluxes do not cease with these measures, clean the patient's head and employ the following regimen (if she suffers dysuria,[10] have her drink ass's milk): boiled vegetables both garden and wild except for garlic, leeks, cabbage, and the long radish; of sea foods the smooth skate, sculpin, conger eel, torpedo, common eel, sole, goby—boil these with onions and coriander in sweet oily brine, and thoroughly; of meats especially pork, but in the second place lamb or mutton, these boiled rather than roasted, and with their sauces; dilute

---

[8] See chs. 83 and 90 below.   [9] See ch. 96 below.
[10] With the text of MV: "if she is without fever."

---

[39] ἐλ. Θ: γένηται MV.   [40] Add. οὖ MV.
[41] δυσουρέῃ Θ: ἄπυρος ᾖ MV.

μοῖσιν· οἴνῳ λευκῷ μελιχρῷ ὑδαρεῖ· λουτροῖσιν πλὴν
τῆς κεφαλῆς, μὴ λίην θερμοῖσι μηδὲ πολλοῖσιν. ἢν
δὲ πρὸς ταύτην τὴν δίαιταν τῆς μὲν ἑλκώσιος καὶ τῆς
φλεγμασίης ἀπαλλάσσωνται, ὑγραὶ δὲ ὦσιν αἱ ὑστέραι, ἀντὶ μὲν λουτρῶν ἀλουσίῃσιν, ἀντὶ δὲ κιρρῶν
οἴνων μέλασιν, ἀντὶ δὲ ὑδαρεστέρων ἀκρητεστέροι-
σιν, ἀντὶ δὲ ἀλφίτων ἄρτοισιν, ἀντὶ δὲ ἰχθύων κρέασιν ὀπτοῖσι καὶ σιτίοισι πᾶσι τοῖσι ξηραντικοῖσιν,
οἷσιπερ ἐπὶ τῇσι διαρροίῃσι χρώμεθα· κλυσμῶν
ἀπηλλάχθαι πάντων, πλὴν οἴνου καὶ ὕδατος· θυμιῆ-
σθαι δὲ τοῖσι στυπτικοῖσιν. ἄριστον δ᾽ ἐν γαστρὶ
ἔχειν. ἢν δέ τις νεᾶνις ᾖ, ἐμέτους νήστιας, πυκνὰ δ᾽
ἀπεμέειν, καὶ μετὰ τοὺς ἐμέτους[42] ἀριστίζεσθαι σμικρόν. αὕτη τῶν ῥόων τῶνδε δίαιτα.

7. (116 L.) Ῥόος λευκός· ῥέει[43] ὡς ὄνου οὖρον, καὶ
ἐν τῷ προσώπῳ οἰδήματα, καὶ τὰ ὑποφθάλμια οἰδέει
ἄμφω, ὑδρωποειδέα τε καὶ οὐ πάνυ εὐειδέα ὁρώμενα,
καὶ τὸ λαμπρὸν ἄπεστι, καὶ οἱ ὀφθαλμοὶ γλαμυροὶ
ἀμβλυώσσοντες, καὶ τὸ χρῶμα ἀφυῶδες καὶ φλυκται-
νοειδές, καὶ ἡ γαστὴρ ἐπανοιδέουσα ἡ νειαίρη, καὶ ἐν
τῇσι γνάθοισι κατὰ σμικρὸν ἐρυθροειδέα τε καὶ σμικρὰ καὶ ὑδαρέα καὶ πονηρά, καὶ ἐν τοῖσι σκέλεσιν
252 οἰδήματα, καὶ ἢν πιέζῃς τῷ | δακτύλῳ, ἐμπλάσσεται
ὡς ἐν σταιτί, καὶ στόμα πτυάλου ἐμπίπλαται· καρδιωγμός τε ὅταν νῆστις ᾖ, καὶ ἐμέουσιν οἷον ὕδωρ ὀξύ·
καὶ ἢν πρὸς ἄναντες πορευθῇ, θᾶσσον ἆσθμα ἔχει
καὶ πνίξ, καὶ σκελέων ἀπόψυξις, καὶ γουνάτων ἀκρα-
σίη, καὶ ἐν τῷ στόματι ἄφθαι, καὶ ὑστέρη πολλὸν[44]

with white honeyed wine; apply baths except to the head, but not too hot or too frequent. If as a result of this regimen the uterus is cured of its ulceration and swelling, but it then becomes moist, cancel the baths and replace yellow wines with dark ones, diluted ones with pure ones, groats with breads, fish with roasted meats, and use all the kinds of drying foods that we employ for diarrhea. Do not apply any enemas except of wine and water, and fumigate with astringent agents. Best would be to become pregnant. If it is a young woman, have her vomit while fasting, repeating this frequently, and after the emesis taking a small meal. This is the regimen for such fluxes.

7. White flux: it flows out like ass's urine, the patient's face swells, as do the areas under both her eyes, which take on a dropsical, not very pleasant appearance, and the brightness of the eyes disappears and they become bleary and dim of vision. The patient's skin is whitish and covered with blisters, her lower belly is swollen, and on her jaws lesions gradually develop that are reddish, small, watery, and troublesome. Swellings arise in her legs, and if you press one with your finger it forms an imprint as if in dough. The patient's mouth is full of saliva, she has heartburn when she goes without food, and she vomits up a kind of acidic liquid. If she walks against a grade, she soon becomes short of breath and feels suffocation, her legs become cold and she has a weakness in her knees, and aphthae form in her mouth. The patient's uterus has a great[11]

---

[11] With the reading of MV: "abnormally great."

---

[42] μ. τ. ἐμ. Θ: μετέπειτα MV.   [43] Add. λευκὸν MV.
[44] π. Θ: παρὰ λόγον MV.

ἀνεστόμωται, καὶ ἐμπέπτωκεν ἐν τῷ στόματι βαρέη
ὥσπερ μόλιβδος· καὶ διὰ τῶν μηρῶν διατείνουσιν
ὀδύναι, καὶ ἀποψύχεται πάντα τὰ κάτω ἀπὸ νειαίρης
γαστρὸς ἀρχόμενα μέχρι ποδῶν, καὶ τὰ θέναρα τῶν
ποδῶν ναρκῶσι, καὶ ἐπιβαίνειν οὐ δύνανται.

Τὴν τοιαύτην χαλεπὸν ἀπαλλάσσειν τῶν νοσημά-
των· αἵ τε γὰρ ἡλικίαι προβεβήκασιν, οἵ τε κάματοι
συγκαταγηράσκουσιν, ἢν μή τι εὐτύχημα τῶν αὐτο-
μάτων λύσῃ γενόμενον. ταύτῃσι χρὴ ἀπαρύσαι, ὅταν
πλεονάζῃ, φαρμάκοισιν ἅσσα χολὴν ξανθὴν μὴ
καθαίρει, ἀλλὰ τὰ οὐρητικὰ ταύτῃσι ξυμφέρει πινό-
μενα, καὶ κεφαλῆς καθάρσιες, καὶ ἀλουσίη, καὶ τὸ
λευκὸν ἐπίθυμον πίνειν,[45] καὶ περίπατοι, καὶ πάσῃ
ξηρασίη ἐν τῇ διαίτῃ. ταῦτα δρῶσα ὑγιὴς μὲν παν-
τελῶς οὐ γίνεται, εὐπετέστερον δὲ διάγει.

8. (117 L.) Ῥόου λευκοῦ θεραπείη· καθαίρεται λευκὸν ὑπόχλωρον, καὶ ὅταν οὐρέῃ, δάκνει καὶ ἀμύσσει,
καὶ ἑλκοῖ τὴν ὑστέρην, καὶ πυρετὸς ἔχει ὀξύς, καὶ
θέρμη πολλή, δίψα, ἀγρυπνίη, καὶ ἔκφρονες γίνονται,
καὶ ὅταν σπουδάσῃ, ἀσθμά μιν ἔχει, καὶ τὰ γυῖα λύ-
ονται. ταύτην μήκωνα πιπίσκειν λευκήν, καὶ κνίδης
καρπόν, ἄμεινον καὶ ῥοιῆς γλυκείης ῥίζαν καὶ φύλλα
καὶ ῥόον καὶ κηκίδα· ταῦτα ἐν οἴνῳ στρυφνῷ πιπίσκειν, καὶ ῥοιῆς χυλόν, καὶ ξυμμίσγειν τυρὸν αἴγειον·
254 | ὑποκάπνιζε δὲ ζειὰς καὶ ὀλόνθους χειμερινοὺς καὶ
ἐλαίης πέταλα καὶ ψώραν, καὶ σικύης λέμματα τρίτον
μέρος, τὰ δ᾽ ἄλλα ἴσον· καὶ τῶν ἑψανῶν ῥυφεῖν, καὶ

dilation and feels a weight in its mouth like a piece of lead. Pains move through her thighs, and all the lower parts become chilled beginning from the lower belly and as far down as her feet; the soles of her feet are numb, she is unable to walk.

Such a patient is difficult to relieve from these diseases since her age is advanced, and the illness is likely to grow old together with her, unless some fortunate spontaneous event frees her of it. In such cases you must draw off fluid when the patient is overfull, by employing medications that do not remove yellow bile. Diuretic potions are helpful here, as are cleanings of the patient's head, avoidance of the bath, drinking white dodder of thyme, walks, and anything in her regimen that tends to dry. A woman who does these things may not become completely healthy, but she will get through the disease more tolerably.

8. Treatment of a white flux: a white, somewhat greenish fluid is expelled and when the patient passes urine, she feels scratching and irritation; her uterus is ulcerated, and she suffers acute fever, much heat, thirst, insomnia, and loss of her senses; when she hurries, she becomes short of breath and loses control over her limbs. Give such a patient white poppy juice to drink, stinging nettle seed, or even better roots and leaves of the sweet pomegranate, sumac, and oak gall: administer these in astringent wine; also give pomegranate juice with added goat's cheese. Fumigate below with rice-wheat, winter wild figs, leaves and scab of the olive, and a third portion of skin peeled from a bottle gourd—the rest in an equal amount. Give a

---

45 πίνειν MV: ποιέει Θ.

τὸ δι' ἀμυγδάλων καὶ σησάμου ῥυφήματα, καὶ αἰω-
ρέεσθαί τε καὶ ὀχέεσθαι καὶ μὴ ἠρεμεῖν.

9. (118 L.) Ῥόος ἄλλος· καθαίρεται οἷον προβάτου
οὖρον πολύ, χροιὴ λευκή, καὶ οἰδέει πᾶσα, καὶ ἐν
τῇσι κνήμῃσι πόμφοι ἀνίστανται, καὶ ἢν ἐπαφήσῃ
τῷ δακτύλῳ, τῇσι κνήμῃσι καὶ τοῖσι ποσὶν ἐμπλάσ-
σεται βοθροειδέα· καὶ ἤν τι φάγῃ, ἐμπίμπλαται, καὶ
φλεγμαίνει, καὶ ἐπειδὰν ὁδοιπορήσῃ καὶ ἔργον τι
δράσῃ, ἆσθμά μιν λαμβάνει καὶ πόνος, καὶ ἡ χροιὴ
λευκή, ἐνίοτε⁴⁶ ὑπόχλωρος.

Ταύτῃ, ἢν ἰσχύῃ καὶ ᾖ νέη, καὶ τἆλλα φαίνηται,
καταρχὰς δίδου ἄνω φάρμακα καὶ κάτω· καὶ τὴν κεφαλὴν καθαίρειν, ἢν μὲν φλεγματώδης ᾖ, ὑφ' ὧν
φλέγμα καθαίρεται, ἢν δὲ χολώδης, ὑφ' ὧν χολή· καὶ
ἢν ὥρη ᾖ τοῦ ἔτεος καὶ ἢν μὴ φύσει ᾖ σπληνώδης,
καὶ ἤν τι τῶν εἰρημένων ᾖ, ὀρὸν διδόναι, ἀφέψειν δὲ
καὶ πίνειν ὡς πλεῖστον χρόνον· καὶ ἐς ἑσπέρην γλυκὺν οἶνον ὑδαρέα, ἢν δέῃ, σιτίου δὲ μὴ ἅπτεσθαι, ἢν
μὴ ἀσθενήσῃ,⁴⁷ ῥυφήματι ὡς ἐλαχίστῳ, καὶ ἀλουτε-
έτω· ἐπειδὰν δέ σοι καιρὸς δοκέῃ εἶναι, τῆς πόσιος
παυέσθω, καὶ μετὰ τὴν κάθαρσιν σιτίοισι χρήσθω,
ἀπεχομένη λιπαρῶν⁴⁸ καὶ γλυκέων, καὶ ἁλμυρῶν, καὶ
λαχάνων δριμέων, χρῆσθαι δ' ἰχθύσι πετραίοισι καὶ
κρέασι μηλείοισιν, ἢ ὀρνιθίοισιν, ἢ λαγῴοισιν, ἄρτῳ
σποδίτῃ,⁴⁹ καὶ λαχάνοισιν ἐφθοῖσι χρήσθω, καὶ ἀγρί-

⁴⁶ ἐνίοτε MV: λίην Θ.
⁴⁷ μ. ἀ. Θ: δὲ ἀσθενὴς ᾖ M: δὲ ἀσθενήσῃ V.

soup of boiled vegetables, and gruel made with almonds and sesame. The patient should be suspended and carried, and never let rest.

9. Another flux: it is cleaned out in a large amount like cattle urine; the patient's skin is white and she has swelling throughout her body; on her calves blisters rise up, and if you press their surface with a finger, there is pitting on the calves and legs. If the patient eats anything, she becomes full and swollen, and when she takes a walk or does any work, she becomes short of breath and labors. Her skin is white or sometimes somewhat greenish.

If such a patient is young and strong, and everything else seems (sc. favorable), at the outset give medications to act both upward and downward. Also clean her head, if it is suffering from phlegm with agents that clean phlegm, if from bile with ones that clean bile. If it is the proper season of the year and the patient is not naturally disposed to disease of the spleen, give whey if any of the signs mentioned is present: boil it off and give it to drink over a long interval. Toward evening give a diluted sweet wine, if this is necessary, but do not let the patient touch food unless she is weak, and in that case employ a very little gruel and forbid bathing. When you think the right time has come, have her stop drinking and after a cleaning employ foods, but avoiding those that are oily, sweet, or salty, and acrid vegetables. Give fish that live in the rocks, meats of sheep, fowl, or rabbits, and bread baked in ashes. Have the patient take boiled vegetables and raw foods both wild and

---

48 Add. καὶ δριμέων M.
49 σποδίτῃ Θ: σιτώδει ἢ ἐρικτοῖσι MV.

οισι τρωκτοῖσι καὶ ἡμέροισιν ἄτερ δριμέων· καὶ περι-
256 πατείτω πρωῒ καὶ | ἀπὸ τοῦ σίτου. ἐπὴν δέ σοι δοκέῃ
ταῦτα ποιέοντι ξηροτέρη εἶναι, κλύζειν τὰς ὑστέρας
τῇ τρυγί· τρεῖς δὲ ἢ τέσσερας ἡμέρας διαλιπὼν, μετέπειτα στρυφνοῖσι κλύσαι, καὶ ἐπισχόντα, ἢν μὲν ᾖ
ξηρή, ἡσυχίην ἄγειν, καὶ ἢν φύσει ᾖ χολώδης ἢ
φλεγματώδης, ὑπὸ δὲ τῆς διαίτης καὶ τῆς φαρμακο-
ποσίης λεπτυνθεῖσα, ἢν μὴ δύνηται ἀναλαβεῖν, γάλα
πινέτω τεσσεράκοντα ἡμέρας βόειον, θερμὸν ἀπὸ
βοός. τὴν δὲ φλεγματώδεα ἄμεινον σιτίοισιν ὡς ἐλαχίστοισι χρῆσθαι, ἕως ἂν γαλακτοποτῇ, ἔστω δὲ μέτρον ὅσον ἐξ κοτύλαι Ἀττικαί, ἄρχεσθαι δ' ἀπὸ δύο,
καὶ προστιθέτω κοτύλην ἑκάστης ἡμέρης, ἄχρι ἂν ἐξ
γένωνται, καὶ ἐκ τοῦ κατ' ὀλίγον ἐπὶ τὸ ἔλασσον, καὶ
μετὰ τὴν γαλακτοποσίην ἀνακόμιζε σιτίοισι καὶ διαίτῃ. καὶ μετὰ τοῦ γάλακτος τὴν πόσιν πινέτω πρωῒ
νῆστις ἀδίαντον, ξηρήνασα καὶ διασήσασα διὰ κρη-
σέρης, τοῦτο διδόναι ἐπ' οἴνῳ μέλανι οἰνώδει[50] κεκρημένῳ.

Ἢν δὲ ὑποστρέφῃ ἡ νοῦσος, πυριῆσαι αὐτὴν ὅλην,
καὶ αὖτις[51] φαρμάκοισι καθῆραι κάτω, καὶ μετὰ τοῦτο
κλύσαι τὰς ὑστέρας, ἢν μὲν φύσει ᾖ φλεγμαώδης,
τῷ κόκκῳ τῷ Κνιδίῳ, ἢ τῇ ῥίζῃ τῆς θαψίης, ἢν δὲ
χολώδης ᾖ, τῆς σκαμμωνίης τῷ ὀπῷ, ἢ τῇ κολοκυν-
θίδι τῇ ἀγρίῃ, κόψας, δύο ἐπιχέας κοτύλας ὕδατος,
ἀφέψειν τὸ ἥμισυ, τούτῳ ξυμμίσγειν μέλι καὶ ἔλαιον
ναρκίσσινον ἢ ἄνθινον· ἔστω δὲ τοῦ μὲν μέλιτος τεταρτημόριον κοτύλης, τοῦ δὲ ἐλαίου μέτρον ἡμιόλιον

cultivated, except pungent ones. Also have her take walks at dawn and after her meals. When, after you have done these things, the woman seems to be quite dry, flush her uterus with wine lees; then rest for three or four days, and after that inject astringent agents, desist, and if she seems dry leave her alone. If the woman is bilious or phlegmatic by nature, and she is losing weight due to her regimen and the potions she is taking, if she is unable to gain flesh, she should drink cow's milk for forty days, taking it warm right from the cow. For a phlegmatic woman it is better to consume as few foods as possible while she is drinking the milk: let the amount be six Attic cotylai, beginning from two and then increasing one each day until the six is reached, and then gradually decreasing to the lower amount. After the program of drinking milk, strengthen the patient by means of foods and regimen; also after drinking the milk have her drink in the fasting state at dawn a maidenhair potion: she should dry the maidenhair and pass it through a straining cloth; then give this sprinkled over diluted vinous dark wine.

If this disease relapses, foment the patient's whole body and clean her again with purgative medications, and then if she is naturally phlegmatic flush her uterus with Cnidian berries or thapsia root, or if she is bilious with scammony juice or wild gourd: pound these, pour on two cotylai of water, boil off half, and mix into this honey and narcissus or lily oil: there should be one quarter cotyle of honey, and half as much oil as honey. Flush again with the

---

50 οἰνώδει Θ: εὐώδει MV.
51 αὖτις om. Θ.

τοῦ μέλιτος· μετακλύζειν δὲ τῷ οἴνῳ καὶ τῷ μέλιτι καὶ
τῷ ἐλαίῳ μούνῳ, καὶ θυμιῆσαι, καὶ ἐπισχεῖν ἡμέρας
τρεῖς ἢ τέσσερας· ἀδίαντον δὲ πινέτω. καὶ ἢν μὴ[52] ἐν
258 γαστρὶ λάβῃ, | ὑποστρέφει ἡ νοῦσος καὶ ἀπόλλυνται·
ὅσας γεραιτέρας λαμβάνει ἡ νοῦσος αὕτη, καὶ κατα-
σήπονται αἱ ὑστέραι, ἐκφεύγουσι δ' ὀλίγαι.

10. (119 L.) Ῥόος ἄλλος· καθαίρεται οἷόν περ ἐξ
ᾠοῦ ὠμοῦ, χλωρὸν ὑπόλευκον, καὶ ἑλκοῖ τὸ αἰδοῖον,
καὶ οἰδίσκεται τούς τε πόδας καὶ τὰς κνήμας, καὶ τὰ
κοῖλα τῶν ὀφθαλμῶν ἐπανοιδέει, καὶ οἱ ὀφθαλμοὶ
ὑγροὶ γλαμυροί, καὶ ἢν βαδίζῃ, ἆσθμά μιν λαμβάνει,
καὶ ἀσθενείη γίνεται. ἡ δὲ νοῦσος φύσει φλεγματώδης, καὶ[53] ἢν μὴ καθαρθῇ, πυρετήνη δὲ χολῆς κινη-
θείσης, φλαῦρον· γεραιτέρας δὲ ἡ νοῦσος αὕτη λαμβάνει ⟨μᾶλλον⟩[54] ἢ νεωτέρας.

Ὁκόταν ὧδε ἔχῃ, ἢν μὲν ἀνοιδέῃ σφόδρα, διδόναι
κάτω φάρμακον πιεῖν ὅ τι φλέγμα καὶ χολὴν ἰνήσε-
ται·[55] ἢν δὲ μὴ ἰσχυρῶς οἰδέῃ καὶ τὸ φλέγμα αὐτὴν
πιέζῃ, ἄνω δοῦναι φάρμακον· καὶ ἢν μὲν ᾖ δυνατή,
ἐλλεβόρῳ· ἢν δὲ μή, ὅ τι χολὴν καὶ φλέγμα ἄγει·
μετὰ δὲ τὰ φάρμακα ὀρὸν ἐφθὸν διδόναι πίνειν ὡς
πλείστας ἡμέρας ξὺν ἁλὶ ὀλίγῳ· ἐπιτρωγέτω δὲ ἡδύοσμον, ἐς ἑσπέρην δὲ σίτου μὴ ἁπτέσθω,[56] ῥυφείτω δὲ
ὀλίγον, καὶ ἐπιπινέτω οἶνον γλυκύν· ἢν χρῄζῃ· ἢν δὲ
μὴ παρῇ, ὀρός· γάλα ὄνειον ἀφεψῆσαι, καὶ διδόναι

_____

[52] μὴ om. MV.   [53] ἡ δέ . . . καὶ Θ· ἢν δὲ φυγῇ ἡ
νοῦσος φλεγματώδης ᾖ· (καὶ M) MV, om. καὶ V.

oil and honey, and with the oil alone, fumigate, leave off for three or four days, and have the woman take the maidenhair potion. If she fails to become pregnant, the disease will relapse and these patients perish. In older women befallen by this disease, the uterus also becomes putrid, and few of them escape.

10. Another flux that, as it is cleaned, has the appearance of what comes out of a raw egg, being yellow and whitish: the vagina becomes ulcerated, the legs and calves swell up, the hollows of the eyes are edematous, the eyes are moist and bleary, and if the woman takes a walk she becomes short of breath and weak. This disease is by nature associated with phlegm, and if the patient is not cleaned, she develops fever when bile is set in motion, and this is an indifferent sign. The disease befalls older women more than younger ones.

When the case is such, if the edema is severe, give a purgative potion that will evacuate phlegm and bile; if however the edema is not so severe but the patient is oppressed with phlegm, give her an emetic. Also, if she is able to bear it, employ hellebore; if not that, then any agent that moves bile and phlegm. After the medications, give boiled whey with a little salt to drink for many days. Have the patient eat some mint, and toward evening abstain from food[12] but take a little gruel, and after that drink sweet wine, if it is needed: if wine is not available, let it be whey; also boil off ass's milk, and give this for four days.

---

[12] With the reading of Θ: "not avoid food."

---

54 Add. recc., Calvus *magis*: om. ΘMV.
55 Littré: ἰήσεται codd.   56 ἁπτέσθω MV: ἀπεχέσθω Θ.

τέσσερας ἡμέρας· ἐς ἑσπέρην δὲ ταῦτα ποιεῖν, ἢν μὴ
πῦρ ἔχῃ νύκτωρ.

Ἢν δὲ ᾖ φύσει σπληνώδης, πνευματώδης, λίφαιμος, μήτε ὀρὸν μήτε γάλα διδόναι [ἢν φυσῇ]·[57] ἢν δὲ
μή, κατωτερικοῖσι φαρμάκοισι καθαίρειν. καὶ ἐπειδάν
σοι δοκέῃ καιρὸς εἶναι, κλύσαι τὰς μήτρας, πρῶτον
μὲν τῷ ἀπὸ τῆς τρυγὸς δὶς ἢ τρίς, μετὰ δὲ τῷ ὀπῷ

260 τῆς σκαμμωνίης, οἶνον δ' ἐπιχέαι γλυκὺν | ὅσον κο-
τύλην Ἀττικὴν καὶ ἐλαίου τέταρτον μέρος κοτύλης,
μάλιστα μὲν ναρκισσίνου, εἰ δὲ μή, ἀνθίνου·[58] μετα-
κλύσαι δὲ τῇ ὑστεραίῃ οἴνου κοτύλῃ, καὶ μέλιτι τε-
τάρτῳ, καὶ ῥητίνῃ τρίτην μοῖραν μέλιτος, ἔλαιον δὲ
ἴσον μέλιτι. ἢν δὲ τὰ ἀπὸ τοῦ κλυσμοῦ ὑποχωρεῦντα
ἴῃ φλεγματώδεα, κλύσαι αὖθις ἐπισχόντα ἡμέρας
τρεῖς ἢ τέσσαρας· ἐκλέψαντα[59] τοῦ κόκκου δύο πόσιας, κλύζειν δὲ τοῖς αὐτοῖς· ἢν δὲ μὴ[60] παρῇ κόκκος,
τῇ ῥίζῃ τῆς θαψίης πόσι, καὶ μετακλύζειν τοῖσιν
αὐτοῖσιν· ἐπὴν δὲ καθαρθῇ τὰ ὑποχωρέοντα πρὸς
τούτους τοὺς κλυσμούς, καὶ ἢν γίνηταί τι αἱματῶδες,
οἷον ἀφ' ἕλκεος, μετακλύζειν τοῖσι στρυφνοῖσι.

Καὶ ἢν μετὰ ταῦτα αὐτὴ φῇ ἑωύτῃ τὸ στόμα τῶν
ὑστερέων σκληρὸν καὶ ὀδύνην ἐνεῖναι, τὸν αὐτὸν τρόπον τῷ σὺν τῷ πικερίῳ, ἔστ' ἂν ἀποξηρανθέωσιν αἱ
ὑστέραι καὶ δοκέῃ ὑγιὴς εἶναι· τὰς δὲ μεταξὺ ἡμέρας
τῶν κλυσμῶν πινέτω ἀκτῆς καρπὸν καὶ λαγωοῦ πυτίην καὶ μήκωνος τὸ κέλυφος καὶ κνίδης καρπόν,

---

[57] Del. Ermerins.   [58] εἰ . . . ἀνθίνου om. MV.

Do these things toward evening unless there is nocturnal fever.

If the patient has a natural tendency to suffer in her spleen, has a breathing disorder, and lacks blood, give neither whey nor milk; if she does not (sc. have these signs), clean with purgative medications. When you think the right time has arrived, flush her uterus, first two or three times with a douche of wine lees, next with scammony juice, and then with an Attic cotyle of sweet wine mixed with a quarter cotyle of oil—best narcissus oil, but if not that, lily oil. Flush again on the next day with a cotyle of wine, honey to a quarter of that amount, resin one third of the amount of the honey, and olive oil equal to the honey. If what is discharged with the douche resembles phlegm, do nothing for three or four days, and then flush again after removing the coverings from two quantities of (sc. Cnidian) berries; reinject with the same things as above. If the berries are not available, use one quantity of thapsia root, and inject again with the same things. When what has passed down is cleaned by these douches in the form of a bloody fluid resembling the discharge from an ulcer, flush again with astringent agents.

If after this has taken place, the woman says that the mouth of her uterus is hard and painful, flush in the same way with a douche containing butter, until the uterus dries up and the patient seems to have recovered. On the days between the injections, have her take a potion made with elderberries, rennet from a hare, poppy capsules, and Cnidian berries, as well as peel from sweet pomegranates,

---

[59] ἐκλέψαντα Littré: ἔπειτα ἐκλέψαι codd.

[60] μὴ om. MV.

ροιῆς γλυκείης τὸν φλοιόν, τρίβων ἴσον ἑκάστου· ἄλφιτον δὲ συμμίσγειν καὶ ἀδίαντον, ἐν οἴνῳ μέλανι
εὐώδει νῆστις· σιτίοισι δὲ χρήσθω μαλθακοῖσι μήθ᾿
ἁλμυροῖσι μήτε δριμέσι· κρέα δὲ ἀμείνω ἰχθύων, ἢ
ὀρνίθια ἢ λαγῷα, καὶ λούσθω θερμῷ μὴ πολλῷ. ἢν
δὲ μὴ λωφήσῃ ὁ ῥόος, ἀλλ᾿ ὑπολίπηται, καὶ ὑγράζω-
νται αἱ ὑστέραι, θυμιᾶσαι τῷ σὺν τῷ σιδίῳ, καὶ
ἔπειτα παρὰ τὸν ἄνδρα ἴτω· καὶ ἢν ἐν γαστρὶ ἔχῃ,
ὑγιὴς γίνεται. αἱ δὲ γεραίτεραι οὐ δύνανται ἀναφέ-
ρειν, ἀλλ᾿ ἀπόλλυνται ὑπ᾿ ἀσθενείης.

11. (120 L.) Ῥόος ἄλλος· ῥέει ὑπόχλωρος οἷον ἐξ
262 ᾠοῦ, καὶ γλίσχρον, | καὶ ἡ γαστὴρ ἀνίσταται ἡ νει-
αίρη, καὶ σκληρὴ γίνεται, καὶ ἢν ψαύσῃς, ἀλγέει, καὶ
βρύχει, καὶ πῦρ ἔχει, καὶ ὀδύνη ἐς τὸ ἐπίσιον καὶ τὰ
αἰδοῖα, καὶ ἐς τὴν νειαίρην γαστέρα, καὶ ἐς τὰς ἰξύας·
καὶ λιποθυμίη, ἀλυσμοί τε καὶ περιψύξιες, ἱδρὼς πολύς, σφυγμοὶ πρὸς χεῖρα ψαίροντες, βληχροί, ἐκλεί-
ποντες. καὶ αὐτίκα ἀπόλλυνται· ἢν δὲ περιῇ, καὶ ἡ
χροιή οἱ τρέπεται, καὶ γίνεται οἷόν περ κηρίον, καὶ ὁ
χρὼς τῷ δακτύλῳ πιεζόμενος μαλθάσσεται, καὶ ἐμ-
πλάσσεται οἷόν περ ἐν σταιτί, καὶ οἰδέουσιν οἱ πόδες
καὶ τὰ σκέλεα. ἡ δὲ νοῦσος γίνεται μάλιστα, ἢν ἐν
αὐτῇ τι διακναισθῇ ἢ ἐνσαπῇ ἐν τόκῳ· αἱ δ᾿ ἀφηλι-
κέστεραι μᾶλλον πάσχουσιν, οὐ πάνυ δ᾿ ἡ νοῦσός
εὐήθης.

12. (121 L.) Ἄλλος ῥόος· καθαίρεται οἷον ἀπὸ κρεῶν
ὀπτῶν χυμός, καὶ ἑλκοῦται τὸ αἰδοῖον καὶ ὅπη ἂν
ἄλλη τοῦ χρωτὸς ἐπιστάξῃ· καὶ ῥῖγος καὶ πῦρ ὀξύ,

squeezing out an equal portion of each of these. Mix into this barley meal and maidenhair, and have the patient take it in the fasting state together with fragrant dark wine. She should employ emollient foods that are neither salty nor pungent: meat is better than fish—either fowl or hare. She should bathe in hot water, but not much. If the flux fails to abate, but continues, and the patient's uterus becomes moist, fumigate it with pomegranate peel, and then have her go to her husband. If she becomes pregnant, she recovers. Older women cannot recover, but tend to succumb due to their weakness.

11. Another flux that flows out yellowish like the contents of an egg, and is sticky. The patient's lower belly becomes raised and hard, and if you touch it she feels pain; she clenches her teeth, becomes febrile, and suffers pain in her pubes and genitalia, lower belly, and loins. She is subject to syncope, restlessness, generalized chills, copious sweating, and a weak intermittent pulsation that flutters when felt. Such patients die at once; if one survives, her skin becomes altered and waxy, and when it is pressed with a finger, it is soft and receives an imprint the way dough does; her feet and legs also swell up. This disease arises in most cases when something lacerates or putrefies in a woman during childbirth. Those who are older suffer greatly, and the disease is far from benign.

12. Another flux that, as it is being cleaned, has the appearance of juice from roasting meat; it ulcerates the vagina and anywhere else on the skin it drips. Rigor and

πυκνόν, μέγα, καὶ φρίκη ὁμοῦ λάζεται· ὀδύναι δὲ αἱ[61]
ἐν πᾶσι τοῖσι ῥόοισιν· αὕτη πᾶσα ἀνοιδίσκεται καὶ
τὰ κάτω τοῦ ὀμφαλοῦ, καὶ τὰ σκέλεα, καὶ ἡ χροιὴ
ἰκτερώδης γίνεται. γίνεται δὲ ὁ ῥόος οὗτος ἀπὸ τοῦδε,
ἐπειδὰν τὸ αἷμα ἐξεραθὲν ὑπόχολον γένηται καὶ μὴ
καθαρθῇ.

Ὁκόταν ὧδε ἔχῃ, φάρμακον πίνειν, ἢν μὲν ἰσχυρὴ
ᾖ, ἐλλέβορον, ἢν δὲ ἀσθενής, ῥίζαν . . . τε καὶ ἐλατήριον, καὶ ὑστέρῳ χρόνῳ γάλα αἰγός, ἢ κάτω ὅ τι
χολήν τε καὶ φλέγμα καθαίρει· κλύσαι δὲ τὰς ὑστέρας τῷ σὺν τῷ κραμβίῳ· πίνειν δὲ ἐλελίσφακον, ὑπερικόν, λίνου σπέρμα, ἴσον ἑκάστου ἐν οἴνῳ μέλανι
αὐστηρῷ, διδόναι νήστει. ἢν δὲ ἡλκωμένον ᾖ τὸ αἰ-
δοῖον, μετακλύσαι τῷ πικερίῳ, ἔπειτα φάρμακον |
264 πῖσαι, καὶ ἐπιχρίειν τὰ ἕλκεα πικέριον, ῥητίνην,
σμύρναν, ἀργύρου ἄνθος· διανιζέσθω δὲ ὕδατι ἀπὸ
μυρσίνης καὶ ἐλελισφάκου χλιερῷ· σιτίοισι δὲ χρήσθω μήθ᾽ ἁλμυροῖσι μήτε δριμέσιν, ὡς μὴ δάκνῃ τὸ
οὖρον, καὶ τῶν θαλασσίων εἴργεσθαι, ὡς καὶ κρεῶν
βοείων, καὶ οἴων, καὶ χοιρείων· τοῖσι δ᾽ ἄλλοισι κρέασι χρῆσθαι ἑφθοῖσι, καὶ ἄρτον σιτείσθω, καὶ οἶνον
οἰνώδεα[62] παλαιὸν πινέτω μέλανα.

Ἢν δὲ ταῦτα ποιέῃ καὶ μὴ ὑγιὴς γίνηται, πυριῆσαι
ὅλην καὶ φάρμακον δοῦναι τῇ ὑστεραίῃ ἄνω, ἔπειτα
διαλιπὼν αὖτις καὶ κάτω· καὶ ἢν μὲν ᾖ ὀρός, μετὰ τὰ
φάρμακα ἀφεψήσας δοῦναι πιεῖν ἑκάστης ἡμέρης, ἐς

---

[61] αἱ om. MV.   [62] οἰνώδεα Θ: εὐώδεα MV.

fever—violent, continuous and severe—accompanied by a
simultaneous shivering, attack; also, the other pains present in all fluxes. The patient has generalized swelling, including in the region below her navel and in her legs, and
her skin becomes jaundiced. This flux arises whenever
blood that has been released becomes bilious and fails to
be expelled.

When the case is such, have the patient, if she is strong,
drink hellebore, but if she is weak, . . . root[13] and squirting
cucumber juice, and later goat's milk or anything that
cleans bile and phlegm downward. Flush her uterus with
a douche containing cabbage, and have her take a potion
of salvia, hypericum, and linseed—an equal amount of
each—in dry, dark wine: give this while she is fasting. If
the vagina ulcerates, give another douche made with butter, and then give a purgative to drink and anoint the ulcers with butter, resin, myrrh, and flower of silver. Have
the woman wash herself out thoroughly with a warm injection made from myrtle and salvia. She should employ
foods that are neither salty nor pungent, in order that her
urine does not become corrosive, and avoid sea foods, as
well as beef, mutton, and pork; other meats should be
prepared by boiling. Have her take bread, and drink aged,
dark, vinous wine.

If on doing these things, the woman fails to recover,
foment her whole body, give an emetic on the succeeding
day, and after waiting a time purge her too. If whey is
available, after the medication boil some off and give it to

[13] Galen suggests that a particular plant is meant with "root,"
but it may well be that a qualifier has been lost from the text here.

ἑσπέρην δὲ ῥοφήμασιν, οἴνῳ δὲ γλυκεῖ λευκῷ· ἢν δὲ
μὴ ᾖ ὀρός, γάλα ὄνου ἑφθὸν πινέτω ἐπὶ τέσσερας
ἡμέρας, ἑσπέρην δὲ τοῖσιν αὐτοῖσι χρήσθω· μετέπειτα δὲ πινέτω ἐπὶ τεσσεράκοντα ἡμέρας γάλα βοὸς
θερμόν, καὶ τῆς ἡμέρης μηδὲν ἐσθιέτω ὡς ἔπος εἰπεῖν,
ἄριστον γὰρ τοῦτο, καὶ γὰρ καθαίρεται καὶ τρέφεται[63] καὶ ἀμβλύνεται· ἐς ἑσπέρην δὲ δειπνείτω κρέας
ὄρνιθος ὀπτὸν ὀλίγον, καὶ ἄρτον σμικρὸν ἐγκρυφίην·
ἐπιπίνειν δὲ οἶνον μέλανα παλαιὸν οἰνώδεα, ἔστ᾽ ἂν
τὸ γάλα πίῃ, ἢν πολλὰ ἴῃ καὶ ταράσσηται τὸ δριμύ.
καὶ ἢν ταῦτα ποιήσασα ἐν γαστρὶ ἔχῃ, ὑγιὴς γίνεται.
ὅσαι δὲ γεραίτεραί εἰσιν, ἐνίοτε ὑποστρέφει ἡ νοῦσος,
καὶ ἀπόλλυνται· τῇσι δὲ νέῃσιν οὐ θανατώδης· χρονίη δέ.

13. (122 L.) Ῥόου ἰχωροειδέος θεραπείη· ῥεῖ ὕφαιμον, οἷόν ἅπερ ἀπὸ κρεῶν ὀπτῶν χυμός,[64] καὶ δάκνει
ὡς ἄλμη, καὶ ἐσθίει καὶ ἐξελκοῖ τὰ αἰδοῖα, καὶ ἡ
ὑστέρη ἀνελκοῦται, καὶ τὰ πέριξ καὶ τοὺς μηροὺς καὶ
τἄλλα· ἐπειδὰν ἐπιστάξῃ ἐπὶ τὰ ἱμάτια, βάπτεται, καὶ
δύσπλυτα ἐμμένει· καὶ ἡ γαστὴρ ἐπαείρεται καὶ
266 σκληρὴ γίνεται, | καὶ ἀλγέει ἢν ψαύσῃ, καὶ θέρμη
ἔχει, καὶ ἐς τὰ αἰδοῖα καὶ ἐς τὴν ἕδρην ὀδύνη καὶ ἐς
τὴν νειαίρην γαστέρα καὶ κενεῶνας καὶ ἰξύας, ἀδυναμίη ψυχρή, καὶ ἡ χροιή οἱ τρέπεται ὡς ἰκτερώδης. ἢν
δὲ ὁ χρόνος μηκύνῃ καὶ ἡ νοῦσος, ταῦτα πάντα πολὺ
μᾶλλον ἐπιλαμβάνει, καὶ τὰ κῦλα ἐπανίσταται, καὶ οἱ
πόδες οἰδέουσι καὶ τὰ σκέλεα ἀπὸ τῶν ἰξύων. ἡ δὲ
νοῦσος λάζεται, ἐπὴν διακναισθῇ τι τοῦ ἐμβρύου ἐν

drink daily; toward evening employ gruels and sweet white
wine. If whey is not available, have the patient drink boiled
ass's milk for four days, and in the evening do the same
things. After that she should drink warm cow's milk for
forty days, and during the day eat nothing, so to speak.
This is the best regimen, since with it she will be cleaned
and well nourished, and her disease will be blunted. Toward evening have her sup on a little roasted fowl and a
small loaf of bread baked in ashes, and after that drink
some aged, dark, vinous, wine up to the time when she
takes her milk, if her flux is copious and irritating because
of its acridity. If after doing these things the woman becomes pregnant, she recovers. In women that are older,
the disease sometimes relapses and they die; in young
women it is not mortal, but it is chronic.

13. Therapy for a serous flux that appears somewhat
bloody like the juice from roasting meat, burns like brine,
and corrodes and ulcerates the genitalia; the uterus becomes ulcerated, as do the area around it, the thighs, and
other regions. Wherever drops fall on to the bedding, they
produce stains which are difficult to wash out. The belly
becomes raised and hard, and is painful when touched;
fever sets in, pain radiates to the genitalia and the seat, as
well as to the lower belly, flanks, and loins, a cold paralysis
develops, and the skin becomes jaundiced. As the disease
extends in time, all its features become more severe, and
the areas below the eyes become edematous, as do the feet
and legs from the loins down. This disease arises when
some part of a fetus is injured during childbirth or an abortion. It must be treated with fomentations, fumigations,

---

[63] καὶ τρέφεται om. Θ.   [64] χυμός om. Θ.

τόκῳ ἢ τρωσμῷ. χρὴ δὲ πυρίην καὶ θυμιῆν καὶ πιπί-
σκειν· ταῦτα πάντα ἀεὶ ἔχει τὸν ῥόον· καὶ ὄνειον γάλα
καὶ τἆλλα, ἢν δέῃ καὶ ἐμέειν· ἢν δὲ ἄπυρος μὴ[65] ᾖ καὶ
βληχρῶς ἔχῃ, ἄμεινον ξηρὴ πυρίη.

14. (123 L.) Ὁκόταν ὡς[66] ἐς τὴν κεφαλὴν τραπῶσιν
αἱ ὑστέραι καὶ τῇδε λάβῃ[67] ὁ πνιγμός, κεφαλὴν βαρύνει, ἄλλῃσι δὲ ἄλλῃ πη τέκμαρ ἴσχεται. σημεῖον δὲ
τόδε· τὰς φλέβας τὰς ἐν τῇ ῥινὶ καὶ τὰς ὑπὸ τοῖς
ὀφθαλμοῖσιν ἀλγέειν φησί, καὶ κῶμα ἴσχει, καὶ
ἀφρίζει ὅταν ῥαΐσῃ. ταύτην χρὴ λούειν θερμῷ πολλῷ·
ἢν δὲ μὴ ἐνακούῃ, ψυχρῷ, καὶ κατὰ κεφαλῆς, δάφνην
τε καὶ μυρσίνην ἐνέψων ἐν τῷ ὕδατι καὶ ψύχων· καὶ
ῥοδίνῳ μύρῳ τὴν κεφαλὴν χριέσθω· καὶ ὑποθυμιάσθω τὰ εὐώδεα, τὰ δὲ κακώδεα ὑπὸ τὰς ῥῖνας· καὶ τὴν
κράμβην ἐσθιέτω, καὶ τὸν χυλὸν ῥυφείτω.

15. (124 L.) Ἢν δὲ πρὸς τὴν καρδίην προσιστάμε-
ναι πνίγωσιν αἱ ὑστέραι, | καὶ ἀνάσσυτος ἴῃ ὁ ἠὴρ
βιώμενος, ἀλησθύει καὶ εἰλέει·[68] καὶ ἔστιν ᾗσιν αὐτίκα
ἐλυσθεῖσα κάτω χωρέει καὶ φῦσα ἔξεισιν, ἢ καὶ ἐμέει
ἀφρώδεα, ἡ δὲ παῦλα ἤδε γίνεται. ᾗσι δὲ οὐκ ἀφίστανται, πράσου τὸν καρπὸν καὶ μήκωνα τρίψας, διεὶς
ὕδατος κυάθῳ δίδου πίνειν· καὶ ὄξος λευκὸν ἀρήγει
πόσις κύαθος· ἢ ἀρκεύθου καρπὸν καὶ ἐλελίσφακον,

268

---

65 Om. μὴ MV.   66 Om. ὡς MV.

67 λάβῃ Potter (cf. Calvus' *strangulant, praefocantve*, Littré's
et que là se fixe la suffocation, and Fuchs' *und schliesslich dort
Erstickungsanfälle auftreten*): λήγῃ ΘV: ληγει M.

68 εἰλέει Θ: ἐμέει MV.

and drinks, all of which contribute to controlling the flux; also ass's milk and the like, if necessary, and an emetic. If the patient is not completely afebrile, but has a mild fever, favor a dry fomentation.

14. When the uterus seems to turn toward the head, and suffocation takes hold there, the patient feels heavy in her head, and various signs appear in one part of her body or another: she says that the vessels in her nose and under her eyes are painful, she loses consciousness, and then as she is improving she has foam in her mouth. You must bathe such a patient in copious hot water; if, however, she fails to respond, pour cold water over her, also down on to her head: boil laurel and myrtle in the water, and then cool it. She should also apply rose unguent to her head, fumigate herself from below with fragrant substances, and place ill-smelling agents beneath her nostrils. Have her eat cabbage, and take the juice as a soup.

15. If the uterus moves close to woman's heart and provokes suffocation, and her breath rushes upward under force, she will be restless and convulsed.[14] In some cases the coiled up uterus immediately moves downward, and wind is expelled, or the patient vomits frothy material and the disease ends. But for women in whom the uterus does not move away (sc. from the heart), grind leek and poppy seeds, dissolve in a cyathos of water, and give this to drink; also of benefit is a cyathos draft of white vinegar. Or take berries of Phoenician juniper and salvia seeds together

---

[14] MV read "vomit" in place of Θ's "convulsed."

ὄξος σὺν τοῖσδεσιν ἢ οἶνον· ἀλεαίνεσθαι δὲ χρή·
ἄλειφα χηνός, κηρωτὴν ἐρρητινωμένην, καὶ πίσσαν
ἐν αὐτῷ τῆξαι, καὶ προσθετὰ ποιέειν.

16. (125 L.) Ὅταν δὲ ὡς πρὸς τὰ ὑποχόνδρια προσπέσωσι, πνίγουσιν· ἐπὴν ἐνθάδε τὸ τέρθρον ᾖ τοῦ
πάθεος, καὶ ἐπιλαμβάνει ἔμετος πυρώδης δριμύς, καὶ
ῥᾴων γίνεται ὀλίγον χρόνον, καὶ ἐς τὴν κεφαλὴν καὶ
ἐς τὸν τράχηλον ὀδύνη διαμπερής. χλιάσματα προστιθέναι, ἢν ἄνω πνίγωσιν· ὑπὸ δὲ τὰς ῥῖνας θυμιᾶν
τὰ κάκοδμα ἐκ προσαγωγῆς, ἢν γὰρ ἀθρόα ᾖ, μεθ-
ίστανται αἱ ὑστέραι ἐς τὰ κάτω καὶ ὄχλος γίνεται·
εὔοδμα δὲ κάτω· καὶ πιεῖν διδόναι τὸν κάστορα καὶ
τὴν κόνυζαν· ἐπὴν δὲ κάτω ἑλκυσθῶσιν, ὑποθυμιῆν
τὰ εἰδεχθέα, ὑπὸ δὲ τὰς ῥῖνας τὰ εὐώδεα. ἢν δὲ αἱ
ὀδύναι παύσωνται, φάρμακον πῖσαι κάτω, καὶ μετα-
πιπίσκειν γάλα ὄνου ἢ ὀρόν, ἢν μὴ σπληνώδης ᾖ ἀπὸ
γενέσιος ἢ λίφαιμος ἢ ἄχροος, ἢ τὰ οὔατα ἠχώδεα
270 διὰ ξυγγενείην, | ἢ[69] ᾖσιν ἠθάδες ἀπὸ νεότητος αἱ
νοῦσοι· τὴν δὲ ἄνω κοιλίην μὴ κινέειν, ὅσαι ἀμβλυώσσουσιν ἢ ᾖσιν[70] ἀμφὶ τὴν φάρυγγα ὄχλοι καὶ
τἄλλα· διδόναι δὲ πτισάνης χυλόν· ἢν σφόδρα εὐη-
μὴς ᾖ, καὶ ἐμεέτω· κλυσμὸς δ᾽ ἄριστος ὁ διὰ ναρκίσ-
σου· προσθετόν, τὸ διὰ κανθαρίδων.

17. (126 L.) Ὅταν αἱ ὑστέραι προστῶσιν πρὸς τὰ
ὑποχόνδρια, πνίγεται ὡς ὑπὸ ἐλλεβόρου, καὶ ὀρθό-
πνους γίνεται, καὶ καρδιωγμοὶ σθεναροί· αἱ δὲ καὶ
ἐμέουσιν ἐνίοτε σίαλον ὀξύ, καὶ τὸ στόμα ὕδατος

with vinegar or wine; warm the patient; melt goose grease, a wax salve mixed with resin, and pitch, and form this into suppositories.

16. When (sc. the uterus) falls against patients' hypochondria, it provokes suffocation. At the time of the disease's crisis, acrid, burning vomiting takes place, for a short time there is relief, and then pain invades the whole head and neck. If the suffocation is located in the upper regions, make warm applications; apply ill-smelling fumigations to the nostrils by intermittent increases—since if these are made all at once the uterus will rush downward and cause trouble—and fragrant fumigations from below. Give castoreum and fleabane to drink. When the uterus is drawn downward, fumigate below with fetid agents and under the nostrils with fragrant ones. If the pains remit, give a purgative potion and after that ass's milk or whey, unless the patient has some congenital condition of the spleen, or is bloodless or pale, or there is an echoing of her ears from birth, or in cases where the diseases are habitual from youth. In patients with dullness of vision, do not move the upper cavity, or in those with sore throats and the like, but give barley gruel. If a patient vomits especially easily, have her vomit as well. The best douche is made from narcissus, the best suppository is the one with blister beetles.

17. When a woman's uterus approaches her hypochondria, she suffocates as if she had taken hellebore, and orthopnea sets in along with violent heartburn. Sometimes such patients vomit acrid saliva and their mouth fills with

---

[69] Om. ἢ MV.   [70] ὅσαι . . . ἧσιν Littré: οσσαι ἀμβλυώσσουσιν η ησιν Θ: ἀμβλυώσσουσιν ἢ MV.

ἐμπίμπλαται, καὶ τὰ σκέλεα ἀποψύχονται. αἱ τοιαῦται,
ἢν μὴ τάχα ἀφιστῶνται ἀπὸ τῶν ὑποχονδρίων αἱ
ὑστέραι, ἄναυδοι γίνονται, καὶ τὰ ἀμφὶ τὴν κεφαλὴν
καὶ τὴν γλῶσσαν νάρκη ἔχει.

Τὰς τοιαύτας ἢν ἀναύδους καταλάβῃς καὶ τοὺς
ὀδόντας συνηρεισμένας,[71] πρὸς μὲν τὰς ὑστέρας
προσθεῖναι εἴριον πρὸς αὐλόν, ὡς ὠθεῖν μάλιστα, τοῦ
πτεροῦ περιελίξας, βάψας ἢ λευκῷ Αἰγυπτίῳ ἢ μυρ-
σίνῳ ἢ βακκάρει ἢ ἀμαρακίνῳ· ἐς δὲ τὰς ῥῖνας, τοῦ
φαρμάκου τοῦ μέλανος, τοῦ τῆς κεφαλῆς, λαβόντα τῇ
μήλῃ ἐμπλάσαι· ἢν δὲ μὴ ᾖ τοῦτο, τῷ ὀπῷ διαλεί-
ψαι,[72] ἢ πτερὸν ὄξει βάψαι καὶ καθεῖναι, διαλεῖψαι δὲ
τὰς ῥῖνας, ἢ τοῦ πταρμικοῦ προσθεῖναι· ὅταν δὲ κλεισθῇ[73] τὸ στόμα καὶ ᾖ ἀναυδίη, δοῦναι πιεῖν τοῦ κάστορος ἐν οἴνῳ· τὰς δὲ ῥῖνας διαλεῖψαι βάψας τὸν
δάκτυλον ἔλαιον φώκης· τὸ δ᾽ εἴριον ἐᾶν προσκεῖσθαι,
μέχρι οὗ καταστέωσιν· ὅταν δὲ παύσωνται, ἀφελέ-
σθαι χρή. ἢν δὲ ἀφαιρεθέντος αὖθις ἀναχωρήσωσι,
272 τὸ εἴριον αὖθις | προσθεῖναι τὸν αὐτὸν τρόπον, ὑπὸ δὲ
τὰς ῥῖνας θυμιᾶν, κέρας μέλαν αἰγὸς ἢ ἐλάφου κνή-
σας, ἐπιπάσσων ἐπὶ σποδιὴν θερμήν, ὅπως μάλιστα
θυμιάσεται, καὶ εἰρυσάτω τὴν ὀδμὴν ἄνω διὰ τῶν ῥινῶν, ὡς ἂν δύνηται μάλιστα· ἄριστον δὲ[74] θυμιῆν φώκης ἔλαιον, ἐπ᾽ ὄστρακον ἐπιτιθέντα[75] ἄνθρακας περι-
καλύψαι, καὶ τὴν κεφαλὴν ὑπερίσχειν, ὡς μάλιστα ἡ

---

71 Τὰς τοιαύτας . . . συνηρεισμένας om. MV.
72 Add. τὰς ῥῖνας MV.

fluid; their legs become cold. Unless their uterus soon moves away from the hypochondria, they lose their speech and experience numbness around their head and tongue.

If you come upon such cases when they are already speechless and clenching their teeth, apply a piece of wool wound around a feather, by means of a pipe, as a suppository against the uterus, pressing it as far in as possible: prepare this by soaking it with white Egyptian oil, myrtle oil, bacchar, or sweet marjoram oil. Into the nostrils fill some black medication[15] for the head, picking it up with a spatula. If this is not available, anoint them with (sc. silphium) juice; or soak a feather in vinegar, and introduce it to anoint the nostrils; or administer some sternutatory. When the patient's mouth is closed and she cannot speak, give castoreum to drink in wine. Anoint the nostrils by dipping a finger in seal's oil. Leave the piece of wool in place until the uterus moves downward, but when it stops moving, remove the wool. If when this is removed the uterus moves up again, apply the piece of wool again in the same way, and foment under the nostrils by grating black horn of a goat or a deer and sprinkling it on to hot ashes, so that it will produce a maximum of smoke, and have the woman draw the fumes up through her nostrils as forcefully as she can. It is best to fumigate the nostrils with seal's oil by placing burning coals on top of potsherds, covering the patient—but leaving her head free in order

---

[15] For this prescription see *Diseases of Women I* 96.

---

[73] κλεισθῇ Foes in note 61: καυθῇ Θ: κλυσθῇ MV.
[74] ἄριστον δέ om. MV.
[75] Add. τοὺς δέ Θ.

ὀδμὴ ἐσίη, καὶ ἐπιστάζειν τοῦ λίπεος, καὶ ἄνω ἑλκέτω
τὴν ὀδμήν· τὸ δὲ στόμα συμμεμυκέναι χρή. ἢν ἄνω
προσπίπτωσι, ταῦτα χρὴ ποιέειν.

18. (127 L.) Ἢν αἱ μῆτραι πρὸς τὸ ἧπαρ κλιθῶσιν,
ἄφωνος ἐξαπίνης γίνεται, καὶ τοὺς ὀδόντας συνερή-
ρεισται, καὶ ἡ χροιὴ πελιδνὴ γίνεται· ἐξαπίνης δὲ
ταῦτα πάσχει, ὑγιὴς ἐοῦσα. γίνεται δὲ μάλιστα παρ-
θένοισι παλαιῆσι καὶ χήρῃσιν ὅσαι λίην νέαι ἐοῦσαι
καὶ τοκήεσσαι χηρεύουσι· γίνεται δὲ μάλιστα καὶ
τῇσιν ἀφόροισι πάμπαν καὶ στείρῃσιν, ὅτι ἐκ τῶν
τόκων εἰσίν· οὐ γὰρ γίνεται ἡ λοχείη κάθαρσις, οὐδ᾽
ἀνοιδίσκεται ἡ ὑστέρη, οὐδὲ μαλθάσσεται, οὐδ᾽ ἐμέει.

Ὅταν ὧδε ἔχῃ, τῇ χειρὶ ἀπώσασθαι ἀπὸ τοῦ ἥπατος παρηγορικῶς τὸ οἶδος ἐς τὸ κάτω, καὶ ἀποδῆσαι
ταινίῃ τὰ ὑποχόνδρια, καὶ τὸ στόμα διανοίγειν, οἶνον
δ᾽ ὡς εὐωδέστατον κεκρημένον ἐγχεῖν, ὅτε χρή, καὶ
προσέχειν πρὸς τὰς ῥῖνας τὰ κάκοδμα· καὶ ὑποθυμιᾶν, πρὸς δὲ τὰς ὑστέρας τὰ εὐώδεα καὶ ἄσσα θυώ-
ματα· καὶ ἐπὴν ἰήσῃ, κάθαιρε,[76] φάρμακον δὲ πῖσαι
κάτω χρή, ἢν μὲν χολώδης ᾖ, ὅ τι χολὴν καθαίρει,
ἢν δὲ φλεγματώδης, ὅ τι φλέγμα· κἄπειτα πιπίσκειν
γάλα ὄνου | ἐφθόν, καὶ τὰς ὑστέρας πυριᾶσαι εὐώ-
δεσι, καὶ προστίθεσθαι τὸ σὺν τῇ βουπρήστει· τῇ δ᾽
ὑστεραίῃ νέτωπον, διαλείπων δ᾽ ἡμέρας δύο κλύσαι
τὰς ὑστέρας εὐώδεσιν· ἔπειτα γλήχωνα,[77] διαλιπὼν δὲ
μίην ἡμέρην, θυμιᾶσαι τοῖσιν ἀρώμασι. ταῦτα ποιέ-

[76] κάθαιρε om. MV.   [77] γλ. om. MV.

that the fumes can enter as easily as possible—and drizzling the oil on to the fire for her to draw up its fumes; she should keep her mouth closed. If the uterus falls against the hypochondria, this is what you must do.

18. If a woman's uterus leans against her liver, she suddenly becomes speechless, clenches her teeth, and her skin becomes livid—she suffers these things suddenly while in a healthy state. This happens most often to old maids and widows who have been widowed while still young and fertile. It also occurs frequently in women who have no children at all and are barren, due to their exclusion from childbirth, since they have never had any lochial cleaning, swelling and softening of their uterus, or regurgitation from it.

When the case is such, use your hand to press the swelling gently down and away from the liver, and bind the hypochondria with a bandage; open the mouth (sc. of the uterus) and infuse diluted very fragrant wine, when it is indicated, and apply ill-smelling agents against the nostrils; also fumigate below against the uterus with fragrant kinds of vapors. Once the patient is cured, clean her by giving a purgative potion—if she is suffering from bile with one that evacuates bile, if she is suffering from phlegm with one that evacuates phlegm. Then give her boiled ass's milk to drink, fumigate her uterus with fragrant agents, and apply a suppository made with buprestis. On the following day give oil of bitter almond, let two more days pass, and flush the uterus with fragrant substances; then fumigate with pennyroyal,[16] and after waiting one further day, with aromatics. Do this for the

16 MV omit "pennyroyal."

ειν τὴν χήρην· ἄριστον δὲ ἐν γαστρὶ ἔχειν· τὴν δὲ
παρθένον ξυνοικέειν·[78] πρὸς δὲ τὰς ὑστέρας ἀείρειν
⟨μηδέν⟩,[79] μηδὲ τὸ φάρμακον πίνειν. νῆστιν τὸν κάστορα, καὶ κόννυζαν δὲ ἐν οἴνῳ ὡς εὐωδεστάτῳ ὡς ἐς
εἴκοσιν ἡμέρας, καὶ τὴν κεφαλὴν μὴ ἀλείφεσθαι εὐώδει μηδενί, μηδ᾽[80] ὀσφραίνεσθαι εὐωδέων.

19. (128 L.) Ἢν δ᾽ αἱ μῆτραι φλεγμήνωσι παρὰ τὸ
πλευρόν, ἢν ψαύσῃς, σκληρὸν φαίνεται, καὶ ὅταν
προσπέσωσι πρὸς[81] τὰ ὑποχόνδρια, πνίγουσι, καὶ
ἐμεῖ φλέγμα ὀξύ, καὶ τοὺς ὀδόντας αἱμωδεῖν ποιέει,
καὶ ἐπειδὰν ἐμέσῃ, ῥᾷον ἔχειν δοκέει. ὅταν δὲ κάτω
ὁρμήσωσιν, ἀφίστανται ἀπὸ τῆς γαστρὸς ἄλλοτε
ἄλλῃ, μάλιστα δὲ ἐς τοὺς κενεῶνας, ἔστι δ᾽ ὅτε ἐμπίπτουσιν ἐς τὴν κύστιν, καὶ στραγγουρίη ἐπιλαμβάνει, καὶ ἐς τὴν ἕδρην, καὶ δοκέει ἀποπατέειν· καὶ τὰ
ἐπιμήνια πρότερον ἢ ὕστερον τοῦ μεμαθηκότος γίνεται, ἢ οὐκ ἐπιφαίνεται.

Ταύτῃ αὐτίκα τῆς νούσου, ἢν τὰ ἄνω πνίγωσι,
χλιάσματα προστιθέναι, καὶ ὑποθυμιᾶν· καὶ προστιθέναι[82] τὰ κακοδμα πρὸς τὰς ῥῖνας, πρὸς δὲ τὰς ὑστέρας τὰ εὐώδεα, καὶ πίνειν κόννυζαν τὴν θηλείην καὶ
κάστορα ἐν οἴνῳ νῆστιν· ἐπὴν δὲ τῇ φύσει καθιστε-
ῶσι, θυμιᾶσαι ἠρεμέως· πινέτω δὲ διουρητικά. ταῦτα
276   ποιέειν ἔστ᾽ ἂν αἱ ὀδύναι | ἔχωσιν· ἐπὴν δὲ παύσων-
ται, πυριᾶσαι ὅλην, ἔπειτα πῖσαι φάρμακον κάτω, ἢν

---

[78] ἀνδρί add. MV.   [79] Add. Littré: ad nares nihil adhibeto
Cornarius.   [80] μηδ᾽ om. MV.

widow—but best would be for her to become pregnant; as far as the young unmarried woman is concerned, she should cohabit. Hold nothing up against the patient's uterus,[17] nor give any purgative potion, but with her in the fasting state give castoreum and fleabane in very fragrant wine for twenty days; do not anoint her head with any aromatic agent, nor let her smell anything of that sort.

19. If the uterus swells up against the side, on palpation it seems hard, and when it falls against the hypochondria, it provokes suffocation and the patient vomits acrid phlegm, which sets her teeth on edge; after vomiting, she feels better. When the uterus rushes back down, it leaves the belly at one time in one direction, at another time in another direction, but moves especially toward the flanks; sometimes it falls upon the bladder provoking strangury or upon the seat causing the patient to feel she has to pass stools. The menses appear earlier or later than they are accustomed to, or fail to appear at all.

To such a patient, if her suffocation is in the upper regions, apply warming treatments right at the beginning of the disease, and fumigate her below; apply evil-smelling substances beneath her nostrils and fragrant ones against her uterus; give her female fleabane and castoreum to drink in wine while she is fasting. When the uterus has settled in its natural location, fumigate it gently; have the patient drink diuretics. Do these things as long as the pains are present: when they cease, foment the whole

---

[17] MV read "nostrils" (ῥῖνας) in place of Θ's "uterus" (ὑστέρας).

---

[81] πρὸς om. MV.   [82] καὶ προστιθέναι om. MV.

μὲν χολώδης ᾖ, ὅ τι χολὴν καθαίρει, ἢν δὲ φλεγματώδης, ὅ τι φλέγμα ἄγει· καὶ πιπίσκειν γάλα ὄνειον
ἢ ὀρὸν αἴγειον ἀφέψων· ἢν δὲ σπληνώδης ᾖ, μὴ
πιπίσκειν τὸ γάλα μηδὲ τὸν ὀρόν· καὶ ἐν τοῖσι καθαρ-
μοῖσι σιτίοισι χρῆσθαι μαλθακοῖσι καὶ ὑποχωρη-
τικοῖσιν· ἰχθύες δ' ἀμείνους κρεῶν· καὶ πυριᾶν τὰς
ὑστέρας, καὶ καθαίρειν προσθετοῖσι, καὶ κλύσαι
αὐτάς· καὶ ὑποθυμιασαμένη παρὰ τὸν ἄνδρα ἴτω· λύσις δὲ τῆς νούσου, ἐπὴν λάβῃ ἐν γαστρί.

20. (129 L.) *Ην αἱ μῆτραι πρὸς τὰς πλευρὰς προσπέσωσι, βὴξ ἴσχει, καὶ ὀδύνη ὑπὸ τὸ πλευρόν, καὶ
προσίσταται σκληρίη ὡς σφαῖρα, καὶ ἁπτομένη πονέει ὡς ἕλκος· καὶ καταφθίνει, καὶ δοκέει οἱ περιπλευ-
μονίη εἶναι, καὶ εἰρύεται, καὶ κυφὴ γίνεται· καὶ τὰ
ἐπιμήνια οὐ φαίνεται, ἐνίῃσι δὲ ἀπόλλυται προφανέντα, τότε γενόμενα ἀσθενέα τε καὶ ὀλίγα καὶ κακήθεα, ἢν ἴδῃς·[83] καὶ ἡ γονὴ οὐ γίνεται τούτου τοῦ
χρόνου.

Ὅταν ὧδε ἔχῃ, φάρμακον χρὴ πῖσαι κάτω ἐλατήριον, καὶ λούειν πολλῷ θερμῷ, καὶ τῶν χλιασμάτων ὅ τι ἂν προσδέχηται προσίσχειν, καὶ προστιθέναι, ὑφ' ὧν καθαίρεται τὸ αἷμα· καὶ λίνου σπέρμα
φῶξαι καὶ κόψαι καὶ σῆσαι, καὶ μήκων ὁ λευκὸς ὠφε-
λέουσι καὶ ἐλελίσφακος σὺν ἀλφίτοισι λεπτοῖσι, καὶ
τυρὸν αἴγειον περιξύσας[84] τὴν ἅλμην, ταῦτα μίσγε,
μίαν μοῖραν τυροῦ καὶ τῶν ἀνάλτων ἀλφίτων μοῖραν,
νήστει δίδου πίνειν ἐν οἴνῳ· ὅταν δ' ἔσπερος ἔῃ, κυκεῶνα παχὺν μέλι ἐπιχέων δίδου· καὶ τῶν ποτημάτων

body, and then give a purging potion—if the patient is suffering from bile, one that evacuates bile, if she is suffering from phlegm, one that draws phlegm. Give boiled down ass's milk or goat's whey to drink, but if the patient has some condition of the spleen, do not give milk or whey. During the cleaning employ foods that are emollient and laxative—fish are better than meats. Foment the uterus, clean it with suppositories, and flush it. After the woman has been fumigated below, send her to her husband: resolution of this disease arrives with pregnancy.

20. If the uterus falls against the ribs, there are coughing and pain in the side, and a hardness shaped like a ball forms, which on palpation is painful like an ulcer. The patient wastes away and seems to have pneumonia; she is contracted and becomes hunchbacked. The menses fail to appear, although in some women they first appear and then disappear again, but when they do appear they are weak, scanty, and unhealthy when you look at them. No (sc. female) seed is generated during this time.

When the case is such, give squirting cucumber juice as a purge, wash with copious hot water, apply the warming treatments that are likely to be accepted, and suppositories that clean blood. Toast linseed, grind it, and sieve it; white poppy is also beneficial, as is salvia together with fine barley meal: grate the rind of goat's cheese with this and mix them together, one portion of the cheese and one of the unsalted barley meal: give this to the patient in the fasting state to drink in wine. When evening arrives, give a thick cyceon over which you have poured honey.

---

[83] καὶ . . . ἴδης Littré: καὶ κακειθεανιδοις Θ: κακίωνα M: κακίονα V.   [84] περιξύσας Θ: ἐπιξύσας M: ἐπιξέσας V.

ἃ δεῖται πιπίσκειν, καὶ πυριᾶν πυκινά,[85] καὶ τῷ θερμῷ

278 | δὲ αἰονᾶν·[86] καὶ τῇ χειρὶ ἡσύχως καὶ μαλθακῶς[87] καὶ
ὁμαλῶς ἀπωθέειν ἀπὸ τοῦ πλευροῦ τὰς μήτρας, καὶ
ἀναδεῖν τὸ πλευρὸν ταινίῃ πλατείῃ, καὶ γαλακτοποτέ-
ειν βόειον γάλα ὡς πλεῖστον ἐπὶ ἡμέρας τεσσερά-
κοντα· σιτίοισι δὲ χρήσθω ὡς μαλθακωτάτοισιν. ἡ δὲ
νοῦσος σπερχνή τε καὶ θανατώδης, καὶ ὀλίγαι ἐκφεύ-
γουσιν ὧδε μελεδαινόμεναι.

21. (130 L.) Ἢν δ᾽ ἐν τῇ ὀσφύϊ αἱ ὑστέραι ἐνῶσιν
ἢ ἐν τῷ κενεῶνι, καὶ ἴῃ πνεῦμα μετάρσιον, καὶ πνιγμὸς ἔχῃ, ἆσθμά μιν λάζυται πυκινόν, καὶ οὐκ ἐθέλει
κινέεσθαι· τρίψας θεῖον ἢ ἄσφαλτον ἢ κώνειον ἢ
σμύρναν, μέλι ἑφθὸν παραχέας, ποιέειν βάλανον μα-
κρὴν πάχετον, καὶ ἐς τὴν ἕδρην ἐντιθέναι.

22. (131 L.) Ἢν αἱ μῆτραι εἰλέωσι σφᾶς ἐς τὸ
μεσηγὺ τῶν ἰξύων, ὀδύνη ἔχει τὴν νειαίρην γαστέρα,
καὶ τὰ σκέλεα εἴρυαται, καὶ τὰς κοχώνας ἀλγέει· καὶ
ὅταν ἀποπατήσῃ, ὀδύναι ἴσχουσιν ὀξέαι, καὶ ὁ ἀπόπατος προέρχεται ὑπὸ βίης,[88] καὶ τὸ οὖρον στρύζει,
καὶ λιποψυχίη λαμβάνει. ὅταν ὧδε ἔχῃ, αὐλίσκον
προσδῆσαι πρὸς κύστιν, καὶ ἐμφῦσαν πρὸς[89] τὰς
ὑστέρας, καὶ πυριᾶν ἢ λοῦσαι πολλῷ καὶ θερμῷ
ὕδατι, καὶ ἐς ἔλαιον καὶ ἐς ὕδωρ καθίζειν· ὑποθυμιῆν
ὑπὸ τὰ αἰδοῖα κακώδεα, ὑπὸ δὲ τὰς ῥῖνας εὐώδεα·
ἐπὴν δὲ ἡ ὀδύνη παύσηται, φάρμακον διδόναι ἄνω,
τὴν δὲ κάτω κοιλίην οὐ χρὴ ταράσσειν. ἐπὴν δὲ αἱ

85 πυκινά om. Θ.   86 δὲ αἰονᾶν Θ: καταιονᾶν (-ῆν V) MV.

Also give the drinks that are requisite, foment frequently, and moisten with warm water. With your hand press the uterus cautiously, gently, and evenly away from the side, and bind the side with a broad bandage; have the patient drink as much cow's milk as she can for forty days, and employ the most emollient of foods. The disease is violent and deadly, and few escape it even if they are treated in this way.

21. If a woman's uterus occupies her lower back or a flank, if her breathing is shallow, and if she suffocates, she will frequently suffer from shortness of breath, and she will not want to move. Grind some sulfur, asphalt, hemlock, or myrrh, and add boiled honey: form this into a long, broad suppository, and insert it into the patient's seat.

22. If a woman's uterus twists upon itself in her loins, pain occupies her lower belly, her legs are contracted, and she aches in her buttocks. When she is at stool, she feels sharp pains, her feces pass with force, her urine squirts out with a noise, and she loses consciousness. When the case is such, tie a little pipe to a bladder and blow air[18] into the uterus; also foment or wash it with copious, hot water, and give the patient a sitz bath of oil and water. Fumigate beneath the genitalia with evil-smelling agents and under the nostrils with fragrant ones. When the pain ceases, give an emetic, but you must not disturb the lower cavity.

---

[18] MV read "infuse warm olive oil" in place of Θ's "blow air."

---

[87] χειρὶ . . . μαλθακῶς Littré: κηρωτῃ ἡσυχῃ καὶ μαλθακῃ Θ: κηρωτῆ ἡσύχα καὶ μαλθακῶς M: χειρὶ ἥσυχα καὶ μαλθακῶς V.   [88] Add. σμικρός MV.

[89] ἐμφ. πρὸς Θ: ἐγκλύζειν ἔλαιον θερμὸν ἐς MV.

ὑστέραι καταστῶσι, φάρμακον πῖσαι κάτω καὶ γάλα
ὄνου, ἢν μὴ σπληνώδης ᾖ· ἔπειτα πυριῆσαι καὶ κατ
αἰονᾶν τὰς ὑστέρας τῷ σὺν τῇ δάφνῃ, καὶ προσθέναι
280 | προσθετὸν καθαρτήριον ὃ μὴ δήξεται· κἄπειτα θυ-
μιωμένη τοῖσιν ἀρώμασι, παρὰ τὸν ἄνδρα ἴτω, καὶ ἢν
ἐν γαστρὶ ἔχῃ, ὑγιὴς γίνεται· ἄτοκοι δὲ πολλαὶ καὶ
πηραὶ τὰ σκέλεα ὡς τὸ πολὺ[90] γίνονται.

23. (132 L.) Ὅσῃσι δὲ τὸ στόμα κλίνεται ἑτέρωσε
καὶ προσπίπτει τῷ ἰσχίῳ, γίνεται γὰρ καὶ τοιαῦτα
διακωλύοντα τὴν ὑστέρην καθαίρεσθαι καὶ τὴν γονὴν
δέχεσθαι καὶ οὐ παιδοποιεῖ·[91] πυριᾶν χρὴ τοῖσιν εὐώδεσι, καὶ μετὰ τὴν πυρίην, ἢν παρασημήνῃ,[92] τῷ δακτύλῳ ἀποστῆσαι ἀπὸ τοῦ ἰσχίου· κἄπειτα ἐξιθύνειν
τοῖσι δαιδίοισί τε καὶ τῷ μολίβδῳ· οὐ γὰρ βιήσεται
θοῶς, ὡς εἴρηται. ὅταν δὲ κατὰ φύσιν ᾖ καὶ ἀνεστομωμέναι γίνωνται, προσθέτοισι μαλθακοῖσι καθαίρειν, καὶ τὰ λοιπὰ[93] ποιέειν.

24. (133 L.) Ἧσιν αἱ ὑστέραι προσπίπτουσι πρὸς
τὸ ἰσχίον, ἢν μὴ τάχα ἀφιστέωνται καὶ πάλιν[94] ἐς
χώρην καθιστῶνται, προσαυαίνονται πρὸς τὸ ἴσχιον,
καὶ ἀνάγκη τὸ στόμα ἀπεστράφθαι καὶ ἀνωτέρω οἴ-
χεσθαι· ὅταν δὲ ἀποστραφῇ, ξυμμῦσαι, ἐκ δὲ τοῦ
ἀπεστράφθαι τε καὶ ξυμμῦσαι, σκληρὸν γενέσθαι,[95]

[90] ὡς τὸ πολὺ om. MV.   [91] Add. ταύτην MV.
[92] παρασημήνῃ Θ: παρασπασαμένη MV.
[93] τὰ λοιπὰ Θ: τὰ ἄλλα MV.   [94] ἀφιστέωνται . . .
πάλιν om. Θ.   [95] ὅταν δὲ ἀποστραφῇ . . . γενέσθαι om. Θ.

When the uterus has settled down, give a purgative potion and ass's milk, unless the patient has some disorder of her spleen. Then foment and moisten the uterus with a preparation of laurel, and apply a cleaning, nonirritating suppository. Then fumigate the woman with aromatics and have her go to her husband: if she becomes pregnant, she recovers. Such women are barren, and generally become lame in their legs.

23. In women whose (uterine) mouth inclines in one direction and falls against a hip—this happens also to be an impediment to the uterus being cleaned and the seed being received, so that sterility results: foment with fragrant agents, and after the fomentation, if it is indicated,[19] remove the uterus from the hip with a finger. Then straighten it out with pine sticks and a lead sound, since, as has been said,[20] it cannot be forced rapidly. When it is in its natural position and its mouth is open, clean with emollient suppositories, and do whatever else is to be done.

24. When a woman's uterus falls against a hip—unless it quickly moves away and settles back in its normal location—it withers next to the hip, and its mouth of necessity turns away and moves upward; when it has turned away and closed, this compels it to become hard and closed, and

[19] Fuchs (n. 48) argues that the MV text ("drawing aside") refers here to the midwife and that Θ's "if it is indicated" represents an early scholarly conjecture.

[20] Emerins explains the absence of any such statement in *Diseases of Women* by the compilatory nature of this book.

καὶ μύειν, καὶ πεπηρωμένον τὸ στόμα τῶν ὑστερέων[96]
εἶναι· καὶ ἀποκλεισθέντα ἐπαναπέμπει τὰ ἐπιμήνια ἐς
τοὺς μαζούς, καὶ βεβρίθασιν οἱ τιτθοί·[97] καὶ ἡ γαστὴρ ἡ νειαίρα ἐπήρται, καὶ δοκέουσιν αἱ ἄπειροι ἐν
282 γαστρὶ | ἔχειν· πάσχουσι γὰρ τοιαῦτα οἷά περ καὶ αἱ
κύουσαι μέχρι μηνῶν ἑπτὰ ἢ ὀκτώ· ἥ τε γὰρ κοιλίη
ἐπιδιδοῖ κατὰ λόγον τοῦ χρόνου, καὶ τὰ στήθεα ἐπαίρεται, καὶ γάλα δοκέει ἐγγίνεσθαι· ὅταν δὲ οὗτος ὁ
χρόνος ὑπερπέσῃ, οἱ τιτθοὶ συνισχναίνονται καὶ γίνονται ἐλάσσονες, καὶ ἡ κοιλίη τωὐτὸ πάσχει· καὶ τὸ
γάλα ἀποδέδρακεν ἄδηλον, καὶ ἡ κοιλίη ἐπ᾽ ἐκεῖνον
τὸν χρόνον, ὃν χρὴ τίκτειν, ἐπειδὰν ἔλθῃ, ἀπόλωλε
καὶ ξυμπίπτει. τοιούτων δὲ γινομένων, αἱ ὑστέραι
ἰσχυρῶς χρόνον ὀλίγον συνέρχονται, καὶ τὸ στόμα
ἐξευρεῖν οὐχ οἷόν τε, οὕτω πάντα ξυνειρύονταί τε καὶ
ξυναναίνονται.

Καὶ ἐν τοῖσι τιτθοῖσι φυμάτια γίνεται σκληρά, τὰ
μὲν μείζω, τὰ δ᾽ οὔ· καὶ οὐκ ἐκπυοῦνται, σκληρότερα
δὲ αἰεί, καὶ[98] ἐξ αὐτῶν φύονται καρκίνοι κρυπτοί. μελ-
λόντων δὲ καρκίνων ἔσεσθαι, πρότερον τὰ στόματα
ἐκπικραίνονται, καὶ ὅ τι ἂν φάγωσι πάντα δοκεῦσι
εἶναι πικρά, καὶ ἤν τις πλέονα δῷ, ἀναίνονται·[99] καὶ
σχέτλια δρῶσι, παράφοροι δὲ τῇ γνώμῃ, καὶ οἱ
ὀφθαλμοὶ σκληροί, καὶ βλέπουσιν οὐκ ὀξέα· καὶ ἐκ
τῶν τιτθῶν ἐς τὰς σφαγὰς ὀδύναι διαΐσσουσι καὶ ὑπὸ
τὰς ὠμοπλάτας· καὶ δίψα ἴσχει, καὶ αἱ θηλαὶ καρφα-

___

[96] τὸ στόμα . . . ὑστερέων om. Θ.

the mouth of the uterus to become incapacitated. It sends
the obstructed menses up into the breasts, and they become heavy. The lower belly rises up, and inexperienced
women believe they are pregnant, because they are expe-
riencing the kind of things pregnant women do until their
seventh or eighth month: the cavity increases gradually as
time passes, the chest rises, and milk seems to be being
formed. When this time is reached, the breasts dry out and
decrease in size, and the belly does the same. Any milk
vanishes, and the belly—when the time is reached when
it should give birth—is undermined and collapses. As
these events occur, the uterus closes forcefully in a short
time, and its mouth cannot be made out because it is
completely drawn together and dried up.

In the breasts hard growths form, some quite large and
others not so large: these do not expel pus, but they remain
very hard, and out of them hidden cancers develop. As
these cancers are developing, patients first have a bitter
taste in their mouths, and everything they eat seems to be
bitter; if someone gives them very much to eat they refuse
it.[21] They do rude things, they become deranged in their
mind, their eyes become hard, and their vision is unclear.
Pains shoot up from their breasts to their throats, and
also under their shoulder blades; they suffer thirst, and

---

21 With the Θ text: "they vomit it up."

---

97 βεβρίθασιν . . . τιτθοί Θ: βεβηότας τοῦ στήθους ποιέει
M: βεβηότας τοὺς τιτθοὺς ποιέει V.
98 τὰ δ᾽ οὐ καὶ . . . καὶ Θ: τὰ δὲ ἐλάττω· ταῦτα δ᾽ οὐ
γίνονται ἔμπυα, ἀλλ᾽ αἰεὶ σκληρότεραι γίνονται· εἶτ᾽ MV.
99 ἀν. MV: ἀνάγονται Θ.

λέαι, καὶ πᾶσα λεπτύνεται,[100] καὶ αἱ ῥῖνες ξηραί τε
καὶ ἐμπεπλασμέναι εἰσίν, οὐκ ἀειρόμεναι· πνεῦμα μι-
νυθῶδες, ὀδμῶνται δ' οὐδέν, καὶ ἐν τοῖσιν οὔασι πόνος μὲν οὐκ ἐγγίνεται, πῶρος δ' ἐνίοτε. ὁπόταν οὖν ἐς
τόσον προΐωσι τοῦ χρόνου, οὐ δύνανται ὑγιέες γίνε-
σθαι, ἀλλ' ἀπόλλυνται ἐκ τούτων | τῶν νοσημάτων·
ἢν δὲ πρότερον ἢ ἐς τόσον ἀφικέσθαι θεραπεύηται,
λύεται τὰ ἐπιμήνια καὶ ὑγιὴς γίνεται.

Θεραπεύειν δὲ τὰς τοιάσδε ὧδε χρή· πρῶτον μὲν
ἀπάντων, ἢν ἔτι ἰσχυρὴν ἐοῦσαν λαμβάνῃς, ἀποσκεψάμενος[101] ἐς τὸ ἄλλο σῶμα φαρμακεύειν ὁποίης ἂν
τινος δοκέῃ δεῖσθαι καθάρσεως· ὁκόταν δὲ καταστή-
σῃς τὸ σῶμα,[102] ὧδε ἰέναι ἐπὶ τὴν τῶν ὑστερέων θεραπείην. ἢν δὲ μηδὲν δοκέῃ τὸ πᾶν σῶμα κινητέον
εἶναι, μηδὲ αἱ προφάσιες ἐντεῦθεν ὡρμῆσθαι, ἀλλ' αἱ
ὑστέραι ἀφ' ἑωυτῶν τὸ νόσημα ἔχωσι, τῆς θεραπείης
ἐνάρχεσθαι ὧδε· πυριὴν πρῶτον τὰς ὑστέρας ὧδε·
χύτρινον λαβόντα ὅσον δύο χοέας[103] χωρέοντα, κάνειον ἐπιθεῖναι ἀπαρτίζων ὅπως μὴ παραπνεύσῃ.[104]
ἔπειτα δ' ἐκκόψαι τοῦ κανείου τὸν πυθμένα, καὶ ποιε-
ειν ὀπήν· ἐς δὲ τὴν ὀπὴν ἐνθεῖναι κάλαμον, ὅσον
μῆκος πηχυαῖον· ἐνηρμόσθαι δὲ χρὴ τὸν κάλαμον τῷ
κανείῳ καλῶς,[105] ὅπως μὴ παραπνεύσεται μηδαμῶς·
ὁπόταν δὲ ταῦτα σκευάσῃς, ἐπίθες τὸ κάνειον ἐπὶ τὸν

---

[100] πᾶσα λεπτύνεται Θ: αὗται πᾶν τὸ σῶμα λελεπτυσμέ-
ναι εἰσί MV.   [101] χρή· πρῶτον . . . ἀποσκεψάμενος MV:
ἀποσκεψάμενος ἢ χρὴ καὶ Θ.

their nipples become parched. Such patients become thin through their whole body, and their nostrils are dry, blocked, and contracted. Breathing decreases, the sense of smell is lost, and although there is no pain in their ears, sometimes a stone forms there. Now when this stage has been reached, sufferers are unable to regain their health, but perish from these conditions. If a woman is treated before she has gone so far, her menses will be released, and she will recover.

Such cases must be treated as follows. First of all, if the patient is still strong when you attend her, while keeping an eye on the rest of her body give a medication to bring about whichever cleaning you judge necessary. When you have restored the body (sc. as a whole), proceed to treat the uterus. If it happens that the whole body shows no signs at all that it requires being set in motion, and the causes of the disease do not seem to have operated from there, but the disease seems to be confined to the uterus itself, then initiate treatment by first fomenting the uterus thus: take an earthenware pot holding two choes and set a lid on top of it, adjusting it so that no vapor can escape from it. Then remove the edges of the lid, make a hole in it, and into the hole insert a pipe about a cubit in length: you must fit the pipe precisely to the lid in order to prevent any vapor at all from escaping. When this is prepared,

---

[102] καθάρσεως . . . σῶμα MV: φαρμακείης· καὶ Θ.

[103] δύο χοέας Θ: δύο ἐκτέας M: ἐκταίας δύο V.

[104] ἀπαρτίζων . . . παραπνεύσῃ Θ: καὶ ξυναρτῆσαι ὅκως παραπνεύσεται (-σηται V) μηδέν MV.

[105] τὸν κάλαμον . . . καλῶς MV: τῷ κανείῳ Θ.

χύτρινον, περιπλάσαι πηλῷ, καὶ ὀρύξαι βόθρον,[106]
ὅσον δύο ποδῶν βάθος, μῆκος δὲ ὅσον χωρέειν τὸν
χύτρινον· ἔπειτα χρὴ καίειν ξύλοις, ἕως τὸν βόθρον
διάπυρον ποιήσῃς· ὅταν δὲ διάπυρος γένηται, ἐξελεῖν
τὰ ξύλα καὶ τοὺς ἄνθρακας, οἳ δὴ ἁδροὶ[107] ἔσονται καὶ

286 διάπυροι, τὴν δὲ σποδιὴν καὶ τὴν μαρίλην ἐν | τῷ
βόθρῳ καταλιπεῖν· ὁπόταν δὲ ὁ χύτρινος ζέσῃ καὶ
ἀτμὶς ἐπανίῃ, ἢν μεν εἴη λίην θερμὴ ἡ πνοιή, ἐπισχεῖν, εἰ δὲ μή, καθίζεσθαι ἐπὶ τὸ ἄκρον τοῦ καλάμου, καὶ ἐνθέσθαι ἐς τὸν στόμαχον, ἔπειτα πυριῆσαι·
ἢν δὲ ψύχηται, ἄνθρακας διαπύρους παραβάλλειν,
φυλασσόμενον, ὡς μὴ ὀξέη ἡ πυρίη· ἢν δὲ φαίνηται,[108] ἀφαιρέειν τῶν ἀνθράκων· τὴν δὲ πυρίην χρὴ
κατασκευάζειν ἐν εὐδίῃ τε καὶ νηνεμίῃ, ὡς μὴ ψύχειν.[109] ἀμφικεκαλύπτεσθαι δὲ χρὴ ἀμφιέσμασιν, ἐς
δὲ τὸν χύτρινον ἐμβάλλειν σκορόδια τῶν αὔων, καὶ
ὕδωρ ἐπιχέαι, ὥστε δύο δακτύλους ὑπερίσχειν, καὶ
καταβρέξαι ἄριστα, ἐπιχέαι δὲ καὶ[110] φώκης ἔλαιον·
καὶ οὕτω θερμαίνειν, πυριῆν δὲ πολλὸν χρόνον.

Μετὰ δὲ τὴν πυρίην, ἢν ᾖ δυνατή,[111] λουσάσθω τὸ

106 καὶ . . . βόθρον Θ: ὅταν δὲ ταῦτα ποιήσῃς, βόθρον
ὄρυξον MV.   107 ἁδρότατοι MV.

108 φυλασσόμενον ὡς μὴ . . . φαίνηται Θ: φυλασσόμενος,
ὅκως μὴ ὀξείην ποιήσῃς τὴν πυρίην· ἢν δὲ παραβαλλομέ-
νων τῶν ἀνθράκων ὀξείη γένηται ἡ πυρίη μᾶλλον τοῦ δέοντος MV.

109 εὐδίῃ . . . ψύχειν Θ: εὐδίῃσιν ὅκου ἄνεμος μὴ προσ-
πνεύσῃ μηδὲ προσψύξει (-ψύξῃ V) MV.

place the lid on the pot and seal it all around with mud, and dig a pit in the earth about two feet deep and just wide enough to hold the pot. Then make a fire in the pit with pieces of wood, continuing until the pit is thoroughly heated; when this has been achieved, remove the remaining wood and the solid, glowing coals, but leave the ashes and dust in the pit. When the pot is boiling and vapor is rising from it, if the vapor is too hot, pause, but if it is not too hot, sit the patient down over the end of the pipe, insert it into her orifice, and foment. If it cools down, add new burning coals, taking care that the fomentation does not become violent, but if it appears already to be so,[22] remove some of the coals. You should carry out the fomentation during good weather and in a protected location where no cooling will take place, and cover the patient all around with cloths. Into the pot put dried garlic germander, pour in water to cover this by two inches in order to steep it well,[23] and add seal's oil; heat in this way, and continue for a good while.

After the fomentation, if the woman is able, have her

---

[22] "the fomentation . . . to be so": MV read "you do not make the fomentation violent; but if when coals are added the fomentation becomes more violent than it should be."

[23] "pour in water . . . steep it well": MV read "break into pieces much of what has been burned and pour on water in order that it will be steeped by the water that is covering it by three inches."

---

[110] ὕδωρ ἐπιχέαι . . . καὶ Θ: πλείονα μὲν καταθρύπτειν τῶν (κατα- M)κεκαυμένων καὶ ὕδωρ ἐπιχέαι ὥστε αὐτὰ καταβρέξ(-ετ- M)αι καὶ ὑπερέχειν τὸ ὕδωρ τούτων οἷον τρεῖς δακτύλους καὶ ἐπιχεῖν MV.   [111] ἢν ᾖ δυνατή om. Θ.

μὲν ὅλον σῶμα πρὸς ἡδονήν, τὴν δὲ ὀσφῦν καὶ τὰ
κάτω τοῦ ὀμφαλοῦ πλέονι·[112] δειπνεῖν δὲ μᾶζαν ἢ ἄρτον καὶ σκόροδα ἐφθά, τῇ δὲ ὑστεραίῃ, ἢν μὲν διαλε-
λυμένη ᾖ πρὸς τὴν πυρίην, διαλιπεῖν τὴν ἡμέρην
ταύτην· ἢν δὲ μή, πυρίην ὀπίσω·[113] πυριωμένη δέ, ἢν
δύνηται σκέψασθαι,[114] κελεύειν ψαῦσαι τοῦ στόματος.
ἡ πυρίη αὕτη φύσης ἐμπίμπλησι τὰς ὑστέρας καὶ ἐς
ὀρθὸν μᾶλλον ἄγει καὶ ἀναστομοῖ.[115] ὁπόταν δὲ πυριήσῃς, ἐπεμβάλλειν χρὴ τῶν σκοροδίων, καὶ τῆς
φώκης τοῦ ἐλαίου ἐπιχεῖν· δρᾶν δὲ ταῦτα, ἄχρι ἂν
δοκέωσιν αἱ ὑστέραι πεφυσῆσθαι, καὶ τὸ στόμα ἄνω
εἱλκύσθαι ἰσχυρῶς· πρὸς γὰρ τὴν πυρίην ταύτην τοῖα
288 ἔσται· διαιτῆν δὲ μετὰ τὰς πυρίας, | ὡς ὅτε τὸ πρότε-
ρον ἐπυριήτο. ὅταν δ' ἀνακινηθέωσιν αἱ ὑστέραι καὶ
εὔλυτοι γένωνται, πυριῆν μαράθου ῥίζῃσιν, ἕτερον
χύτρινον κατασκευάσας τὸν αὐτὸν τρόπον, τὰς δὲ ῥίζας τῶν μαράθων ἀμφιπλύναντα φλάσαι, καὶ ἐς τὸν
χύτρινον ἐμβαλεῖν, καὶ ἐπιχέαι ὕδωρ καὶ τὸν αὐτὸν
τρόπον πυριῆν. μετὰ δὲ λούειν, καὶ ἐς ἑσπέρην δειπνεῖν μᾶζαν· ἢν δὲ ἄρτον βούληται καὶ βολβία καὶ
σηπίδια τῶν σμικρῶν, ἐν οἴνῳ δὲ πάντα ἑψεῖν καὶ
ἐλαίῳ. ἐν δὲ τῇ πυρίῃ πειρᾶσθαι τοῦ στόματος ψαύειν·

112 πλέονι Littré: πλεῖον εἶ Θ: πλέον Μ: πλέον εἰ V.
113 ὀπίσω om. Θ.   114 σκέψασθαι om. Θ.
115 ἡ πυρίη αὕτη . . . ἀναστομοῖ Θ: ἢν ἡ πυρίη αὐτέη
φύσης ἐμπιπλᾷ τὰς ὑστέρας, ἐμπιπλάμεναι δὲ φύσης τὸ
σῶμα ἐκ τοῦ λίην ἀπεστράφθαι καὶ προσπεπτωκέναι πρὸς
τῷ ἰσχίῳ, ὡς ὀρθὸν μᾶλλον καθεστᾶσι καὶ ἀναστομοῦνται.

wash her whole body as much as she pleases, and use a good amount (sc. of water) on her lower back and the region below her navel. Have her dine on barley cake or bread with boiled garlic, and on the next day cease treatment for a day if she has been relaxed by the fomentation; if not, foment again. After the patient has been fomented, if she is able to examine herself ask her to palpate the orifice of her uterus. These fomentations fill the uterus with wind, straighten it, and dilate it.[24] After you have fomented, inject some garlic and after that seal's oil: continue until the uterus seems to be inflated and its mouth is drawn forcefully upward; for the fomentation will achieve this. After the fomentations, return to the same regimen as before them. When the uterus has moved up and is free, prepare another earthen pot in the same way and foment with fennel roots: wash the fennel roots thoroughly, pound them, and place them in the pot; pour water over them, and then foment in the same way. Afterward have the woman take a bath and toward evening dine on barley cake. If she wants to have bread with small eledones and cuttlefish, boil these all together in wine and olive oil. During the fomentation, try to palpate the mouth of the

[24] "These fomentations fill . . . dilate it": MV read "If the fomentation fills the uterus with wind, and as it is being filled her body turns away due to the excess and her uterus falls against a hip, it will become straight and dilate; since fomentation is thus natural and able to bring these things about, you must foment."

ὡς οὖν τοιαύτης ἐούσης τῆς πυρίης καὶ ταῦτα ποιέειν δυναμένης, οὕτω πυριῆν χρή. MV.

αὕτη ἡ πυρίη ἄγει τὰς ὑστέρας ἄγχιστα· ταῦτα δὲ
χρὴ ποιέειν ἐφ' ἡμέρας πέντε ἢ ἕξ, σκεπτόμενον ἐς
τὴν ἄνθρωπον· ἢν μὲν διάλυτός τε[116] καὶ ἀσθενὴς γίνηται, διαλαμβάνων ὁπόσον ἄν σοι δοκέῃ χρόνον· ἢν
δὲ μὴ διαλύηται, πυρίην ἅπασαν ἡμέρην.

Μετὰ δὲ τὰς πυρίας πειρᾶσθαι προστιθέναι τῶν
προσθέτων τῆς σιάλου δαιδὸς τῆς πιοτάτης· χρίμα δὲ
λίπα ἔστω, ποιέειν δὲ ὧδε[117] μῆκος μὲν δακτύλων ἕξ,
πλῆθος δὲ πέντε ἢ ἕξ, εἶδος δὲ ἔξουρα· εἶναι δὲ θάτερον θατέρου σμικρῷ παχύτερον· τὸ δὲ παχύτερον εἶναι ὁπόσον δάκτυλος ὁ λιχανός, καὶ τὸ εἶδος ὅμοιον
δακτύλῳ ἐξ ἄκρου λεπτότατον, ἀγόμενον δὲ παχύτε-
ρον· ποιέειν ὡς λειότατόν τε καὶ στρογγυλώτατον,
φυλασσόμενον ὅπως σχινδαλαμὸς μηδεὶς ἐνέσται·
προστιθέναι δὲ πρῶτον τὸ λεπτότατον καὶ[118] ἀναπαυ-
έσθω ὡς μὴ ἐκπέσῃ φυλάσσουσα· προστιθέτω δὲ
πρῶτον τὸ ἄκρον, εἶτα αἰεὶ μᾶλλον, ὁμοῦ τε ἐπιστρέ-
φειν καὶ ἀπωθέειν κυκλόσε τὸ δαίδιον· καὶ ὅταν σμικρὸν προσδέξηται, ἐπισχεῖν ἐπὶ τῷ σμικρῷ τούτῳ,
290   φυλάσσουσα ὡς μὴ ἐκπέσῃ· εἶτα | αὖθις ἀπωθέειν τὸν
αὐτὸν τρόπον, ἄχρις οὗ τεσσέρων δακτύλων τὸ ἔσω
τοῦ στόματος τῶν ὑστερέων γένηται· ὅταν δὲ τοῦτο
προσδέξηται, τὸ μετὰ τοῦτο προστιθέναι, ὁμοῦ τὸ
προσκείμενον[119] ἀφαιρέουσα πρὶν ξυμπεσεῖν τὸ
στόμα, ὡς προσκείσεται θάτερον δαίδιον ὀρθοῦ ἔτι
ἐόντος καὶ ἀνεῳγμένου· οὕτω δὲ τοῦτο ἔσται, ἢν τὸ
μὲν ἕλκηται,[120] τὸ δὲ προστιθέαται.

uterus: this fomentation draws the uterus very close. You must foment for five or six days, and pay careful attention to the patient: if she becomes exhausted and weak you should stop treatment for as long as you think correct; if she does not become exhausted, foment every day.

After the fomentations, attempt to introduce (sc. into the uterus) pessaries made from very oily pine sticks: anoint these with a fatty unguent and prepare five or six of them six inches long and shaped like a cone; each one should be slightly thicker than its predecessor, and the thickest one should be the size of your index finger; shape them like a finger and make them narrower at one end and progressively thicker; make them very smooth and rounded, and be careful that they do not have splinters. First introduce the thinnest stick, and have the woman pause to make sure that it does not fall out; she should first insert the extremity of the stick and then more and more of it, simultaneously twisting it around in a circle and pressing it forward. When the stick is admitted a little, first stop with this small step to have the woman check that it has not fallen out, and then push it forward again in the same way until it is four inches into the uterine mouth. After this stick has been admitted, apply the one after it—while at the same time the woman is removing the previous one—before the mouth of the uterus closes, so that the new stick will be introduced with the mouth still remaining straight and open. This is how one is removed and the next one inserted.

---

116 διάλυτός τε Θ: διαλύηται MV.   117 ὧδε om. MV.
118 καὶ Θ: ὅταν δὲ προσθῆται MV.   119 προσκείμενον MV: πρὶν Θ.   120 ἕλκηται Θ: ἐξερέηται M: ἐξαιρέηται V.

Χρὴ δὲ καὶ μολύβδιον ἐξελάσαντα ἴκελον εἶδος
ποιῆσαι τῷ δαιδίῳ τῷ παχυτάτῳ,[121] κοῖλον δὲ ὅπως
ξυνέξει·[122] τοῦ δὲ μολυβδίου εἶναι τὸ εὖρος ἐληλασμέ-
νον οἷον ἐπὶ τὰ ἕλκεα·[123] ὅπως δὲ τὸ στόμα τοῦ μοτοῦ
λεῖον ἔσται καὶ μὴ τρώσει, ποιέειν ταῦτα ὡς καὶ τῶν
δαιδίων· ὁπόταν δὲ ποιηθῇ ὁ μοτὸς ὁ μολύβδινος,
στέατος αὐτὸν ἐμπλῆσαι οἶος τριπτοῦ· ὅταν δὲ παρ-
εσκευασμένος ᾖ, τὸ μὲν δαίδιον ὑφελεῖν, τὸ δὲ μολύβδιον ἐνθεῖναι. ἢν καῦμα δὲ παρέχῃ προστεθὲν τὸ
μολύβδιον, ἀφαιρέειν, τὴν δὲ δαῖδα ὀπίσω προστιθέναι, καὶ αὖτις[124] τὸ μολύβδιον ἀποβάψαι ἐς ὕδωρ ψυ-
χρόν· ἔπειτ' ὀπίσω προσθεῖναι, τὴν δὲ δαῖδα ἀφελεῖν·
προσκεῖσθαι δὲ χρὴ αἰεί τι· τὴν ἡμέρην ἄμεινον τὴν
δαῖδα ἔχειν, νύκτα τὸ μολύβδιον. ἢν δὲ ἀναστῆναι
θέλῃ, φυλασσομένη τοῦτο δράτω, ὅπως ὁ μοτὸς ἀτρε-
μέῃ· εἰ δὲ μή, αὐτίκα κατόπιν προστιθέναι. ἢν δὲ τῶν
δαιδίων, ἃ δὴ προσέκειτο, μηδὲν προσδέχηται, λεπτό-
τερον ποιέειν, μέχρις εὐπιθὲς ᾖ.[125] ἢν δὲ τὸ στόμα μὴ
δύνηται ἀνοίγεσθαι, μηδ' αἱ ὑστέραι ἀγχοῦ προσίω-
292 σιν, | αὖτις ἐπαναχωρέειν ἐπὶ τὴν ἀρχαίην πυρίην·[126]
ἀπὸ δὲ τῆς πρώτης ἐπὶ τὴν δευτέρην, μέχρις ὅτου
μαλαχθῶσι καὶ πελαστάτω προσίωσι.

Καὶ ἐπειδὰν ἀναστομωθῶσι, προσθεῖναι προσθετὰ
ποιήσαντα δύο,[127] εἶναι δὲ τὸ μέγεθος ὅσον ἐλαίη,

---

[121] παχυτάτῳ MV: κοιλοτάτῳ Θ.
[122] ξυνέξει MV: δὴ ἁρμονίη καλῶς ἕξει Θ.
[123] Add. ἐξελαύνεται MV.   [124] αὖτις om. MV.

Next hammer out a lead sound in the same way to make it like the thickest stick in shape, but hollow so that it can hold something.[25] The width of the lead sound after it has been hammered should be the same as the sound used for ulcers. In order that the mouth of the sound will be smooth and cause no damage, form it like the pine sticks. When the lead sound is fashioned like this, fill it with kneaded sheep's fat: then with this prepared pull the pine stick out and insert the lead sound. If on being inserted the lead sound provokes burning, remove it, put the pine stick back in place, and wash the sound off with cold water: then remove the stick and put the lead back in—there must always be something in place: during the day it is best for the patient to have the stick, and at night the sound. If she wishes to stand up, have her do this carefully so that the sound will not be disturbed: if it is disturbed, immediately reinsert it. If none of the sticks applied is admitted, make one thinner, until it does what you wish. If the mouth is impossible to open and the uterus does not come close enough, revert to the fomentation you used in the first instance, and then go from the first procedure to the second until the uterus becomes soft and approaches very close.

When the uterus has been dilated, make two suppositories and apply them—they should be the size of an olive.

---

[25] With the Θ text: "there will be a true harmony."

---

[125] εὐπιθὲς ᾖ Θ: οὗ προσδέξηται ἐκ τούτων ἐς ἐκεῖνα καθιστᾶναι (καθεστάναι V) καὶ εὐπιθῆσαι (εὐπειθήσει V) MV.
[126] πυρίην Θ: ἰητρείην MV.
[127] δύο om. MV.

τούτων δὲ τὸ ἕτερον προσκεῖσθαι, μέχρις ὅτου ἐκτακῇ,
καὶ ἔπειτα θάτερον προστιθέναι· ὅταν δὲ προστίθεσθαι μέλλῃς, βάπτειν[128] δὲ τὸ προσθετὸν τὸ ἐκ τῆς
ῥητίνης, ἀφίστησι γὰρ ἀπὸ τοῦ στόματος τῶν ὑστερέων, ἐς ἔλαιον ῥόδινον ἢ ἴρινον.[129] πρὶν δὲ τὸ μολύβδιον προσκεῖσθαι χρή, ὅπως ἐν ὀρθῷ τῷ στόματι
ἐρηρείσεται ἐς τὸ ἔσω τοῦ σώματος· ἔπειτα μαλθακτήρια προστιθέναι, λοῦσθαι δὲ θερμῷ καὶ πρὸ τῶν
προσθετῶν καὶ ὕστερον, καὶ δεῖπνον διδόναι τῶν θαλασσίων.[130] ὅταν δὲ ἡμέραι γένωνται δύο ἢ τρεῖς ἀπὸ
τῆς προσθέσιος, σκεψάμενος, ἢν μὲν τὸ στόμα καλῶς
ἔχῃ τῶν ὑστερέων καὶ καθαρὸν ᾖ, παύεσθαι,[131] μετὰ
δὲ τοῦτο μαλθακτήρια· ὅταν δὲ ἀποφλεγμήνῃ, αἰεὶ
προσκείσθω τὸ δαίδιον ἢ τὸ μολύβδιον· ἢν δὲ μήπω
καθαρὸν ᾖ[132] τὸ στόμα τῶν ὑστερέων, αὖτις προστιθέναι[133] τὰ φάρμακα, καὶ τἆλλα ποιέειν,[134] ὡς εἴρηται.
καὶ τὸ στόμα ἀναφυσᾶν χρὴ καὶ τὰς ὑστέρας ὧδε,
προσθετὸν ποιήσαντα τὸ σὺν τῷ σύκῳ προσθεῖναι·
ποιέειν δὲ δύο καὶ ταῦτα· πρὸ δὲ τοῦ προσθεῖναι πυριῆσαι[135] καὶ τῇ πυρίῃ τῇ ἐκ τοῦ μαράθου τῇ πρόσθεν
εἰρημένῃ·[136] μετὰ δὲ τὴν πυρίην τῇ ὑστεραίῃ προστιθέναι τὸ φάρμακον λούσαντα· λούειν δὲ καὶ ἐς

---

128 εἶναι . . . μέλλῃς (μέλης V) ἄψαι MV: ὡς (blank space
equivalent to five or six letters) κοτινάδα πυριάσαι δὲ πρὶν
θάτερον τούτῳ προσκεῖσθαι μέχρι ὅτου ἐκτακῇ· ὅταν δ'
ἐκτακῇ, καὶ θάτερον προστίθεσθαι· βάπτειν Θ.

129 ἢ ἴρινον om. Θ.   130 Add. τι τῶν εἰρημένων MV.

131 Add. τούτου τοῦ προσθετοῦ MV.

Apply one of these and wait until it melts, and then apply the second one: when you are about to make the application, immerse the suppository—which is to be made from resin—in rose oil or iris oil, in order that it will not stick in the mouth of the uterus. First, however, you must insert a lead sound so that the suppository will sit straight in the mouth (sc. of the uterus) and extend into its body. Then use emollient suppositories: have the patient bathe in hot water both before and after the application of the suppositories, and give her a dinner of sea foods. Two or three days after the application of the suppositories make an assessment, and if the mouth of the uterus is in a good state and clean, stop this treatment and after that use emollients; after the inflammation has disappeared, have the patient always retain a pine stick or a lead sound. If, on the other hand, the mouth of the uterus is not yet clean, apply medicinal suppositories again and take the other measures that have been described. You should also inflate the mouth and the uterus by making a suppository out of a fig and inserting it: make two of these. Before the insertion, foment the patient with the fennel fomentation described above; on the day after the fomentation, give the patient a bath and then apply the medicinal suppository; also wash her again toward evening. Immerse this sup-

---

132 αἰεὶ προσκείσθω . . . ῇ om. MV.

133 Add. τὸν αὐτὸν τρόπον MV.

134 Add. κατὰ τὸν ὑφηγημένον τρόπον· ὅταν δὲ καλῶς ἔχῃ τὸ σῶμα, ἀναφυσῆν MV.

135 πρὸ δὲ τοῦ . . . πυριῆσαι om. Θ.

136 τῇ πρόσθεν εἰρημένῃ MV: πυριῆσαι (blank space equivalent to fifteen or sixteen letters) Θ.

294 ἑσπέρην· | βάπτειν δὲ καὶ τοῦτο ἐς ἔλαιον, ὅταν προσ-
τίθηται· μετὰ δὲ τὴν κάθαρσιν τῇ ὑστερον[137] ἡμέρῃ
μαλθακτήρια ἀρήγει ἄχρι ἂν φλεγμήνωσι· μετὰ δὲ
τὴν φλεγμασίην ποιέειν τὴν δαΐδα[138] κατὰ τὸν ἔμπροσθεν λόγον· ποιέειν δὲ καὶ περὶ τοῦ προσθέτου τούτου
τοιαῦτα, οἷα καὶ περὶ τοῦ προτέρου. καὶ ἢν μὲν ἅπαξ
προστεθὲν ἀρκεόντως δοκέῃ ἀναφυσῆσαι τὰς ὑστέρας, πεπαῦσθαι· ἢν δέ τι χρήζῃ, αὖτις προστιθέναι
τὸν αὐτὸν τρόπον ὅνπερ τὸ πρότερον.

Μετὰ δέ, ὅταν δοκέῃ σοι καιρὸς εἶναι, πυρίην
κατασκευάσαι τὴν ἐκ τῶν θυμιημάτων· σκευάσαι δὲ
ὧδε·[139] κύπαιρον κόψαντα κατασῆσαι ὅσον σκαφίδα,
καὶ κάλαμον μυρεψικὸν ἕτερον τοσοῦτον, καὶ σχοίνου
τῆς μυρεψικῆς ἴσον, καρδαμώμου[140] τε ἴσον, καὶ κυμίνου Αἰθιοπικοῦ, καὶ ἀννήσου,[141] καὶ πηγάνου ξηροῦ,
καὶ ὑπερικοῦ, καὶ μαράθου σπέρμα· ὅταν δὲ ταῦτα
παρασκευάσῃς, ἐγχέας ἐς τὸν χύτρινον οἴνου αὐστη-
ροῦ κοτύλας ἓξ ὡς εὐωδεστάτου λευκοῦ, καὶ ἐπιπάσαι
τῶν κεκομμένων ὅσον τεταρτημόριον, καὶ ἀναταρά-
ξαι· εἶτα ἐπιχέαι ὅσον τριώβολον ὁλκὴν μύρου Αἰγύπτιον κράτιστον,[142] ἢ ἀμαράκινον, ἢ ἴρινον ἄκρον·
ὅταν δὲ ἐπιχέῃς, ἀναταράξαι·[143] πυριᾶν δὲ μαλθακῇ
πυρίῃ πλέονα χρόνον ἡμέρας δύο·[144] λούειν δὲ πρὸ
τῆς πυρίης·[145] ἐπὴν δὲ παύσηται πυριωμένη, ἐπίθημά

---

137 λούσαντα· λούειν . . . ὕστερον Θ: ὅταν δὲ μέλλῃ προστιθέναι πρὸ πάντων τῶν προσθετῶν λούσαντα, οὕτως προ-
στιθέναι· τῇ δὲ ὑστεραίῃ MV.

pository in olive oil when you insert it. On the day after
the cleaning, emollients are beneficial, and for as long as
the inflammation lasts; after the inflammation ends, use
the pine stick as indicated above, and do the same things
along with this application as you did with the ones above.
If one application seems to inflate the uterus sufficiently,
cease the treatment, but if more inflation is still needed,
apply treatment again in the same way as before.

Afterward, when you think it the right time, prepare a
fomentation made from fumigants as follows: grind and
sift a bowl of galingale, the same of aromatic reed, a like
amount of aromatic rush, the same of cardamom, and also
of Ethiopian cumin, anise, dry rue, hypericum, and fennel
seed. With these prepared, pour six cotyles of dry, very
fragrant, white wine into an earthen pot, sprinkle one
quarter this amount of the powdered ingredients over it,
and stir together. Then add three obols of Egyptian unguent of the strongest kind, or of sweet marjoram, or of
fine iris oil, and stir. Apply a gentle fomentation of this for
an extended time over two days, after bathing the patient
before the fomentation. When the fomentation is completed, you should place a lid over the preparation in order

---

138 τὴν δαῖδα Θ: τὰ ἄλλα MV.

139 σκευάσαι δὲ ὧδε Θ: σκευάζεται δὲ τόνδε τὸν τρόπον·
κατασκευάζειν δὲ τὴν ἐκ τῶν θυμιημάτων πυρίην ὧδε (τόνδε
τὸν τρόπον V) MV.   140 καρδαμώμου ΘV: καρδάμου M.

141 ἀννήσου MV: ἀννήθου Θ.

142 κράτιστον Θ: ὡς βέλτιστον MV.

143 ὅταν δὲ . . . ἀναταράξαι om. Θ.

144 ἡμέρας δύο om. MV.

145 Add. πυριῆν δὲ ἡμέρας δύο ἢ τρεῖς MV.

τι χρὴ ἐπικεῖσθαι ἐπὶ τῇ πυρίῃ, ὅπως μὴ ἀποπνέῃ·
μετὰ δὲ τοῖσι δαιδίοισι καὶ τῷ μολύβδῳ χρῆσθαι.
ὅταν δὲ πυριήσῃς ἡμέρας δύο ἢ τρεῖς, ἑτέρας διαλιπεῖν δύο ἢ τρεῖς· τὰς μεταξὺ ἡμέρας λούσασθαι δὶς
τῆς ἡμέρης.[146] | ἐσθίειν δὲ πράσα ἑφθὰ καὶ ὠμὰ καὶ
σίσυμβρον[147] καὶ ῥαφανῖδας καὶ κάρδαμον καὶ σκόροδα ἑφθὰ καὶ ὀπτά, καὶ τὸ πῶμα ἄκρητον, καὶ τοῖσι
θαλασσίοισι σιτίοισι χρῆσθαι· καὶ αὖτις[148] πυριᾶν
χρὴ ἐπιπάσαντα ὀλίγιστα ἑκάστου τῶν κεκομμένων
ἐς τὴν ὑπάρχουσαν πυρίην, καὶ οἶνον ἐπιχέαι· καὶ
ἔλαιον,[149] ἢν δοκέῃ δεῖσθαι, τούτῳ πυριᾶν. ὅταν δέ σοι
δοκέῃ καλῶς ἔχειν τῆς πυρίης, καὶ τὸ στόμα μαλθακὸν εἶναι καὶ ἀνεστομωμένον ὡς χρή, καὶ αὖται ἐγγὺς
αἱ ὑστέραι, μετὰ τὴν πυρίην τῇ ὑστέρῳ ἡμέρῃ προστιθέναι τὸ φάρμακον τὸ ξὺν τῇ σμύρνῃ· προσθετὰ δ'
εἶναι δύο.

Μετὰ δὲ τὴν κάθαρσιν τῇ ὑστεραίῃ τὰ μαλθακτήρια προστιθέναι, μέχρι ὅτου ἀποφλεγμήνωσιν αἱ
ὑστέραι· καὶ[150] τοῖσι δαιδίοισι[151] χρῆσθαι· τὴν δὲ[152]
δύναμιν τεκμαιρόμενον τοῦ σώματος, αὖτις ἀναπυ-
ριᾶν ὡς γέγραπται· ἢν δὲ δοκέῃ ἡ πυρίη ἔχειν
φλαύρως,[153] ἑτέρην ἀρχῆθεν ποιέειν. ὅταν δὲ προπυ-
ριάσῃς αὖτις τὸν αὐτὸν τρόπον, ὅνπερ πρότερον προ-
επυρίασας, προστιθέναι τὸ φάρμακον τὸ σὺν τῷ στέ-
ατι· ποιεῖν δὲ δύο τὰ προσθετά, καὶ προστιθέναι τὸν

[146] ὅταν δὲ πυριήσῃς . . . ἡμέρης om. Θ.
[147] σίσυμβρον om. MV.

that no vapor will be given off; then make use of the pine sticks and the lead sound. When you have fomented for two or three days, then take a break for another two or three days: during these intermediate days have the patient wash twice daily, eat leeks both boiled and raw, watercress, radishes, cress, garlic both boiled and baked, take an undiluted wine, and employ sea foods; you must foment again by sprinkling very small amounts of each of the powdered ingredients on to the fomentation already prepared, and adding wine; if you think oil is needed as well, foment with this. When you think the fomentation is working well, the mouth is as soft and open as it should be, and the uterus is close, on the day after the fomentation apply medicinal suppositories of myrrh: let there be two of these.

On the day after the cleaning apply emollient suppositories and continue until the inflammation of the uterus ends, and also use the pine sticks. Estimate the strength of the patient's body, and foment again as indicated. If the fomentation seems to being going badly, go through a new fomentation from the beginning: when you have carried out a preliminary fomentation again as before, apply a medicinal suppository made with fat: make two of these

---

148 σιτίοισι . . . αὖτις Θ: τοῖσι προειρημένοισι χρέεσθαι. ὅταν δὲ αἱ ἡμέραι διαλιπόμεναι παρέλθωσι MV.

149 καὶ ἔλαιον transf. MV after δεῖσθαι.

150 καὶ Θ: μετὰ δὲ τοῦτο MV.   151 Add. καὶ μολυβδίοισι MV.   152 τὴν δὲ Θ: μετὰ δὲ ταῦτα διαλιπεῖν ἡμέρας ὡσὰν (ὅσας V) ἂν δοκέῃς εἶναι τὰς φλεγμασίας ἀποσκηπτόμενον τῶν ὑστερέων καὶ τὴν MV.   153 ἡ πυρίη ἔχειν φλαύρως Potter: ἡ πυρίη εἶναι φλαῦρος Θ: om. MV, see next note.

αὐτὸν τρόπον, ὅνπερ πρότερον· μετὰ δὲ τὰς προσθέσιας τῇ ὑστέρῃ ἡμέρῃ κατὰ τὸν ἔμπροσθεν λόγον
τὰ μαλθακτήρια προστιθέναι· τούτῳ δὲ τῷ φαρμάκῳ
χρῆσθαι δι᾽ ἡμέρας τετάρτης προπυριῶντα·[154] αἰεὶ[155]

298 δὲ | νεοχμὸν ποιέειν τὸ φάρμακον ἐφ᾽ ἑκάστῃ προσ-
θέσει, μέχρι ἂν ὕφαιμα μὴ[156] καθαίρηται· ὅταν δὲ
τοιαῦτα καθαρθῇ, πεπαύσθω· ὅταν δὲ παύσηται,[157]
διαιτᾶν δὲ τούτῳ τῷ τρόπῳ· ἀνερωτᾶν ἐν ὁκοίῃσιν
ἡμέρῃσιν αὐτῇ ἐγίνετο τὰ ἐπιμήνια· ἀπὸ τούτων ἀρ-
ξάμενος διαιτᾶν τῇδε τῇ διαίτῃ· πρῶτον μὲν λούσθω
πολλῷ θερμῷ, πλὴν τῆς κεφαλῆς· ὅταν δὲ παύσηται,
δοῦναι τυρὸν δριμὺν καὶ πήγανον οἴνῳ διέντα μέλανι,
ἴσον ἴσῳ κεκρημένον· ἐπὶ τοῦτον ἄλφιτα ἐπιπάσαντα
πιεῖν δοῦναι ἐκ τοῦ λουτροῦ εὐθέως· εἶτ᾽ ἄριστον δοῦναι ἐν καιρῷ, μᾶλλον[158] μᾶζαν ἢ ἄρτον καὶ πράσα
ἑφθὰ καὶ ὠμά, καὶ τῶν ὁμοιοτρόπων δριμέων πάντων
τῶν προγεγραμμένων,[159] καὶ ἔτνος ἄλλοτε καὶ ἄλλοτε,
τὸν δὲ ἀφρὸν μὴ ἀφαιρέειν, καὶ σίλφιον πολὺ ἐγ-
κλῶντα, καὶ σκόροδα ἐγκαθεψῶντα πολλά· ὄψοισι δὲ
χρῆσθαι τοῖσι σελάχεσι πᾶσιν ἐφθοῖσιν ὀξυγλύκεσι,
βολβιδίοισι, πουλυποδίοισι, σηπιδίοισιν[160] ἐν οἴνῳ

154 ἡ πυρίη (330, 21) . . . προπυριῶντα Θ: ἔτι ὑπάρχουσα
πυρίη ἱκανὴ εἶναι, ἐς ταύτην ἐκβαλὼν τῶν θυμιημάτων
ὥσπερ τὸ πρότερον, καὶ οἶνον ἐπιχέον (-χέων V) καὶ ἔλαιον·
ἢν δοκέῃ αὕτη ἄχρηστος εἶναι ἡ πυρίη, ἑτέρην ἐξ ἀρχῆς
κατασκευάζειν MV.   155 αἰεὶ Θ: δεῖ MV.
156 μὴ om. MV.   157 ὅταν . . . παύσηται om. MV.
158 μᾶλλον Θ: ἐσθίειν δὲ MV.

and apply them the same as the ones before. On the day after the treatment, apply emollients according to the method above, using this medication for four days after the preliminary fomentation.[26] Always make the medication fresh for each application, and continue the series until the discharge is no longer[27] bloody. After the cleaning has taken place, have the patient stop the treatment and then employ a regimen like this: ask her on which days her menses occurred, and beginning from these establish the regimen. First have her bathe in copious hot water except for her head; when she has finished this, take strong cheese with rue mixed in dark wine that has been diluted equally with water: sprinkle barley meal over this, and give it to her to drink immediately on coming from the bath. Then give her breakfast at the appropriate time, having her eat barley cake rather than bread, boiled and raw leeks, and all the same pungent agents indicated above, and also from time to time a soup on which you have left the foam: whip into this much silphium and boil into it copious garlic. As main dish have the patient employ all the selachians boiled in sweet and sour sauce, and eledones, octopuses, and small cuttlefish in wine and oil; she

[26] "If the fomentation seems . . . preliminary fomentation": MV read "If the fomentation seems to be adequate, add to it some of the fumigants used above, and pour on wine and olive oil, but if this fomentation seems to be ineffective, prepare another one from the beginning."

[27] MV omit "no longer."

---

[159] πάντων τῶν προγεγραμμένων om. Θ.
[160] σηπιδίοισιν om. MV.

καὶ ἐλαίῳ· φύλλον ὑποτετριμμένον πίνειν ὡς πλεῖστα,
καὶ ἐσθίειν ὡς πλεῖστα καὶ ἐπ᾽ ἀρίστῳ καὶ ἐπὶ δείπνῳ·
λοῦσθαι δὲ μετὰ τὸ δεῖπνον, ὅταν μέλλῃ ἀναπαύεσθαι.

Αὕτη ἡ δίαιτα μέχρι ἡμερῶν πέντε ἢ ἕξ· μετὰ δὲ
ταύτας τὰς ἡμέρας[161] πρωὶ μὲν διδόναι νήστει ἀκτῆς
καρπὸν ὅσον πυρῆνας ἓξ ἐν οἴνῳ ἀκρήτῳ καὶ σηπίης
ᾠὰ ὅσον δέκα ἢ δυοκαίδεκα· ταῦτα τρίψαντα ὁμοῦ
λεῖα πρωὶ διδόναι νήστει πίνειν.[162] καὶ μετὰ τὴν πόσιν
ἐπισχοῦσαν λοῦσθαι, καὶ πιεῖν τὸ πήγανον καὶ τὸν
τυρόν, καὶ οὕτως ἀριστᾶν τι τῶν προγεγραμμένων·
δειπνεῖν δὲ ὀψιαίτερον, λούεσθαι δὲ χρὴ δὶς τῆς ἡμέ-
ρης· αὕτη ἡ δίαιτα ἡμερέων τρισκαίδεκα ἢ τεσσερεσ-
καίδεκα. ὅταν δὲ αὗται αἱ ἡμέραι διέλθωσι, καὶ κατά-
300 ποτα ποιέειν ἐκ τοῦ ὀποῦ τοῦ | σιλφίου, ὅσον κύαμον,
καὶ διδόναι τοῦτο τὸ πρῶτον· μετὰ δὲ τἆλλα τὰ προ-
γεγραμμένα ποιέειν. ἐπὴν δ᾽ ἡμέραι γένωνται πέντε
καὶ εἴκοσιν ταύτῃ τῇ διαίτῃ, τὰ μὲν ἄλλα ποιέειν
κατὰ ταὐτά, πρὸ δὲ τοῦ ἀρίστου, ὅταν μέλλῃ ἀριστᾶν,
προτρίψαι[163] σκορόδου ἄγλιθας ὅσον τέσσερας, καὶ
τυρὸν δριμὺν ὅσον ἀστράγαλον, καὶ ἄλφιτα παραμῖξαι ὀλίγα, καὶ ποιῆσαι μαγίδα· ταύτην δὲ πρῶτον
καταφαγεῖν, πρὸ δὲ τοῦ δείπνου ἄκρητον ἐπιρροφεῖν,[164] καὶ οὕτω σιτεῖσθαι· τὰ δ᾽ ἄλλα τὴν αὐτὴν διαιτᾶν. ὅταν δ᾽ ἐννέα ἡμέραι ἢ δέκα[165] γένωνται διάλοι-

---

[161] ὅταν μέλλῃ ἀναπαύεσθαι. Αὕτη ἡ δίαιτα (Ταύτῃ τῇ
διαίτῃ M) . . . ἡμέρας MV: ἔπειτα Θ.

should drink as many macerated leaves (sc. of silphium) as possible, and eat as much as she can at breakfast and dinner. After her dinner have her bathe when she is about to take a rest.

This regimen until the fifth or sixth day; after these days give the patient at dawn before she has broken her fast six elderberries in undiluted wine and ten or twelve cuttlefish eggs: grind all these together smooth, and give them to her fasting at dawn. Have her wait a while after taking the potion, and then bathe, drink the rue and cheese, [28] and breakfast on some of the foods indicated above; she should dine later, and bathe twice daily. Follow this diet for thirteen or fourteen days. After these days have passed, make a pill from silphium juice to the amount of a bean and first give this; then do the rest of what is indicated above. After following this regimen for twenty-five days, continue the rest of it in the same way, but before breakfast when the patient is about to eat, first crush four cloves of garlic, add strong cheese to the amount of a vertebra, mix this together with a little barley meal, and bake this as a cake; first have the patient eat this, then before her dinner take some undiluted wine, and after that the meal—but otherwise follow the same regimen. During the nine or ten days remaining until the time (sc. of

---

[28] Cf. the preceding paragraph: "strong cheese with rue."

---

[162] Om. νήστει πίνειν Θ.
[163] ὅταν μέλλῃ ἀριστᾶν προτρῖψαι MV: τρίψας Θ.
[164] πρὸ δὲ . . . ἐπιρροφεῖν om. Θ.
[165] ἢ δέκα om. MV.

ποι ἐς τὸν χρόνον, διδόναι[166] καὶ μετὰ τῶν ᾠῶν καὶ
τῆς ἀκτῆς κύμινον Αἰθιοπικόν, καὶ τοῦ κάστορος
ὅσον ὀβολόν. ὅταν δὲ ἡμέραι λοιπαὶ δύο ἔωσι, τούτων
μὲν ἀπαλλαγῆναι πάντων τῶν πωμάτων καὶ τῶν
καταποτίων, τὸ δὲ σὺν τῇ δαιδὶ φάρμακον διδόναι
νήστει λουσαμένη· μετὰ δὲ τοῦ φαρμάκου τὴν πόσιν
διδόναι λινόζωστιν καὶ κράμβην ὁμοῦ ἕψοντα ἐν
ὕδατι, ἡδύναντα ὀξεῖ καὶ γλυκεῖ καὶ ἁλσὶ καὶ σιλφίῳ
καὶ ἐλαίῳ ἐπὶ τῷ ἀρίστῳ· ταῦτα δίδου καὶ αὐτὰ τρώγειν καὶ τὸν χυλὸν ῥοφέειν,[167] καὶ τὸ πῶμα ἀκρητέ-
στερον πίνειν· ὄψοισι δὲ χρῆσθαι πουλυποδίῳ ἑφθῷ,
ἢ σηπιδίοισι. ταῦτα μὲν ἐπὶ τῷ ἀρίστῳ, ἐπὶ δείπνῳ δὲ
κρέας ἢ αἰγὸς ἢ ὅϊος ἢ ἀρνὸς δίεφθον, καὶ πράσα, καὶ
τῶν ἄλλων δριμέων ὅ τι ἂν βούληται· λουσάσθω δ᾽
ἀπὸ τοῦ δείπνου· αὕτη ἡ δίαιτα τὰς δύο ἡμέρας τὰς
ὑστάτας.

Ἢν δὲ πρὸς τὴν δίαιταν ταύτην μὴ κατασπασθῇ
τὰ ἐπιμήνια, τὸν ἐπιόντα μῆνα τὴν αὐτὴν δίαιταν ἐξ
302   ἀρχῆς | διαιτᾶν ἄχρι τῶν ὑστάτων δύο ἡμερέων·
ταύτας δὲ τὰς δύο ἢ τῇ πρότερον ἡμέρῃ τῆς ὑστάτης
προσθετὰ ποιήσας, προσθεῖναι κατὰ τὸν ἔμπροσθεν
λόγον, τοῦ φαρμάκου τοῦ ἐν τῷ ὕδατι ποιευμένου·
προστιθέσθω δὲ προλούσας. σκέψασθαι δὲ καὶ τῶν
ὑστερέων χρὴ ὅπως ἂν ἔχωσιν αἰεὶ παρὰ πάντα τὸν
χρόνον, ὅπως καλῶς ἕξουσι, καὶ τὸ στόμα ὀρθόν τε
καὶ ἀνεστομωμένον. καὶ ἢν δοκέωσι πρὸ τῆς προσθέσιος προπυριατέαι εἶναι· ἢν δὲ καταρραγῇ τὰ ἐπιμήνια, ἢν μὲν συχνά, ἐλάσσοσι τοῖσι λουτροῖσι χρή-

her menses), also have her take along with the eggs and the elder some Ethiopian cumin and an obol of castoreum. When two days remain, cease all the drinks and pills, and give her the medication made from pine wood to take in the fasting state after having bathed. In conjunction with the medication give her as potion at breakfast mercury herb and cabbage boiled together in water and seasoned with vinegar, honey, salt, silphium, and the finest olive oil: give this to eat, along with its juice to drink. She should also drink straight wine; as main dishes, employ boiled octopus or small cuttlefish: this is for breakfast. For dinner give boiled meat of goat, sheep, or lamb, and also leeks and any other of the pungent vegetables the patient desires. After her dinner, she should bathe; this is the regimen for the final two days.

If with this regimen the menses are not drawn down, in the following month use the same regimen, starting from the beginning and continuing until the final two days. On these two days or on the next to final one make suppositories and apply them according to the plan above, using a medication made with water: have the patient apply the suppository to herself after she has taken a bath. You must also pay attention continually to the state of the uterus, checking that it is in a good condition, and that its mouth is straight and open. If it seems to require a fomentation before the suppository, if the menses have already come down and are heavy, employ fewer baths, but if their

---

166 Add. τοῦτο πρῶτον MV.
167 καὶ τὸν . . . ῥοφέειν om. Θ.

σθαι· ἢν δ᾽ ἐλάσσω, πλέοσιν· ἢν δὲ προσημήναντα
μὴ[168] ἴῃ, αὖτις τῇ διαίτῃ τῇ αὐτῇ ἐκθεραπεύειν, ἄχρι
ἂν φανῇ τὰ ἐπιμήνια· ὅταν δ᾽ ἅπαξ ἔλθῃ, τῇσι τοι-
αύτῃσιν ἄριστον ἐν γαστρὶ λαμβάνειν. αὕτη πασέων
τῶν ὁμοιοτρόπων νούσων θεραπείη.

25. (134 L.) Ἢν δ᾽ αἱ μῆτραι ψαύσωσι τοῦ ἰσχίου
καὶ προσκέωνται, στερρὸν γίνεται ὑπὸ τὸν κενεῶνα,
καὶ ὀδύναι νειαίρης γαστρός, καὶ ἐς τὸν κενεῶνα καὶ
ἐς τὰς ἰξύας καὶ ἐς τὰ σκέλεα ἡ ὀδύνη ἐσπίπτει, καὶ
τιταίνεται· καὶ ἐκπυΐσκονται, καὶ ἔμμοτοι γίνονται,
αἵδε ῥεόμεναι ὄλλυνται, ἢν μὴ καύσῃς ἢ τάμῃς. ὅταν
οὕτως ἔχῃ, φάρμακον πῖσαι κάτω, καὶ λούειν πολλῷ
θερμῷ· καὶ πυριῆν τὰς ὑστέρας, καὶ ἐγχέαι οὖρον παλαιὸν ἀναζέον ἐς τὸ κοῖλον τῆς πυέλου, καὶ ἀμφικα-
θίζεσθαι περικαλύψας εἵματι τὴν γυναῖκα, ὡς μὴ
παραπνέῃ· καὶ ἐπειδὰν ἀποψύχηται τὸ οὖρον, ἐμβάλλειν μύδρους διαπύρους,[169] καὶ πυριῆν μέχρι οὗ ἂν φῇ
304   ἀμαυρὰ βλέπειν | καὶ λιποθυμέειν· λοῦσθαι δὲ ἀπὸ
τῆς πυρίης θερμῷ ὕδατι· κἄπειτα ἁψαμένην τῷ δακτύλῳ[170] ἕλκειν τὸ στόμα τῶν ὑστερέων πρὸς τὸ ὑγιὲς
ἰσχίον, καὶ τὰς νύκτας προστίθεσθαι μαλθακτήρια·
ἐπὴν δὲ φῇ κατ᾽ ἰθὺ εἶναι, προπυριάσας τοῖς εὐώδεσιν, αὖτις προστιθέναι τὰ μαλθακτήρια, καὶ τοὺς μο-
λύβδους τρεῖς ἡμέρας, ἕνα ἑκάστης ἡμέρης· μετὰ δὲ
τὴν ἐχέτρωσιν ἢ τὴν σκίλλαν τρεῖς ἡμέρας. μετὰ δὲ
σκεψάμενος ἐν τοῖσιν ἐπιμηνίοισιν, ἤντε χολώδεα ᾖ,
ἤντε φλεγματώδεα, ἤντε αἷμα διεφθορὸς ᾖ καὶ δέῃ
αὐτὴν αἷμα καθῆραι, προστιθέναι, ὅτου ἄν σοι δοκέῃ

flow is less, use more baths. But if in spite of giving signs they are about to appear they fail to do so, rigorously apply the same regimen again until they actually do. If they once pass, it is best for such women to become pregnant. This is the treatment for all diseases of this kind.

25. If the uterus touches the hip and clings to it, a hardness arises below the flank, there are pains in the lower belly, and pain also invades the flank, loins and legs, becoming quite intense. Pus breaks out and requires tents: such women succumb from the fluxes unless you incise or cauterize them. When the case is such, give a purgative medication to drink, and wash with copious hot water. Also foment the uterus: pour boiling stale urine into the hollow of a tub, have the woman sit down over it, and cover her with garments in order that no vapor escapes. When the urine has cooled down, put red-hot stones into it and continue the fomentation until the patient says her vision is unclear and she is about to faint. After the fomentation, bathe her in hot water. Then have the patient take hold with a finger and draw the mouth of her uterus toward her healthy hip, and at night apply emollient agents. When she says the mouth is straight, first foment with fragrant substances and then reapply emollients, and also insert one lead sound a day for three days; after that apply bryony or squill for three days. Next examine whether the menses are bilious or phlegmy, or whether the blood is putrid and the patient needs to have it cleaned out, and apply whichever suppositories you think are most needed, and after

---

168 μὴ om. MV.
169 Add. ἐς τὸ οὖρον MV.
170 τῷ δακτύλῳ MV: τῶν δακτύλων Θ.

μάλιστα δεῖσθαι, καὶ μετακλύζειν τοιούτοισι· προστιθέναι δὲ τὰ προσθετά, ἔστ᾽ ἂν αἷμα καθαρὸν ἄγηται,
καὶ ταῦτα ἐνεργεῖν τρεῖς ἡμέρας. ἐλάφου δὲ προστιθέσθω στέαρ τηκτὸν ἐμβάπτων μαλθακὸν[171] εἴριον,
ἔπειτα γλήχωνα τὴν ἡμέρην· θυμιωμένη τοῖσιν ἀρώμασιν οὕτω παρὰ τὸν ἄνδρα ἴτω.

Ἢν δ᾽ ἐκ τόκου ἡ νοῦσος γένηται, κεκαθαρμένης
πάντα ἐκ τῆς πυρίης τοῦ οὔρου, αὐτίκα ἰέναι ἐς τὰ
εὐώδεα, κἄπειτα λουσαμένην ἄλειφα λευκὸν Αἰγύπτιον προστιθέσθω πρὸς τὸ ὑγιὲς ἰσχίον, καὶ κατακείσθω ἐπὶ τοῦτο. ἢν δὲ μὴ πρὸς ταῦτα μεταστῶσιν
αἱ ὑστέραι, πινέτω νῆστις γλυκυσίδης κόκκους τοὺς
μέλανας πέντε, ἐν οἴνῳ εὐώδει, καὶ ἐπὶ τῷ σίτῳ τρωγέτω σκόροδα ὠμὰ καὶ ἐφθὰ καὶ ὀπτά· ὄψοισι δὲ χρήσθω ὡς λεχώ.[172] ἢν δὲ μὴ γίνηται ὑγιής, ὡσαύτως
καθαίρειν αὐτὴν ὡς ἐπὶ τῆς πρόσθεν. ἐπὶ δὲ ταύτης
τῆς νούσου, ἢν μὴ αὐτίκα σχῇ ἐν γαστρί, ἄτοκος
γίνεται. ἢν δ᾽ ὧδε ἐχούσῃ τὰ ἐπιμήνια μὴ γίνηται, καὶ
πῦρ ἐπιλάβῃ, φαρμάκοισι πρῶτον καθαίρειν ἄνω μά-
306 λιστα, ἢν δὲ ἀσθενὴς ᾖ, | κάτω· καὶ μετὰ τὴν φαρμα-
κοποσίην, ἤν σοι δοκέῃ φλεγματώδης εἶναι, ἐμείτω
καὶ νῆστις καὶ ξὺν τῷ σίτῳ, καὶ ὡς τὰ πολλὰ ὑγιὴς
ἔσται.

26. (135 L.) Ἢν αἱ μῆτραι πρὸς τὸ ἰσχίον λυθῶσι,[173] τὰ ἐπιμήνια παχέα ἐόντα οὐ γίνεται· ὀδύνη δὲ

[171] τηκτὸν ἐμβάπτων (ἐμβάπτον M) μαλθακὸν MV: πηκτὸν
ἐμβάπτων εἰς μαλκὸν Θ.

that an injection of the following kinds—apply the suppositories until clean blood is discharged and then continue for three days—have the patient apply melted deer's fat in which a piece of soft wool has been immersed, and then pennyroyal during the day; after fumigating herself with the aromatic substances, she should approach her husband.

If the disease arises after childbirth, the woman should receive a general cleaning by means of a fomentation with urine; then immediately change over to fragrant substances, and then after she takes a bath have her apply white Egyptian unguent against her healthy hip, and lie down on that side. If in spite of these measures the uterus does not move back, have her drink in the fasting state five black peony grains in fragrant wine, and after her meal take garlic raw, boiled, and baked. As main dishes she should employ the same ones as a woman in childbed. If the patient still fails to recover, clean her the same way you cleaned the preceding patient. If in this disease the patient does not soon become pregnant, she will be sterile. If with her in this condition, no menses appear and fever sets in, first purge her with medications acting mainly upward, but if she is weak, downward. After she takes the potion, if you believe she is phlegmatic, have her vomit both with an empty stomach and with her meal: generally she will recover.

26. If a woman's uterus dislocates toward a hip, her menses become thick and cease to appear. Pain invades

---

172 λεχώ ΘV: ἐλαχίστοισι M.
173 λυθῶσι Θ: εὐανθῶσιν MV.

ἐς τὴν νειαίρην γαστέρα ἀφικνέεται· ἀφικνεῖται δὲ
καὶ ἐς τὸν κενεῶνα, καὶ δάκνεται. ὅταν ὧδε ἔχῃ, λούειν
πολλῷ καὶ θερμῷ, καὶ διδόναι σκόροδα ὡς πλεῖστα
τρώγειν, καὶ γάλα πίνειν ἱκνευμένως, ἔπειτα οἶνον
ἄκρητον· καὶ πυριᾶσαι ὅλην, καὶ φάρμακον δοῦναι
ἄνω· ἢν δ᾽ ἀσθενεστέρη ᾖ, κάτω. ἢν δὲ ἰηθῇ, πυριῆν
τὰς ὑστέρας μαράθῳ, ξύμμισγε δὲ καὶ ἀψίνθιον· ἐπὴν
δὲ νεοπυρίητος ᾖ, ἀφέλκειν τὸ στόμα τῷ δακτύλῳ
ἠρέμα τῶν ὑστερέων πρὸς τὸ ὑγιὲς ἰσχίον, παρηγορικῶς μαλθάσσοντα τὸ σῶμα καὶ τὰ ἀμφιπονεύμενα·
καὶ προσθεῖναί τι μαλθακτήριον, μετὰ δὲ τοὺς μο-
λύβδους, καὶ αὐτίκα σκίλλαν, ἔπειτα ναρκίσσινον
μίαν ἡμέρην διαλιπών. ἐπὴν δέ σοι δοκέῃ καθαρή τις
εἶναι, τὸ νέτωπον προσθέσθω ἐν εἰρίῳ· τῇ δ᾽ ὑστεραίῃ
ἔλαιον ῥόδινον.

Ἐπιμηνίων δὲ ἰόντων, ἄμεινον μὴ προστίθεσθαι·
ἢν δὲ μὴ ἴῃ, κανθαρίδας τέσσερας, ἀπτέρους καὶ ἄπο-
δας καὶ ἄτερ κεφαλῆς, καὶ γλυκυσίδης κόκκους πέντε
τοὺς μέλανας καὶ σηπίης ᾠά, καὶ σελίνου σπέρμα
ὀλίγον ἐν οἴνῳ διδόναι πίνειν· καὶ ἢν ὀδύνη ἐνῇ καὶ
στραγγουρίη ἔχῃ, ἐν ὕδατι θερμῷ ἐγκαθήσθω, καὶ
πινέτω μελίκρητον ὑδαρές· ἢν δὲ μὴ καθαίρηται,
αὖτις τὸ φάρμακον πινέτω· ἢν δὲ ἴῃ, ἀσιτήσασα πινέτω, καὶ ξυνέστω τῷ ἀνδρί. ἢν δὲ μὴ γίνηται, διδό-
308 ναι[174] | ὅ τι κατασπάσει, ὁρῶν πρὸς τὴν δύναμιν τῆς
γυναικός, καὶ τότε ἀσφαλὲς φοιτᾶν πρὸς τὸν ἄνδρα·
ἢν γὰρ ἔχῃ ἐν γαστρί, ὑγιὴς γίνεται. ἐν δὲ τῇ καθάρσει, ἢν ἴῃ πολλή, λινόζωστιν ἐσθιέτω, καὶ πουλύπο-

her lower belly and flank, and it gnaws. When the case is such, wash her with copious hot water, and give her a great amount of garlic to eat, the right amount of milk to drink, and then undiluted wine. Foment the patient's entire body and give her an emetic—but if she is weak, a purge. If she is cured, foment her uterus with fennel, and also add wormwood. Right after the fomentation gently move the mouth of her uterus with your finger in the direction of her healthy hip, and gradually relax her body and the parts next to those involved. Apply an emollient suppository, after that lead sounds, and then immediately squill; after leaving one day free, then narcissus oil. When you think the patient is clean, have her apply as suppository oil of bitter almond on a piece of wool, and on the following day rose oil.

If the menses are flowing, it is better not to use a suppository, but if they fail to do so give four blister beetles without wings, legs and head, five black grains of peony, cuttlefish eggs, and a little celery seed in wine to drink. If pain arises along with strangury, have the patient seat herself in hot water, and drink dilute melicrat. If this fails to clean her, have her drink the medication again; but if her menses do pass, have her fast, take a drink, and cohabit with her husband. If they still do not pass, give the patient another emmenagogue, keeping an eye to her strength, and then she can proceed with confidence to her husband, for if she becomes pregnant, she recovers. If much passes during the cleaning, have the patient eat mercury herb

---

174 ἢν δὲ μή . . . διδόναι MV: διδόναι δὲ καὶ Θ.

δας ἐφθοὺς ἀπαλούς, καὶ σιτίοισι μαλθακοῖσι χρήσθω.

27. (136 L.) Ἢν ἐς τὸ ἰσχίον αἱ ὑστέραι ἢ τὸν κενεῶνα καταστηρίξωσι λεχοῖ, προστιθέναι ἐς θάτερον
ἰσχίον ἔλαιον Αἰγύπτιον λευκὸν ἢ ῥόδινον, ἐπὶ δὲ τὸ
ὑγιὲς ἰσχίον ἄμεινον κατακεῖσθαι· πίνειν δὲ γλυκυσίδης κόκκους πέντε τοὺς μέλανας, καὶ ἀκτῆς καρπὸν
ἐν τοίσδεσσιν ὅσον χηραμύδα, τοῦ δὲ κάστορος ὡς
κύαμον· καὶ σιτίοισι χρῶ μαλθακοῖσι· λινόζωστις
ἐναρμόζοι πρὸ τοῦ σιτίου ἐφθὴ ὡς κράμβη· ῥυφείτω
δὲ καὶ τοῦ ὕδατος· τὰ δὲ δριμέα τρωγέτω, πλὴν ῥα-
φανῖδος, καὶ κρομμύου, καὶ καρδάμου· ἄριστον δὲ
θριδακίνη.

28. (137 L.) Ὁπόσα δ’ ἀπὸ τῶν ὑστερέων συμβαίνει
γίνεσθαι νοσήματα, τάδε λέγω· ὅταν αἱ ὑστέραι ἐκ
χώρης κινηθέωσι, προσπίπτουσιν ἄλλοτε ἄλλῃ· ὅπῃ
δ’ ἂν προσπέσωσιν, ὀδυνήματα καταστηρίζουσιν
ἰσχυρά. καὶ ἢν ἅψωνται[175] τῆς κύστιος, ὀδύνην παρ-
έχουσι καὶ τὸ οὖρον οὐ δέχονται, οὐδὲ τὴν γόνην ἐπὶ
σφᾶς ἕλκουσι· καὶ ἄμφω ἀλγέει, καὶ ἢν μὴ ταχέες
λύσιες γένωνται, διαπυΐσκονται αἱ ὑστέραι χρόνῳ
ὕστερον κατὰ ταὐτὰ τὰ χωρία, ᾗ ἂν προσαυανθῶσι·
γίνεται ταῦτα δὲ κατὰ κενεῶνάς τε καὶ βουβῶνας καὶ
ὑπὲρ τοῦ κτενός. |

310   Χρὴ δὲ ἐν ἀρχῆσιν, ὅταν ἡ ὀδύνη ἔχῃ, ὧδε θερα-
πεύειν· χλιάσματα προστιθέναι, καὶ ἐν ὕδατι θερμῷ
εἶναι, ἀγαθὸν γὰρ καὶ τοῦτο· ἢ σπόγγοισιν ἐν ὕδατι
θερμῷ ἐκπιεζομένοις πυριῆν· καὶ πίνειν τῶν ὑστερι-

and tender boiled octopus, and also employ emollient foods.

27. If, in a woman who has just given birth, the uterus becomes fixed in her hip or flank, apply white Egyptian oil or rose oil on the opposite hip; but it is better for the patient to lie on her healthy hip. She should take five black grains of peony in a potion, and with them a cheramys of elderberries, and castoreum to the amount of a bean; also employ emollient foods. Mercury herb boiled like cabbage would be suitable before she eats, taken along with its fluid, and she should eat pungent vegetables except for radish, onion, and cress; best is wild lettuce.

28. For all the diseases that are wont to arise from the uterus, I state the following: When the uterus moves out of its natural position, it moves sometimes in one direction and at another time in another direction, and wherever it comes to rest it provokes violent pains. If it seizes the bladder, it causes pain and fails to receive urine there, nor does it attract seed to itself. Both parts suffer pain, and if no timely resolution follows, with the passage of time the uterus will putrefy and wither in the parts where it is, i.e., in the flanks and groins, and above the pubes.

At the beginning, when the pain is present, you must treat as follows: Apply warm compresses, and it is also good for the patient to be immersed in hot water; or foment with sponges soaked in hot water and squeezed out. Also have her drink medications appropriate for the

---

175 ἄψωνται Potter: ἄψηται τις codd.

κῶν φαρμάκων· ἢν δὲ μὴ πρὸς ταῦτα λύηται, φαρμακεύειν κάτω, ἢν δέηται, καὶ ἄνω, ὁπότερον ἂν ἁρμόζῃ
μᾶλλον δεῖσθαι. τοῦτο δὲ διαγνώσῃ τῷδε τῷ τρόπῳ·
ἢν μὲν ἐς τοὺς βουβῶνας καὶ κτένα καὶ κύστιν ἐγ-
χρίμψῃ, αὗται χρήζουσιν ἄνω φαρμακείης· ἢν δ' ἐς
τοὺς κενεῶνάς τε καὶ τὰ ὑποχόνδρια, αὗται δέονται
κατωτερικῶν φαρμάκων· μετὰ δὲ ταύτας τὰς καθάρσιας εὐθὺς καθαίρειν τὰς ὑστέρας.[176] τὰ δὲ νοσήματα
πάντα τὰ τοιουτότροπα γεραιτέρῃσι μᾶλλον γίνεται
ἢ νεωτέρῃσι, πρὸς τὰς ἀπολείψιας τῶν ἐπιμηνίων· γίνεται δὲ καὶ νέῃσιν ἐούσῃσιν, ὅταν χηρεύσωσι πολὺν
χρόνον.

Ἢν δ' ἐς τὴν ἕδρην τράπωνται, τὰ ὑποχωρήματα
κωλύονται, καὶ ὀδύναι ἴσχουσι τήν τε ὀσφῦν καὶ τὴν
νειαίρην γαστέρα καὶ τὸν ἀρχόν. ὅταν δὲ ὧδε ἔχῃ,
λούειν χρή μιν τῷ θερμῷ, καὶ πυριῆν τὴν ὀσφῦν, καὶ
ὑποθυμιῆν τὸ κατόπιν κακώδεσι, καὶ προστιθέναι
ἄσσα καθαίρει τε καὶ ἐλαύνει τὰς ὑστέρας, καὶ πιπίσκειν ὅ τι ἂν προσδέχηται τῶν ξυμφερόντων μάλιστα. ἢν δὲ κάτω ἐγκέωνται ἐς τοὺς βουβῶνάς τε καὶ
οὐρητῆρα, ὀδύναι γίνονται ἰσχυραί, καὶ νάρκα ἐν
τοῖσι σκέλεσι, καὶ ὁ οὐρητὴρ ἀποφράσσεται, καὶ τὸ
οὖρον οὐ μεθίησι. τὰς τοιαύτας θεραπεύειν ὧδε· πρὸς
μὲν τὰς ῥῖνας προστιθέναι τὰ εὐώδεα καὶ μύρα, πρὸς
δὲ τὰς ὑστέρας τὰ δύσοδμα θυμιῶντα.

29. (138 L.) Ἅπασα δὲ πρόφασις ἱκανὴ τὰς ὑστέρας
παροτρῦναι, ἢν ἔχωσί τι φλαῦρον· καὶ γὰρ ἀπὸ ῥί-
312 γεος τῶν ποδῶν καὶ ὀσφύος, καὶ | ἀπὸ τοῦ ὀρχέεσθαι

uterus. If the disease is not relieved with these measures, give a purgative medication, or if necessary an emetic, according to which is required. Make the decision thus: if the disease occupies the groins, the area above the pubes, and the bladder, emetics are indicated, but if it attacks the flanks and the hypochondria, a purge downward is needed. After these cleanings, immediately clean the uterus. All diseases like this occur more frequently in older women than in younger ones, and involve a cessation of the menses. They also occur in younger women who have been widows for a longer time.

If these diseases turn in the direction of the seat, they obstruct the stools, and pains occupy the lower back, lower belly, and the anus. When the case is such, bathe the patient in hot water, foment her lower back, next fumigate her from below with ill-smelling substances, apply suppositories that clean and empty the uterus, and give the kind of useful drinks that will be best received. If the uterus encroaches below toward the groins and the urethra, violent pains supervene together with numbness of the legs; the urethra is obstructed, and no urine is passed. Treat thus: to the nostrils apply fragrant agents and unguents, and to the uterus fetid fumigations.

29. Any cause may suffice to displace the uterus if it is diseased, since it can result from a chill of the legs and lower back, from dancing, from winnowing grain, from

---

[176] μετὰ δὲ ταύτας . . . ὑστέρας om. MV.

καὶ πτίσαι καὶ κεάσαι καὶ δραμεῖν πρὸς ἄναντες χωρίον καὶ πρὸς κάταντες, καὶ ἀπ' ἄλλων.[177] ταῦτ' οὖν
χρὴ σκέπτεσθαι ἐς ὅλον τὸ σῶμα καθορῶντα, ὅταν
τὰ παρεόντα συθῇ[178] νοσήματα· τὰ γὰρ τοιάδε πάντα
ἀνάγκη ἐστὶ τῷ πλέονι ἢ τῷ ἐλάσσονι νοσέειν· καθ'
ὃ δ' ἂν ἐκλάμψῃ μάλιστα, ταύτῃ τὰ ἐξαπιναῖα δῆλα
τῶν νοσημάτων· ὁκόταν οὖν τὰ ἐξαπιναῖα ταῦτα
συθῇ,[179] ἀνωτέρω χρὴ λαμβάνεσθαι ἐκ τοῦ παντὸς
ἀνθρώπου. ὅσαι δὲ ἀποψύξιες σκελέων ἢ ναρκώσιες
ψύχει γίνονται ἐν τοῖσιν ὑστερικοῖσι, ταῦτα πάντα
μεταρσιοῖ τὰς ὑστέρας. ἐν τοῖσι τοιούτοισι καταχεῖν
χρὴ θερμὸν ὕδωρ κατὰ τῆς ὑστέρης καὶ τῶν πέριξ
χωρίων, καὶ θερμαίνειν αὐτὰς καὶ τὰ σκέλεα, καὶ ὅταν
προσπεπτωκυῖαι ἔωσιν.

30. (139 L.) Ἐὰν περιστραφῶσιν αἱ μῆτραι ἐκ τόκου κατὰ τὰ δεξιά, τά τε λοχεῖα οὐ γίνονται, καὶ
ὀδύνη ἴσχει τὴν νειαίρην γαστέρα καὶ τὰς ἰξύας καὶ
τοὺς κενεῶνας, καὶ τὸ δεξιὸν σκέλος βαρύνεται, καὶ
νάρκα ἔχει, καὶ[180] τεταρμαίνει· καὶ οὐκ ἂν δύναιτο τοῦ
στόματος θιγεῖν τῶν μητρέων, ἀλλ' ὄψει λείας τε καὶ
ὁμαλὰς ἰσχυρῶς. ὅταν ὧδε ἔχῃ, φάρμακον χρὴ πῖσαι
ὑφ' οὗ καθαίρεται καὶ ἄνω καὶ κάτω, κάτω δὲ μᾶλλον,
καὶ πυριῆν ὅλον τὸ σῶμα καὶ τὰς μήτρας ὡς μάλιστα
προσηνέως, καὶ λούειν θερμῷ δὶς τῆς ἡμέρης, καὶ τῶν
ποτημάτων ὅ τι μάλιστα προσδέχεται πειρώμενος·
καὶ τῷ ἀνδρὶ συνευδέτω θαμινά, καὶ τὴν κράμβην
ἐσθιέτω.

chopping wood, from running up or down a grade, and from other things as well. Now this must be evaluated by observing the entire body when the actual diseases are erupting, since all of the parts will be involved in the disease to a greater or less degree: wherever the disease violently erupts, there its signs suddenly become evident. When this sudden eruption occurs, it must be understood from the former state of the person's whole body. Chilling of the legs or numbness and cold in patients with uterine disorders all elevate the uterus. In such diseases you must pour warm water down over the uterus and adjacent areas, and heat these and the legs; also when the uterus has fallen against some part.

30. If after childbirth the uterus turns toward the right, the lochia will not pass, pain will occupy the lower belly, loins, and flanks, and the right leg will be subject to heaviness, numbness, and trembling, palpation of the mouth of the uterus will be impossible, but you will perceive the uterus to be very smooth and even. When the case is such, give the patient a potion that cleans both upward and downward, although more downward, and foment her whole body and her uterus in a very gentle way, wash her with hot water twice a day, and give her the drinks you find by experience are most acceptable; also have her sleep often with her husband, and eat cabbage.

---

177 Add. ἱκανῶν MV.
178 συθῇ Littré: λυθῇ codd.
179 συθῇ Littré: λυθῇ codd.
180 Add. οὐ MV.

31. (140 L.) Ἢν αἱ ὑστέραι κατὰ τ' ἀριστερὰ κλι-
314 θῶσιν ἢ τὸ ἰσχίον, | ὀδύνη ἔχει ὀξείη τε καὶ σπερχνὴ
τάς τε ἰξύας καὶ τοὺς κενεῶνας καὶ τὸ σκέλος, καὶ
ἐπισκάζει. ὅταν οὕτως ἔχῃ, φάρμακον χρὴ πῖσαι ἐλατήριον, τῇ δ' ὑστεραίῃ ὑποθυμιᾶν· τῶν τε κριθέων
χοίνικας δύο, καὶ ἐλαίης φύλλα κατακνῆσαι σμικρά,
καὶ κηκίδα κατακόψας καὶ σήσας, καὶ ὑοσκυάμου
τρίτον χοίνικος· ταῦτα μίξας, καὶ ἐλαίῳ περιποιήσας
ὅσον ἡμικοτύλιον ἐν χύτρῃ καινῇ, ὑποθυμία τέσσερας ἡμέρας· τῆς δὲ νυκτὸς γάλα βοὸς καὶ μέλι καὶ
ὕδωρ πινέτω,[181] καὶ τῷ θερμῷ λούσθω.

32. (141 L.) Ἢν παραλοξαίνωνται αἱ μῆτραι καὶ
δοχμοὶ ἔωσι καὶ τὸ στόμα σφέων, τὰ ἐπιμήνια ταύτῃ
τὰ μὲν κρύπτονται, τὰ δὲ προφανέντα οἴχονται· καὶ
οὐχ ὅμοια γίνονται, ἀλλ' αἰεὶ[182] κακήθεα καὶ ἐλάσσονα ἢ πρὸ τοῦ. καὶ ἡ γονὴ οὐκ ἐγγίνεται τούτου τοῦ
χρόνου, καὶ ὀδύνη ἴσχει τὴν νειαίρην γαστέρα καὶ
τὰς ἰξύας καὶ τὸ ἰσχίον, καὶ ἐφέλκεται αὐτό. ὅταν
οὕτως ἔχῃ, φάρμακον χρὴ πῖσαι ἐλατήριον, καὶ λούειν θερμῷ, καὶ πυριᾶν. ὅταν δὲ νεοπυρίητος ἢ νεόλου-
τος ᾖ, τὸν δάκτυλον παραφάσασα, ἀπορθούτω, καὶ
παρευθυνέτω τὸ στόμα τῶν μητρέων, καὶ ὑποθυμιάσθω τὰ εὐώδεα. καὶ τῶν ποτημάτων δίδου ὅ τι ἂν
μάλιστα προσδέχηται πειρώμενος· σιτίοισι δὲ χρήσθω μαλθακοῖσι, καὶ σκόροδα ἐσθιέτω καὶ ὠμὰ καὶ
ἐφθά. καὶ τῷ ἀνδρὶ συνευδέτω· καὶ ἐπὶ τοῦ ὑγιέος
ἰσχίου κατακείσθω ἐπὶ θάτερον καὶ πυριήσθω. ἡ δὲ
νοῦσος δυσαπάλλακτος.

31. If a woman's uterus inclines toward her left side or hip, violent acute pain will seize her loins, flanks, and a leg, and she will limp. When the case is such give a potion of squirting cucumber juice, and on the next day fumigate from below: Take two choinixes of barley groats, a small amount of crushed olive leaves, some pulverized and sieved oak galls, and one third chionix of henbane: mix these together, and prepare in a half-cotyle of olive oil in a new pot: apply as a fumigation from below for four days. At night have the patient drink cow's milk, honey and water, and bathe herself in hot water.

32. If a woman's uterus shifts to an oblique position with its mouth at an angle, her menses will at one time be absent, at another reappear, and then be lost again; they are no longer as they should be, but worse in quality and less in amount than before. During this time the woman fails to become pregnant, and pain occupies her lower abdomen, loins, and hip, which is drawn up. When the case is such, give the patient a purge of squirting cucumber juice, bathe her in hot water, and apply a vapor bath. Immediately after bathing or the vapor bath have her insert her finger and straighten and widen the mouth of her uterus, and fumigate herself with fragrant substances from below. Give the patient whichever beverages you know by experience will be most acceptable, and have her employ emollient foods, and eat garlic both raw and boiled. She should sleep with her husband, lie on her healthy hip, and apply vapor baths to the other one. This disease is very hard to allay.

---

[181] πινέτω om. Θ.   [182] αἰεὶ om. MV.

33. (142 L.) Ἢν δὲ ἄγχιστα ἔωσιν, ἐμέειν θαμινά-
κις· τὰς δ' ὑστέρας πυριᾶν δυσόδμοισι, μέχρι ἐς χώ-
ρην ἱδρυθέωσιν· διαίτῃσι δὲ χρῶ μὴ λαπακτικῇσιν. |

316   34. (143 L.) Ἢν αἱ μῆτραι προΐωσιν ἐξωτέρω τῆς
φύσιος, πυρετὸς ἔχει τὸ αἰδοῖον καὶ τὴν ἕδρην,[183] καὶ
τὸ οὖρον στάζει θαμινὰ κατ' ὀλίγον, καὶ δάκνεται
τὸ αἰδοῖον· πάσχει δέ, ἢν ἐκ τόκου ἐοῦσα τῷ ἀνδρὶ
ξυνεύνηται. ὅταν ὧδε ἔχῃ, μύρτα καὶ λωτοῦ πρίσματα
ἑψῆσαι ἐν ὕδατι, καὶ θεῖναι τὸ ὕδωρ ἐς τὴν αἰθρίην·
προσαιονᾶν δὲ ὧδε ὡς ψυχρότατον τὸ αἰδοῖον· καὶ
τρίβων λεῖα πρόσπλασσε· ἔπειτα πίνουσα ὕδωρ φακῶν ξὺν μέλιτι καὶ ὄξει ἐμείτω, ἔστ' ἂν αἱ ὑστέραι
ἀνελκυσθῶσι· καὶ τὴν κλισίην ἀνεκὰς χρὴ ποιέειν τὰ
ἀπὸ τῶν ποδῶν· καὶ ὑποθυμιᾶν τὰ αἰδοῖα τὰ κακώδεα,
καὶ πρὸς τὰς ῥῖνας εὐώδεα· σιτίοισι δὲ χρήσθω ὡς
μαλθακωτάτοισι καὶ ψυχροῖσι, καὶ τὸν οἶνον ὑδαρέα
λευκὸν πινέτω, καὶ μὴ λούσθω, ἢ[184] τῷ ἀνδρὶ συνίτω.

35. (144 L.) Ἢν παντάπασιν ἐκ τοῦ αἰδοίου ἐκπέ-
σωσιν αἱ ὑστέραι, ἐκκρίμναται οἷον ὄσχη, καὶ ὀδύνη
λάζεται τὴν νειαίρην γαστέρα καὶ τὰς ἰξύας καὶ τοὺς
βουβῶνας· καὶ ὅταν ἐπιγένηται χρόνος, οὐκ ἐθέλου-
σιν ἐς χώρην ἰέναι. ἡ δὲ νοῦσος λαμβάνει, ὅταν ἐκ
τόκου ταλαιπωρήσῃ, ὥστε ψαίρειν τὰς ὑστέρας, ἢ τῷ
ἀνδρὶ συνίῃ ἐν τῇ λοχίῃ καθάρσει. ὅταν ὧδε ἔχῃ,
ψύγματα προστιθέναι παρηγορικὰ πρὸς τὸ αἰδοῖον·
καὶ τὸ ἔξω ἐὸν ἀποκαθήρας, σίδην ἐν οἴνῳ μέλανι

---

183 ἕδρην Θ: νειαίρην γαστέρα MV.

33. If the uterus is very close (sc. to the exterior), have the patient vomit frequently, and foment her uterus with ill-smelling substances until it settles back into its normal location. Avoid laxatives in her regimen.

34. If a woman's uterus moves away from its natural position to the outside, heat will occupy her vagina and seat, and her urine will trickle continually a little at a time and irritate her vagina. A woman suffers in this way if after giving birth she sleeps with her husband. When the case is such, boil myrtle berries and nettle tree sawdust in water and set the water out in the open air: then douche the patient's vagina with this as cold as possible; also grind the same ingredients fine and apply them in a poultice. Then have the patient drink lentil juice with honey and vinegar as an emetic, and continue to vomit until her uterus is retracted. You should also raise her bed at the foot end. Fumigate her genitalia with ill-smelling substances and her nostrils with fragrant ones. Have her take very emollient and cold foods, drink diluted white wine, and abstain from bathing or intercourse with her husband.

35. If a woman's uterus falls completely out of her vagina, it will hang there like a scrotum, pain will occupy her loins, lower belly, and groins, and after a certain time has passed it will refuse to return to its proper position. This condition arises when after giving birth a woman exerts herself so that her uterus quivers, or she has intercourse with her husband during the lochial cleaning. When the case is such, apply cooling, soothing compresses to the vagina. The part that protrudes outside should be cleaned off, washed with pomegranate boiled in dark wine, and

---

¹⁸⁴ ἤ Potter: καὶ codd.: μηδὲ Linden.

ἑψήσας, τούτῳ περιπλῦναι, καὶ εἴσω ἀπωθέειν· ἔπειτα
μέλι καὶ ῥητίνην συντῆξαι ἴσον ἑκατέρου, καὶ ἐγχέειν
ἐς τὸ αἰδοῖον, καὶ κατακεῖσθαι ὑπτίην, ἄνω τοὺς πό-
318 δας ἔχουσαν ἐκτεταμένην. κἄπειτα | σπόγγους προσ-
θεῖσαν ἀναδῆσαι ἐκ τῶν ἰξύων. ἔστ᾽ ἂν δ᾽ οὕτως ἔχῃ,
σιτίων ἀπεχέσθω, ποτῷ δ᾽ ὡς ἐλαχίστῳ χρήσθω,[185]
μέχρι ἑπτὰ ἡμέραι παριῶσι.

Καὶ ἢν μὲν οὕτως ἐθέλωσιν ἐνακούειν καὶ ἀπιέ-
ναι·[186] ἢν δὲ μή, ἄκρα περιξέσαι τὰ λέγνα τῆς ὑστέρης καὶ ἀμφιπλῦναι, χρῖσαι δὲ τῇ πισσηρῇ· ἔπειτα
πρὸς κλίμακα δῆσαι τοὺς πόδας, τὴν δὲ κεφαλὴν
κάτω ἔχειν, καὶ τῇ χειρὶ εἴσω ἀπωθέειν· ἔπειτα λύειν,[187] καὶ ξυνδῆσαι αὐτῆς τὰ σκέλεα ἐναλλάξ, καὶ ἐὰν
νύκτα καὶ ἡμέρην οὕτω, καὶ διδόναι ὀλίγον χυλὸν
πτισάνης ψυχρόν, ἄλλο δὲ μηδέν. τῇ δὲ ὑστεραίῃ
κατακλίνας ἐπὶ τὸ ἰσχίον, σικύην προσβάλλειν ὡς
μεγίστην, καὶ ἐὰν ἕλκειν πολὺν χρόνον, καὶ ἐπὴν
ἀφέλῃς, μὴ ἀποσχάσῃς· ἀλλὰ κατακλίνας καὶ μὴ
προσφέρεσθαι μηδὲν ἀλλὰ τὸν χυλόν, ἕως ἑπτὰ ἡμέραι παριῶσιν· ἢν δὲ δίψα ἔχῃ, ὕδωρ ἐλάχιστον πι-
νέτω· ἐπὴν δὲ αἱ ἑπτὰ ἡμέραι παριῶσι, σιτίοισιν ὡς
μαλθακωτάτοισι καὶ ἐλαχίστοισι χρῆσθαι· ὅταν δ᾽
ἀποπατῆσαι θέλῃ, ἀνακειμένη δράτω ἔστ᾽ ἂν τεσσε-
ρεσκαίδεκα ἡμέραι γένωνται· ἔπειτα κλυζέσθω χλι-
αρῇσι πυρίῃσιν· ἄμεινον δὲ θερμῷ, ὡς ἐξ ἡλίου, καὶ
περιπατείτω ὡς ἐλάχιστα, καὶ μὴ λούσθω· κοιλίην δὲ
μὴ λύσῃς· σιτίοισι δὲ ὀλιγίστοισι χρήσθω, καὶ μὴ
δριμέσι μηδ᾽ ἁλμυροῖσι· καὶ θυμιάσθω τὸ αἰδοῖον

then pressed back inside. Next melt equal amounts of honey and resin, and infuse this into the vagina with the patient lying on her back and holding her legs raised and spread apart. Then have the patient insert sponges and suspend them from her waist, and as long as this goes on have her abstain from food and employ as few drinks as she can until the seventh day arrives.

If with these measures the disease can be made to respond and go away, fine. But if not, abrade the outer border of the uterus all around, wash it thoroughly, and anoint it with pitch ointment. Then tie the patient's feet to a ladder with her head facing downward, and with your hand press the uterus back in. Then release her, tie her legs together against one another, and leave her this way for a night and a day; give her a little cold barley gruel, but nothing else. On the next day have her recline on her hip, apply the largest bloodletting cup you have, leave it for a long time, but when you remove it do not scarify. Leave the patient reclining and do not give her anything but gruel until seven days have passed: if she is thirsty, have her drink a very little water. When the seven days are over, have her employ a very small amount of very emollient foods. When she wants to go to stool, have her do so in the reclining position until fourteen days have passed; then have her flush herself with warming fomentations—better yet, with hot ones as if taken out of the sun—walk about a very little, and avoid the bath, and do not give any laxative for her cavity. She should employ the minimum of foods, and none that is pungent or salty; have her fumigate

---

185 χρήσθω om. Θ.   186 Add. ἅλις ἔστω MV.
187 λύειν MV: λούειν Θ.

τοῖσι κακώδεσι· καὶ ἐπειδὰν ἄρχηται περιπατεῖν, τὴν σφενδόνην φορείτω.

36. (145 L.) Ἦν δ' ἔξω τοῦ αἰδοίου τὸ στόμα τῶν
320 μητρέων ἐκπέσῃ,[188] | τοῦ αὐχένος τῶν μητρέων πλη-
σίον τοῦ αἰδοίου κειμένου καὶ ἐόντος εὐρέος· γίνεται
δὲ τοῦτο μᾶλλον τῇσιν ἀτόκοισι, μάλιστα δὲ γίνεται
ἐκ ταλαιπωρίης, ἐπὴν ταλαιπωρήσῃ ἡ γυνή, καὶ αἱ
μῆτραι θερμανθῶσι καὶ ἱδρώσωσιν, ἐκτρέπεται τὸ
στόμα αὐτῶν διὰ τοῦ αὐχένος, ἅτε ἐν ὑγροτέρῳ καὶ
ὀλισθηροτέρῳ καὶ θερμοτέρῳ[189] χωρίῳ γενόμενον ἢ ἐν
τῷ πρὶν χρόνῳ· καὶ ἐπὴν τοῦτο γένηται, θύουσιν[190]
ἔξω πρὸς τὸ ψύχος, καὶ σφέων τὸ στόμα ἔρχεται
ἔξω[191] ἐκτραπέν.

Καὶ ἢν μὲν ἐν τάχει θεραπευθῇ, ὑγιὴς γίνεται,
ἄφορος δὲ πάντως· ἢν δὲ μὴ ἐν τάχει, ἔξω οἱ ἔσται
σκληρὸν τὸ στόμα, καὶ ῥεύσεται[192] ἰχὼρ ἄλλοτε καὶ
ἄλλοτε γλίσχρος καὶ κάκοδμος, καὶ ἢν τὰ ἐπιμήνια
χωρήσῃ, ἢν ἔτι ἐν τῇ ἡλικίῃ εἴη ἐν εὐνῇ εἶναι· χρόνου
δὲ γενομένου, ἡ νοῦσος ἀνίητος γίνεται, καὶ ξυγκατα-
γηράσκουσιν ἔξω αἱ μῆτραι ἔουσαι.

Ἦν αἱ μῆτραι ἐξίσχωσι, περινίψας αὐτὰς ὕδατι
χλιαρῷ, καὶ ἀλείψας ἐλαίῳ καὶ οἴνῳ, πάλιν ἐνθεῖναι
καὶ ἀναδῆσαι· καὶ ὑποθυμᾶν τὰ κακώδεα, ὑπὸ δὲ τὰς
ῥῖνας τὰ εὐώδεα. ἢν δὲ πλείονα χρόνον αἱ ὑστέραι
ἐξίσχωσι καὶ περιψύχωνται ναρκωδέως, καταχεῖν
ὕδωρ θερμὸν πολύ, ὅπως διαπυριηθῶσιν. ἢν δὲ ἤδη
φυσῶνται, καὶ ὄξους μίσγοντα σὺν ὕδατι πυριᾶν ἢ

her vagina with evil-smelling substances; when she begins to take walks, she should wear the sling.

36. If the mouth of a woman's uterus projects out of her vagina with its neck lying against the vagina and gaping, this is most often occurring in women that have never given birth, generally from exertion when they are exerting themselves and their uterus becomes heated and sweats, so that its mouth turns inside out through the neck, because it is in a place that has become moister, slipperier and hotter than it was before. When this happens, the uterus moves outside toward the cold, and its mouth prolapses and is inside out.

If this condition is treated at once, the woman may recover, but she is left completely sterile. If not treated at once, the protruded mouth will become hardened, and a fetid, sticky discharge will flow out of it from time to time, and if the menses pass, if the woman is still in that age, she should be in bed. With the passage of time such a disease becomes incurable, and such women grow old with their uterus prolapsed.

If the uterus is prolapsed, wash it all around with warm water, anoint it with olive oil and wine, place it back inside, and bind it there; fumigate it from below with fetid substances, and fumigate under the nostrils with fragrant ones. If the uterus has been prolapsed for a longer time, and it has become cold and insensitive, pour much hot water down over it in order to thoroughly heat it. If it is already inflated, mix some vinegar together with water,

---

188 Add. οἷα MV.   189 καὶ θερμοτέρῳ om. Θ.
190 θύουσιν Θ: ἰθύουσιν MV.
191 ἔξω om. MV.   192 Add. αὐτόθεν MV.

δάφνης ἢ μυρσίνης ὕδωρ, καὶ ἠρέμα προωθέειν, καὶ
κηρωτῇ ἢ μύρῳ χρίειν· ἢν ἐσακούῃ· ἢν δὲ μή, περι-
κλύσαι τῷ ὕδατι, καὶ ὄξος ὀλίγον χλιήνας καταχέαι,
εἶτα ἁλὶ ψαύειν· ὅταν δὲ συντακέωσι, περινίψας ὡς
λέλεκται, ἐνθεῖναι, καὶ τἆλλα ποιέειν τὰ | προειρη-
μένα· ἔλαιον δὲ μὴ προσφέρειν, μηδ᾽ ἄλλο πῖον, μηδὲ
λίπα ἔχον.

37. (146 L.) Ἢν ὑποπτυχθῇ τι τῶν στομάτων τῆς
ὑστέρης, τὰ ἐπιμήνια οὐ γίνονται, ἢ ὀλίγα τε καὶ πονηρὰ καὶ ἀλγεινά, καὶ ὅταν τῷ ἀνδρὶ συνεύδῃ, ἀλγέει,
καὶ ὅ τι ἂν ὁ ἀνὴρ μεθίῃ,[193] ἔξεισι· καὶ οὐκ ἐθέλει
ψαύεσθαι, οὐδ᾽ ἕλκουσι τὴν γονήν, καὶ ὀδύνη ἴσχει
τὴν νειαίρην γαστέρα καὶ τὰς ἰξύας, καὶ τὸ στόμα
τῶν ὑστερέων ⟨οὐ⟩[194] δῆλον ψηλαφήσει.

Ὅταν ὧδε ἔχῃ, πυριᾶσαι οὔρῳ παλαιῷ· ἔπειτα
ἐμεσάτω τῷ φακίῳ ξυμμίξασα μέλι καὶ ὄξος· ἔπειτα
λούσθω ὕδατι θερμῷ· ἔπειτα ἐγχέασα ἐς φιάλην ἀρ-
γυρέην ἢ χαλκέην ἔλαιον λευκὸν Αἰγύπτιον καὶ ἅλας,
καλυψαμένην δὲ καθέζεσθαι ἀμφὶ τὴν φιάλην· καὶ ἢν
μὲν εἴη ὀδμὴ διὰ τοῦ στόματος τοῦ ἐλαίου, φάναι
αὐτὴν τέξεσθαι, καὶ τὴν ὑστέρην ἔτι ὑγιῆ εἶναι. ἢν δὲ
μὴ ᾖ ὀδμή, θαρσύνειν, καὶ ἐπειδὰν μέλλῃ εὕδειν,
προστιθέσθω τὸ Αἰγύπτιον ἔλαιον ἐν εἰρίῳ· τῇ δ᾽ ὑστεραίῃ σκεψάσθω ἤν τι μᾶλλον κατ᾽ ἰθὺ ᾖ τὸ στόμα
τῶν ὑστέρων· καὶ ἢν φῇ εἶναι, πυριᾶσαι εὐώδεσιν

and foment with fluid from laurel or myrtle, or press the uterus gently back in, and anoint with a cerate or an unguent. If the uterus cooperates, fine; if not, flush it with water and pour a little warmed vinegar down over it, and then coat it with salt: when the uterus has shrunken together, wash it as indicated, press it back in, and do the other things mentioned above. Do not apply olive oil or anything else that is oily, nor anything containing fat.

37. If some part of the mouth of a woman's uterus folds over on itself, her menses will not appear, or they will be scanty, troublesome, and painful, and when she sleeps with her husband, she will feel pain, and whatever the man ejaculates will run back out. She refuses to be palpated, her uterus fails to attract the (sc. male) seed, pain occupies her lower belly and loins, and the mouth of her uterus cannot be clearly felt.

When the case is such, foment with aged urine; then have the patient mix honey and vinegar together with lentil soup as an emetic; then have her bathe in hot water. Next she should pour white Egyptian oil into a pan made of silver or bronze along with salt, and covering herself sit down over the pan. If the odor of the oil is present in her mouth, announce that she will give birth and that her uterus is still healthy. But if there is no odor, still take heart, and when she is about to go to bed have her apply the Egyptian oil in a piece of wool as a suppository; on the following day have her examine herself to see whether the mouth of her uterus is straighter. If she says it is, foment

---

[193] μεθίη om. Θ.   [194] Add. Foes in note 155, after Calvus' *nec tactu vulvarum os comparet* and Cornarius' *et os uterorum ad contactum non comparet*.

ἡμέρας τρεῖς, καὶ προστιθέναι προσθετὰ ὁποῖα μὴ
ἀναδήξεται· καθαίρειν δὲ καὶ μετακλύζειν εὐώδεσι καὶ
μαλθακοῖσιν, ἴσας ἡμέρας τῇσι πρόσθεν· ὅταν δὲ τὰ
ἐπιμήνια γένηται, νηστεῦσαι, καὶ ἀλουτήσασα παρὰ
τὸν ἄνδρα ἴτω, θυμιασαμένη τοῖσιν ἀρώμασιν. ἄτοκοι
δὲ καὶ ἐκ ταύτης τῆς νούσου γίνονται, ἢν μὴ μελεδαν-
θῶσιν.

38. (147 L.) Ἢν ἑλκωθῶσιν αἱ μῆτραι καὶ πρόσω
324 χωρήσωσιν ἐξωτάτω, | ἐλαίῳ χρίων λίπα τὰς χεῖρας
προστιθέναι, καὶ δίδου καταπότιον σμύρνης ἀκρήτου,
τρεῖς σπυράδας καταπιεῖν, καὶ πινέτω δάφνην χλω-
ρὴν τετριμμένην, οἴνῳ[195] διεῖσα, καὶ ὑγιὴς οὕτω γίνεται.

39. (148 L.) Ἢν δ' ἐξορούῃ ᾖ τὰ αἰδοῖα, προσθετόν,
ἀννήσου[196] καρπὸν καὶ σελίνου τρίψας λεῖα, πρόσθες
πρὸς τὸ αἰδοῖον.

40. (149 L.) Ἢν μὴ κατὰ χώρην μένωσιν, ἀλλ' ὁτὲ
μὲν ἔνθα, ὁτὲ δ' ἔνθα ἴωσιν, ὀδύνας παρέχουσιν· αἱ δὲ
ἀφανέες γίνονται, τοτὲ δ' ἐξίασιν ἕως ἕδρης· καὶ ὅταν
μὲν ὑπτίη ᾖ, κατὰ χώρην μένουσιν· ὅταν δ' ἀναστῇ ἢ
ἐξ ὕπνου ἔγρηται ἢ ἐπικύψῃ ἢ ἄλλο τι κινηθῇ, ἐξέρ-
χονται, πολλάκις δὲ καὶ ἡσυχίην ἐχούσῃ.

Ταύτην χρὴ ὡς μάλιστα ἠρεμεῖν τε καὶ ἀτρεμέειν
καὶ μὴ κινεῖσθαι, καὶ τὸν κλισμὸν κεῖσθαι πρὸς ποδῶν ὑψηλότερον· καὶ τοῖσιν ἐμέτοισι χρέεσθαι, χρὴ
γὰρ ἀντισπάσαι ἀνεκάς· καὶ τοῖς στρυφνοῖσιν δὲ αἰονᾶν, καὶ ὑποθυμιᾶν τὰ κακώδεα,[197] ὑπὸ δὲ τὰς ῥῖνας
εὐώδεα· καὶ τῶν ῥοιῶν διὰ τοῦ ὀμφαλοῦ τρήσαντα

her with fragrant substances for three days, and apply non-irritating suppositories. Clean her and then flush her with fragrant, emollient agents for the same number of days as before. When the menses appear, have the woman fast, fumigate herself with aromatics, and without washing go to her husband. Many women become sterile from this disease unless they are cared for.

38. If a woman's uterus ulcerates and moves very far out, anoint your hands generously with olive oil and replace it; also give her a pill of unmixed myrrh—she should swallow three pellets—and have her drink ground green laurel dissolved in wine to make her well.

39. If the genitalia protrude, grind seeds of anise and celery fine and apply this as a pessary in the vagina.

40. If a woman's uterus does not remain in place, but moves at one time in one direction and at another time in another direction, this causes pains. Sometimes it cannot be seen, while at other times it protrudes as far as her anus. As long as she lies on her back it remains in place, but when she stands up or awakes from her sleep or bends over or makes any other movement, it protrudes, and often even when she is at rest.

This patient must remain as still as possible and rest, and never move, and her couch should be inclined with the foot-end elevated. Have her employ emetics, since you must create an upward attraction; she should foment with astringents, and fumigate from below with fetid agents and under the nostrils with fragrant ones. Pierce a pome-

---

195 Om. οἴνῳ Θ.

196 ἀννήσου MV: ἀνήθου Θ.

197 τὰ κακώδεα MV: μαλακώδεα Θ.

μέσην, ἐν οἴνῳ χλιαίναντα—ἥτις ἂν ἁρμόζῃ μάλιστα
καὶ μή τι κωλύῃ λίην,[198]—προστίθει ὡς ἐσωτάτω· εἶτ᾽
ἀναδῆσαι δεῖ ταινίῃ πλατείῃ, καὶ ἀναλαβεῖν, ὡς μὴ
ὀλισθάνῃ, ἀλλὰ μένῃ, καὶ ποιέῃ τὸ δέον· καὶ τῶν μηκώνων σὺν τῷ τυρῷ καὶ τοῖς ἀλφίτοισι πιπίσκειν, ὡς
ἐν τῇ πρὸς τὸ πλευρὸν[199] προσπτώσει γέγραπται· καὶ
τῶν ποτημάτων πειρώμενος ὅ τι ἂν μάλιστα προσδέχηται πιπίσκειν· σιτίοισι δὲ ὡς μαλθακωτάτοισι
326 χρήσθω, καὶ μετ᾽ | ἀνδρὸς οὐ χρὴ κοιμᾶσθαι, ἄχρις
οὗ κατὰ χώρην οἱ ἔωσι.[200] ἢν αἱ μῆτραι ἐκπέσωσι,[201]
κισσὸν ὡς ξηρότατον τρίψας λεῖον, ἐνδήσας ἐς
ὀθόνιον, προσίσχειν, καὶ λιπαρὸν προσφέρειν μηδέν·
πίνειν δὲ διδόναι πυροὺς προκόψας, καὶ μήκωνα
ὀπτήν,[202] καὶ ἐλελίσφακον, καὶ κύπαιρον, καὶ ἄννησον, ταῦτα τρίψας λεῖα, διεὶς οἴνῳ, καὶ τῶν κυρηβίων
τῶν ἀπὸ τῶν κριθέων, διδόναι δὶς τῆς ἡμέρης, ἐφ᾽
ἑκάτερον ἡμικοτύλιον.

41. (150 L.) Ἢν ἐς τὰ σκέλεα καὶ τοὺς πόδας τρα-
πῶσι, γνώσῃ τῷδε· οἱ μεγάλοι δάκτυλοι[203] σπῶνται
ὑπὸ τοὺς ὄνυχας, καὶ ὀδύνη ἔχει τὰ σκέλεα καὶ τοὺς
μηρούς, καὶ ἔγκειται καὶ θλίβει τὰ ἀμφὶ τὸν μηρὸν
νεῦρα. ὅταν οὕτως ἔχῃ, λούειν χρὴ πολλῷ καὶ θερμῷ
θαμινά, καὶ πυριᾶν, ἢν ἀνδάνηται, καὶ ὑποθυμιᾶν τὰ
κακώδεα, καὶ τῷ μύρῳ τῷ ῥοδίνῳ ἀλειφέσθω λίπα.

---

[198] λίην om. MV.   [199] πρὸς τὸ πλευρὸν om. MV.
[200] οἱ ἔωσι Ermerins: οἵη τε ᾖ μαίνειν Θ: οἵη τε ᾖ MV.
[201] ἐκπέσωσι MV: ἐκθέωσι Θ.

granate through the center of its navel, warm it in wine—viz. a pomegranate that fits well and is not going to resist very much—and insert it as far in as possible. Then you must tie it with a flat bandage that holds it up so that it does not slip out, but stays in place and has the required effect. Have the patient drink a potion made from poppies together with cheese and barley meal, as described in the chapter devoted to the uterus falling against the side;[29] also give the drinks you know by experience will be best received. Have her employ very emollient foods, and avoid sleeping with her husband until her uterus has returned to its proper location. If the uterus prolapses, grind some very dry ivy fine, wrap it up in a piece of linen cloth, insert it, and do not employ any fat. Also give a potion made with ground wheat, baked poppies, salvia, galingale, and anise: grind these fine, dissolve them in wine, add barley bran, and give this twice a day, each time a half cotyle.

41. If (sc. a woman's uterus) turns toward her legs and feet, you will know this by the following: her large toes are contracted under the nails, pain is present in her legs and thighs, and the cords around her thighs are pulled tight and cause pressure. When the case is such, you should wash the patient frequently with much hot water, foment her if she finds that desirable, and fumigate her below with ill-smelling substances; she should anoint herself generously with rose unguent.

---

[29] See ch. 19 above.

---

202 ὀπτήν Θ: λεπτήν MV.
203 Add. τοῖν ποδοῖν MV.

42. (151 L.) Ἢν δ' ἄναυδος γίνηται ἐξαπίνης, καὶ τὰ σκέλεα ψυχρὰ εὑρήσεις καὶ τὰ γούνατα καὶ τὰς χεῖρας· καὶ ἢν ψαύσῃς τῆς ὑστέρης, οὐκ ἐν κόσμῳ ἐστί· καὶ καρδίη πάλλεται, καὶ βρύκει, καὶ ἱδρὼς πολύς, καὶ τἆλλα ὅσα οἱ ὑπὸ ἱερῆς νούσου ἐπίληπτοι, καὶ ἀπ' οὔατος δρῶσι. ταύτῃσι καταχεῖν ὕδωρ ψυχρὸν χρὴ πολὺ κατὰ τῶν σκελέων τέως, τὰ δ' ἄλλα ποιέειν, ἢν δέῃ,[204] ὡς πρόσθεν εἴρηται.

43. (152 L.) Ἢν δὲ κινηθεῖσαί που προσπέσωσι καὶ ὀδύνην παρέχωσιν, ἐλαίης ψῶραν καὶ δάφνης καὶ κυπαρίσσου πρίσματα ἑψήσας ἐν ὕδατι, ἐς εἴριον ἐμβαλὼν προστίθει. |

328   44. (153 L.) Ὅταν δὲ γυνὴ ἐκ τόκου ἐοῦσα φορτίον αἴρηται μέζον τι τῆς φύσιος, ἢ πτίσσῃ, ἢ κεάσῃ ξύλα, ἢ δράμῃ, ἢ τοῖα δράσῃ, αἱ ὑστέραι ἐκπίπτουσι πρὸς ταῦτα μάλιστα· ἐνίοτε δὲ καὶ πρὸς πταρμόν, ὃ[205] γὰρ βιᾶται, ἢν βίῃ πταρνυμένη[206] ἐπιλάβηται τῆς ῥινός.

Περιπλύνειν χρὴ τὰς ὑστέρας ὕδατι χλιερῷ, ἔπειτα τεύτλων χυλῷ ἀποζέσαντα ὁμοίως, εἶτα ἀκρήτῳ οἴνῳ μέλανι. ἢν δὲ μὴ ἐσακούῃ, μαλθακτήρια ποιέειν χρή· ταῦτα δὲ ποιέειν πρότερον ἢ ψυχθῆναι, καὶ ἐνθεῖναι εἴσω παρηγορικῶς· εἶτ' ἐκτείνειν τὰ σκέλεα καὶ ἐπαλλάξαι, καὶ ὑπὸ ταῦτα ὑποτιθέναι τι μαλθακόν, ποτοῦ δὲ εἴργειν τὰς τοιαύτας χρὴ ὡς μάλιστα, καὶ τὴν κοιλίην χρὴ φυλάσσειν, ὅπως μὴ ἐκταραχθῇ· πρὸς δὲ τὴν ῥῖνα τῶν εὐωδέων διδόναι τι. τὰς τοιαύτας χρὴ ἐξ ὑστέρου, ἢν μὴ ἀτρεμίζωσιν, ἀλλὰ κινέωνται,

42. If a woman suddenly loses her ability to speak, you will also find her legs are cold, as are her knees and hands; and when you palpate her uterus it is not in order, and her heart palpitates, and she grinds (sc. her teeth), and she sweats copiously, and there are other signs like those of patients with the sacred disease, and they do unheard of things: you should pour copious cold water down over the patients' legs for a time, and apply the other measures indicated above if they are necessary.

43. If a woman's uterus moves in some direction so that it prolapses and provokes pain, boil olive scab with laurel and cypress sawdust in water, place it on a piece of wool, and apply this as a suppository.

44. When a woman who has just given birth then lifts some weight greater than she should, or winnows grain, or chops wood, or runs, or does any other such thing, her uterus is very likely to prolapse from this; sometimes it also happens as the result of sneezing, since this can be very violent if a person forcefully holds their nose when they sneeze.

Wash the patient's uterus with tepid water, then in the same way with boiled beet juice, and finally with undiluted dark wine. If there is no response, make emollient pessaries before the uterus cools down, and gently insert them. Then have her extend her legs and cross them, and place some soft object under them; such women must restrict their drinks as much as they can, and their cavity must be protected to prevent its being disturbed; apply fragrant agents to their nose. To such patients, if later their uterus fails to remain in place but moves around, give hellebore.

---

204 ἢν δέῃ om. MV.   205 ὁ V: οὐ ΘM.   206 μὴ add. M.

ἐλλεβορίζειν· ἢν δὲ μὴ ἐξαρκέωσι, καὶ ἐμέται ποιέειν,
καὶ ἀλουτείτω, σιγᾶν τε καὶ ἠρεμέειν.

45. (154 L.) ῍Ην πρησθῶσιν²⁰⁷ αἱ μῆτραι, ἡ γαστὴρ
αἴρεται καὶ φυσᾶται καὶ σμαραγεῖ, καὶ οἱ πόδες οἰδέουσι καὶ τὰ κοῖλα τοῦ προσώπου, καὶ ἡ χροιὴ ἀειδὴς
γίνεται, καὶ τὰ ἐπιμήνια κρύπτεται, καὶ ἡ γονὴ οὐκ
ἐγγίνεται τούτου τοῦ χρόνου· καὶ ἀσθμαίνει, καὶ
ἀφρίζει τε καὶ ἀλύει, καὶ ὅταν ἐξ ὕπνου ἔγρηται, ὀρ-
θοπνοία μιν ἔχει, καὶ ὅ τι ἂν φάγῃ ἢ πίῃ λυπέει
330 αὐτήν, καὶ στένει τε καὶ ἀθυμέει μᾶλλον ἢ πρὶν | φα-
γεῖν, καὶ πνίγεται, καὶ τὰ νεῦρα ἕλκεται, καὶ αἱ μῆτραι
καὶ αἱ κύστιες ἀλγέουσι, καὶ οὐκ ἔστι ψαῦσαι τῇ
χειρί· οὐδὲ τὸ οὖρον προΐενται, οὐδὲ τὴν γονὴν δέχονται.

Ὅταν οὕτως ἔχῃ, φάρμακον χρὴ πῖσαι κάτω, καὶ
θερμῷ λούειν καὶ καθιννύσθαι, καὶ θαμινὰ οὖλον τὸ
σῶμα· ἔστι δ᾽ ὅτε ἄχρι ὀμφαλοῦ καὶ πυριᾶν διαλείποντα, καὶ προστιθέναι ὑφ᾽ ὧν καθαρεῖται καὶ μὴ
ἀδαξήσεται· ὑποθυμιάσθω δὲ τὰ εὐώδεα ὑπὸ τὰ αἰδοῖα, ὑπὸ δὲ τὰς ῥῖνας τὰ κάκοδμα· καὶ ποτήματα
δίδου, ἃ καθαίρουσιν ὑστέρας καὶ ἐλάσει ἐς χώρην·
καὶ τὴν λινόζωστιν ἐσθιέτω, καὶ τὸ γάλα μεταπινέτω,
ὡς ἐπὶ τοῦ πλευροῦ γέγραπται. ἡ δὲ νοῦσος οὐ χρονίη.

²⁰⁷ πρησθ. Foes in note 172: προσθέωσιν V: ἀγρησθῶσιν
ΘΜ.

If such measures are not effective,[30] also induce vomiting, forbid bathing, and have the patient stay still and quiet.

45. If a woman's uterus becomes inflated, her abdomen will be raised, filled with air, and loud, and her legs will have edema, as will also the hollows of her eyes. Her skin becomes unsightly, her menses are absent, and during this time she does not become pregnant. She suffers from shortness of breath, she foams from her mouth, and she is restless; and when she wakes up from her sleep, she has orthopnea. Anything she eats or drinks irritates her, she moans, and she becomes more dejected than she was before she ate; she suffocates, her cords are drawn tight, her uterus and bladder are painful, and it is impossible to palpate her with the hand. No urine is passed, nor is the (sc. male) seed taken in.

When the case is such, administer a purgative potion and wash the patient with hot water both in a sitz bath and by pouring it often over her whole body; foment her sometimes as far up as her navel, leave intermissions, and apply suppositories that will clean but not irritate her. Have the woman fumigate herself below with fragrant agents under her genitalia, and above with ill-smelling ones beneath her nostrils. Give potions that will clean the uterus and drive it back to its proper location: have the patient take mercury herb, and then drink milk, as prescribed for diseases in the side.[31] This disease does not last long.

---

[30] This is the interpretation of Foes and Gardeil, but Littré and Fuchs understand the inadequacy to be of the woman's strength: "quant les forces ne sont pas suffisantes"; "reichen ihre Kräfte dazu nicht aus."

[31] See ch. 19 above.

46. (155 L.) Ἢν σκιρρωθέωσιν αἱ μῆτραι, τό τε στόμα τρηχὺ γίνεται, καὶ τὰ ἐπιμήνια κρύπτεται· ὅταν δὲ ἴῃ, ὡς ψάμμος φαίνεται τρηχεῖα· ἢν δὲ καθάψηται τῷ δακτύλῳ, τρηχὺ τὸ στόμα ὡς πῶρον εὕροις τῆς μήτρης, ὃ προφύεται αἰεί. ὅταν ὧδε ἔχῃ, τῆς κυκλαμίνου χρὴ τρίψαντα, καὶ ἅλας, καὶ σῦκον ὁμοῦ μίσγειν, καὶ ἀναποιέειν μέλιτι βαλανίδα, καὶ πυριάσαντα κλύσαι τοῖσι καθαίρουσιν· ἐσθιέτω δὲ τὴν λινόζωστιν καὶ τὴν κράμβην ἑφθήν, καὶ τὸν χυλὸν ῥυφείτω καὶ πράσων, καὶ θερμῷ λούσθω.

47. (156 L.) Ἢν αἱ μῆτραι σκιρρωθέωσι, τά τε ἐπιμήνια ἐπηλυγάζονται, καὶ τὸ στόμα αὐτῶν ξυμμύει, καὶ οὐ κυΐσκεται, καὶ στερρὸν[208] ἐστι· καὶ ἢν ψαύσῃς, ὡς λίθος δοκέει κεῖθι εἶναι, καὶ τὸ στόμα τρηχὺ καὶ πολύρριζον καὶ οὐ λεῖον ἰδεῖν, καὶ τὸν δάκτυλον οὐκ ἐσίησιν, ὅς μιν καθορῇ· καὶ πῦρ λαμβάνει 332 περίψυχρον, καὶ βρυγμός, καὶ τὰς | μήτρας ὀδύνη ἔχει καὶ τὴν νειαίρην γαστέρα καὶ τοὺς κενεῶνας καὶ τὰς ἴξύας. πάσχει δὲ ταῦτα, ἤν οἱ διαφθαρέντα τὰ ἐπιμήνια σαπῇ· ἔστι δ' ὅτε καὶ ἐκ τόκου ἢ ψύχεος ἢ πονηρῆς διαίτης καὶ ἄλλως. φάρμακον χρὴ πιπίσκειν, καὶ λούειν πολλῷ τῷ θερμῷ, καὶ ὕδατι καὶ ἐλαίῳ πυριᾶν. ὅταν δὲ νεόλουτος ἢ νεοπυρίητος ᾖ, τὴν μήλην καθεὶς ἀναστομοῦν, καὶ ἀνευρῦναι τὸ στόμα αὐτῶν τῷ δακτύλῳ ὁμοίως, καὶ προστιθέναι μαλθακτήρια ὡς εἴρηται, καὶ τῶν ποτημάτων ὡσαύτως πιπίσκειν καὶ θεραπεύειν.

46. If a woman's uterus forms a scirrhus, its mouth will be rough, and her menses will disappear; when they do pass, they will have the appearance of coarse sand. If the patient is palpated with a finger, you will discover that the mouth of her uterus is rough like a stone that is continually advancing. When the case is such, chop cyclamen, mix it together with salt and a fig, make this into a suppository with honey, and then apply a vapor bath and douche with cleaning agents. Have the patient eat mercury herb and boiled cabbage, drink a decoction from these and from leeks, and bathe in hot water.

47. If a woman's uterus forms a scirrhus, her menses will disappear, the mouth of the uterus will close, and she will fail to become pregnant and will remain sterile. If you palpate her, something like a stone seems to be present, the mouth (sc. of the uterus) will appear to be rough and fibrous rather than smooth and will not admit your examining finger. Fever with concomitant chills and chattering of the teeth sets in, and pain occupies her uterus, lower belly, flanks, and loins. These things are suffered by the woman if her menses are spoiled and putrid; sometimes they also follow child birth, or from coldness, or from an unhealthy regimen, or otherwise. Give a purgative potion, bathe the patient in copious hot water, and foment her with water and olive oil. Immediately after she has been bathed and fomented, dilate the mouth of her uterus by inserting a probe, and widen it with a finger in the same way. Apply emollient suppositories as indicated, give the same potions to drink, and manage your therapy along the same lines.

---

[208] στερρόν Θ: ὡς ἕτερον MV.

48. (157 L.) *Ἢν δὲ σκιρρωθῶσιν*[209] *αἱ μήτραι, τὸ στόμα σκληρὸν γίνεται τῶν ὑστερέων καὶ ξυμμέμυκε, καὶ τὰ ἐπιμήνια οὐ γίνεται, ἀλλ᾽ ἐλάσσω καὶ κακίω· καὶ πῦρ καὶ ῥῖγος λαμβάνει, καὶ ὀδύνη ἐμπίπτει ἐς τὴν νειαίρην γαστέρα καὶ τὴν ὀσφῦν καὶ τοὺς κενεῶνας. ὅταν ὧδε ἔχῃ, λούειν πολλῷ καὶ θερμῷ, καὶ χλιάσματα προστιθέναι ἢν ἡ ὀδύνη ἔχῃ, καὶ πυριᾶν τὰς ὑστέρας βληχρῶς πολὺν χρόνον, τῷ ἀπὸ τοῦ σικύου τοῦ ἀγρίου ὕδατι· ἔπειτα προστιθέσθω μαλθακτήρια, τρεῖς ἡμέρας ταῦτα ποιείτω· καὶ ἢν ψαυούσῃ αὐτῇ τὸ στόμα μαλθακὸν τῶν ὑστερέων φαίνηται, μοτοῦν ὠμολίνῳ καθετῆρι, ᾧ τοὺς ἐμπύους, μοτοῖσι τρισί· τῷ μὲν πρώτῳ λεπτῷ, τῷ δὲ δευτέρῳ ὀλίγον παχυτέρῳ· ὁ δὲ παχύτατος*[210] *ἔστω ὅσον ὁ σμικρὸς δάκτυλος, μῆκος δὲ πέντε δακτύλων· χρίων χηνείῳ ἀλείφατι, προστιθέναι, προπυριήσας τοῖσιν εὐώδεσι, καὶ βάλανον μαλθακτικὴν* | *νίτρου ὅσον ἐπαλείφοντα, ὡς μὴ τρώσῃ, οὐ γὰρ κεντέειν χρή, καὶ προσκείσθω δύο ἡμέρας· ἀφίσταται δὲ οἷον λοπὸς καὶ δέρμα παχύ· διαλιπὼν δὲ τρεῖς ἡμέρας, τὴν κυκλάμινον καὶ τὸ σὺν τῷ ναρκισσίνῳ. ἢν δὲ ταῦτα μὴ καθαρθῇ, κατανοῶν*[211] *πολύ, προστιθέναι τέως τὸ ξὺν τῇ βουπρήστει· προσκείσθω δὲ τὴν ἡμέρην· καὶ ἐπὴν δάκνῃ ἰσχυρῶς, ἕλκειν τὴν βάλανον, καὶ διανίζεσθαι τὸ αἰδοῖον ὕδατι θερμῷ καὶ ἐν ἐλαίῳ ἵζεσθαι· τῇ δ᾽ ὑστεραίῃ λουσαμένη ἐλάφου στέαρ τήξασα ἔριον μαλθακὸν ἀναφυρήσασα προστιθέσθω. καὶ ἢν σοι δοκέῃ ἔτι καθάρ-*

334

48. If a woman's uterus forms a scirrhus, its mouth will be hard and closed, and her menses will no longer pass, or be less in amount and worse in quality. Fever and chills set in, and pain invades her lower belly, sacral region, and flanks. When the case is such, bathe the patient in copious hot water, apply warm compresses if the pain is present, and foment the uterus gently over a long time with juice of the squirting cucumber. Then have her insert emollient suppositories for three days, and if on palpating herself she thinks that the mouth of her uterus is soft, plug it with a pessary of raw flax like the one used for internal suppurations, employing three of these: the first should be thin, the second a little wider, and the third one as large as your little finger and five inches in length; smear goose grease on these and insert them after first fomenting with fragrant substances. Also use an emollient suppository anointed with soda, but only so much that it causes no irritation—for it should not sting—and leave this in place for two days: it debrides by causing a thick layer of skin to peel off. After an intermission of three days, apply cyclamen and a suppository of narcissus. If these do not bring about cleaning, observe the case very closely, and for a time apply a buprestis suppository: it should remain in place for the day: when it is causing severe irritation, remove it, flush the vagina thoroughly with hot water, and have the patient take a sitz bath in olive oil. On the next day have the patient bathe and then melt deer's fat, smear it on to a piece of soft wool, and insert it. If you think that

---

[209] σκιρωθῶσιν Θ: σκληρυνθέωσιν MV.

[210] -τατος MV: -τερος Θ.   [211] κατανοῶν Littré: κατανοὸν Θ: κατανόον M: κατὰ νόον V.

σιος δεῖσθαι, διαλιπὼν τρεῖς ἡμέρας, προστιθεῖναι τὸ
σὺν τῷ ναρκισσίνῳ· τῇ δ' ὑστεραίῃ τὸ νέτωπον·
ἔπειτα πάλιν διαλιπὼν τρεῖς ἡμέρας, κλύσαι τὰς
ὑστέρας εὐώδεσι καὶ λιπαροῖσι· τῇ δ' ὑστεραίῃ τὴν
γλήχωνα προστιθέναι μίαν ἡμέρην· τῇ δ' ἐπομένῃ
θυμιᾶν τοῖσιν ἀρώμασι· σιτίοισι δὲ δριμέσι χρήσθω[212] καὶ θαλασσίοισιν, καὶ κρέασιν. ἐν δὲ τοῖσιν
ἐπιμηνίοισι πινέτω τὸν κάστορα, καὶ ἀσιτέουσα
ἀλουτείτω, θυμιωμένη, καὶ τὸν κυκεῶνα πίνουσα,
παρὰ τὸν ἄνδρα ἴτω.

49. (158 L.) Ἢν δὲ μὴ ἐσδέχηται τὸ στόμα τῶν
ὑστερέων τὴν γονήν, ἀλλὰ στειρῶδες ᾖ καὶ ξυμ-
μεμύκῃ, προστιθέσθω μολύβδιον, ὡς εἴρηται, τρεῖς
ἡμέρας λουσαμένη θερμῷ, καὶ μαλθακοῖσι χρήσθω,
καὶ ἀνακεέσθω ὑπτίη, καὶ ἐς ὕδωρ θερμὸν ἱζέσθω, καὶ
εἴριον ἐς μύρον ἀποβάπτουσα προστιθέσθω· καὶ τὴν
νύκτα παρατιθέναι παρὰ | τὰ ἰσχία ἱμάτια εἰρινέα ἢ
λινέα μαλθακά, ὡς μὴ περιρρηδὲς ᾖ τὸ σῶμα. προσθέτοισι μαλθακοῖσι· σμύρναν πιοτάτην καὶ πίσσαν[213]
καὶ κηρὸν καὶ στέαρ χηνός· ἔστω δὲ τῆς σμύρνης μὲν
τὸ ἥμισυ, τῶν δ' ἄλλων διπλάσιον· ἐν εἰρίῳ δὲ προσ-
τίθεσθαι· ἔστω δὲ δύο· προσκείσθω δὲ λουσαμένη τὴν
ἡμέρην, τὸ δὲ ἐς τὴν νύκτα ἔστ' ἂν μαλθακὸν ᾖ, καὶ
ἐπειδὰν ὑφέληται, τῷ εὐώδει ὕδατι περινιζέσθω. ἢ
ἐκλέψας κόκκους πεντεκαίδεκα—ἔστω δὲ καὶ Ἰνδικοῦ
ποσόν, ἢν δοκέῃ δεῖν—ἐν γάλακτι δὲ γυναικὸς κουρο-

336

---

[212] χρήσθω om. Θ.   [213] καὶ πίσσαν om. Θ.

still more cleaning is needed, wait for three days, and then apply the narcissus suppository again, and on the succeeding day oil of bitter almond; then wait again for three days, and wash the uterus with fragrant and fatty injections. On the day after that, apply a suppository of pennyroyal for one day, and on the next day fumigate the woman with aromatics, and have her employ pungent foods along with sea foods and meats. During her menses, have her drink castoreum; then after going without food, she should abstain from bathing, receive fumigations, drink a cyceon, and go to her husband.

49. If the mouth of a woman's uterus will not admit the (sc. male) seed, but it is closed and she is sterile, have her introduce a lead sound as described above, bathe for three days in hot water, employ emollients, recline on her back, take a sitz bath in hot water, and apply as suppository a piece of wool dipped in unguent; at night lay woolen blankets or soft linen sheets against her hips to prevent her body from turning on to its side. Apply emollient suppositories: very oily myrrh, pitch, wax, and goose grease—the myrrh should be half in amount and the other ingredients double as much; make the application in a piece of wool, or rather two. Have the patient make one application during the day, after having taken a bath, and then another one toward night, inserting the suppository while it is still soft, and after she removes it she should wash herself thoroughly with fragrant water. Or shell fifteen (sc. Cnidian) berries—there should also be a certain amount of the Indian medication if this seems necessary—macerate them in the milk of a woman who is nursing a male

τρόφου τρίβειν, καὶ παραμίσγειν ἐλάφου μυελὸν καὶ
τἆλλα ὅσα εἴρηται, καὶ μέλιτι ὀλίγῳ μίσγειν· τὸ εἴριον μαλθακὸν καθαρὸν ἔστω, καὶ προστίθεσθαι τὴν
ἡμέρην· ἢν δὲ βούλῃ ἰσχυρὸν²¹⁴ ποιέειν, καὶ σμύρνης
μικρὸν παραμίσγειν· ἄριστον δὲ ᾠοῦ τὸ πυρρὸν²¹⁵ καὶ
αἰγὸς στέαρ καὶ μέλι καὶ ἔλαιον ῥόδινον, τούτοισιν
ἀναφυρᾶν, παραχλιαίνειν δὲ²¹⁶ παρὰ τὸ πῦρ καὶ τὸ
ἀποστάζον εἰρίῳ ξυλλέγειν καὶ προστιθέναι. ἢ στέαρ
ἐρυθρὸν τὸ ἡδυντὸν χηνός, μύρον ῥόδινον, ταῦτα
συμμίσγειν, καὶ προστιθέναι εἴριον ἀναδεύσας. ἄμεινον δὲ χηνὸς ἔλαιον, ἢ ὄϊος στέαρ, κηρὸς λευκός, ῥητίνη, νέτωπον, ἔλαιον ῥόδινον, ταῦτα τῆξαι· λουσαμένη δὲ προστιθέσθω χλιαρὰ εἴσω πρὸς τὸ στόμα
τῶν ὑστερέων. ἢ ἐλάφου μυελὸν καὶ στέαρ χηνὸς
τήξας ῥοδίνῳ ἢ ἰρίνῳ ἐλαίῳ ἀναφυρᾶν· εἴριον δὲ μαλθακὸν ἄγαν προστίθει.

50. (159 L.) Γυναικὶ ὅταν αἱ ὑστέραι σκληραὶ γένωνται καὶ ἐς τὰ αἰδοῖα ἐξίωσι, καὶ οἱ βουβῶνες
σκληροὶ γίνωνται, καὶ καῦμα ἐν τοῖς | αἰδοίοισιν ἐνῇ,
καρκινοῦσθαι φιλέει πάντα. ὅταν ὧδε ἔχῃ, σικύης
χρὴ τὸ ἔνδον τρῖψαι καὶ κηρίον, ὕδατος κοτύλην ἐπιχέας, ἐνιέναι ἐς τὴν ἕδρην, καὶ καθαίρεται.

51. (160 L.) Ἢν τὸ στόμα τῶν ὑστερέων σκληρὸν
γένηται ὑπὸ ξηρασίης, καὶ ἐν σχήματι ἑτέρῳ ᾖ ὁ αὐ-
χήν, τῷ δακτύλῳ γνώσῃ παραψαύσας· καὶ ἢν ἄνω ὡς
ἐς τὸ ἰσχίον εἰληθῶσι, μὴ προσφέρειν δριμὺ²¹⁷ μηδέν·
ἢν γὰρ ἑλκωθῇ ἐπὴν φλεγμήνῃ, κίνδυνος τὸ πάμπαν

338

infant, and add to this deer's marrow and the other ingredients indicated, and also a little honey; the wool should be soft and clean, and the application should be made during the day. If you want to make the suppository more powerful, mix in a little myrrh as well; but best take the yolk of an egg, goat's fat, honey, and rose oil, mix these together, warm them near a fire, collect the drops on to a piece of wool, and insert this. Or take the finest red goose grease and some rose unguent, mix these together, sop them up in a piece of wool, and insert it. Still better are goose oil, sheep's fat, white wax, resin, oil of bitter almond, and rose oil: melt these and then have the woman bathe and apply the suppositories warm internally against the mouth of her uterus. Or melt deer's marrow and goose grease, and mix them with rose oil or iris oil: apply in a very soft piece of wool.

50. When a woman's uterus becomes hard and moves out into her genitalia, her groins become hardened, and heat is felt in the genitalia; all such tumors tend to become cancerous. When the case is such, knead the insides of a bottle gourd together with wax, mix this into a cotyle of water, and force this into the seat: cleaning will follow.

51. If the mouth of a woman's uterus becomes hard because it is dry, and its neck is deformed, you will recognize this by palpating with your finger. If the uterus rolls upward toward a hip, do not apply any irritant suppositories, since there is a danger that if the uterus ulcerates when it becomes inflamed the woman will become totally

---

214 -ρὸν Θ: -ρότερον MV.   215 πυρρὸν Θ: λευκὸν MV.
216 παραχ. δὲ Θ: ἔλαιον δὲ παραχ. MV.
217 δριμὺ om. Θ.

ἄτοκον γενέσθαι· προστίθεσθαι δὲ ἄσσα μὴ ὀδάξε-
ται, ὑφ' ὧν καθαρεῖται.

52. (161 L.) Ὅταν δὲ σκληραὶ ἔωσιν αἱ ὑστέραι καὶ
τῇδε ἀλγέωσι, προσθετὰ τιθέναι οἷον ἐλάφου μυελὸν
ἢ χήνειον στέαρ ἢ ὕειον, καὶ ἴρινον μύρον ξὺν μέλιτι,
καὶ μαλθάσσειν ᾠοῦ τὸ πυρρὸν καὶ κηρὸν λευκόν·
ἐπίπλασμα, κρίθινον ἢ πύρινον ἄλητον ξὺν ὕδατι καὶ
πηγάνῳ ἕψειν.

53. (162 L.) Ἢν αἱ μῆτραι ξυμμύσωσι, τὰ στόματα
σκληρὰ γίνεται σφέων, καὶ τὴν γονὴν οὐκ ἔτι δέχονται, ἀλλ' αὐτόθι, ἐπὴν ξυνευνηθῇ τῷ ἀνδρί, καὶ ἢν
κινήσῃ τὰ σκέλεα, χωρέει· καὶ ὀδύναι τὴν νειαίρην
γαστέρα καὶ τὰς ἰξύας καὶ τοὺς βουβῶνας ἔχουσι,
καὶ τὰ ἐπιμήνια παντάπασιν οὐ γίνεται· ἢν δὲ γίνηται, ὀλίγα καὶ πονηρὰ καὶ ἄχροα.

Ὅταν ὧδε ἔχῃ, λούειν πολλῷ θερμῷ· μετὰ δὲ τὸ
λοῦτρον διδόναι τοῦ κάστορος ξὺν τῇ ῥίζῃ γλυκυσί-
340 δης, ξυμμίσγειν χρὴ ἐν οἴνῳ | μέλανι οἰνώδει· σιτί-
οισι δὲ χρήσθω, ὡς λεχώ· καὶ ἢν δῆλα ᾖ τὰ ἐπιμήνια,
ἐπισχεῖν μίην ἡμέρην, καὶ πυρίην ὅλην, καὶ φάρμακον διδόναι ποτόν, εἰ ἠθάδες εἶεν, ἄνω, ἢν δὲ δοκέῃ
δεῖσθαι, ἢ κάτω, ἢ γάλα ὄνου ἢ ὀρόν· καὶ πυριᾶσαι
βληχρῶς, καὶ μαλθακτήρια ὑστέρης προσάγειν, καὶ
κυκλάμινον καὶ ναρκίσσινον. σικύης δὲ λαβεῖν τῶν
μακρῶν²¹⁸ τὴν ἐντεριώνην, καὶ κατακνήσας, ἐξελὼν τὸ
σπέρμα, παραστάζων γάλα γυναικὸς κουροτρόφου,
τρίβειν, παραμιγνύων σμύρναν ἄκρητον, καὶ μέλι ὡς
κάλλιστον, καὶ ἔλαιον λευκὸν Αἰγύπτιον, καὶ ποιεῖ μὴ

sterile; instead apply nonirritant suppositories by which it will be cleaned.

52. Suppositories for when a woman's uterus becomes hard and painful: deer's marrow, goose grease, or lard; iris unguent with honey; soften white wax with the yolk of an egg. Poultices: boil barley or wheat meal with water and rue.

53. If a woman's uterus closes together, its mouth will become hard and no longer admit the (sc. male) seed, which will run out right there when she has slept with her husband, if she moves her legs. Pains will occupy her loins, lower belly, and groins, and her menses will disappear completely; or if they do appear, they will be scanty, troublesome, and of an unhealthy color.

When the case is such, bathe the patient in copious hot water, and after the bath give her castoreum together with peony root: mix these into strong dark wine. She should employ the same foods as a parturient. If her menses appear, wait for one day and then foment her whole body; give her a potion, if she is used to this, to act upward if it seems to be required, or downward: either ass's milk or whey; also use a gentle fomentation, and apply emollient suppositories for the uterus containing both cyclamen and narcissus. Take the insides of a large bottle gourd, scrape it, remove the seeds, and on to them dribble the milk of a woman who is nursing a male child; crush these and add undiluted myrrh, very fine honey, and white Egyptian oil,

---

218 μακρῶν Θ: μικρῶν MV.

ὑγρόν, ἀλλὰ ξηρότερον·[219] τοῦτ᾽ ἐμπλάσαι ἐς εἴριον
μαλθακὸν καθαρόν, καὶ ἐμβάψας ἐς ἔλαιον λευκὸν
Αἰγύπτιον, λουσαμένη[220] προστιθέσθαι· ὑποδείσθω δὲ
καὶ ἐν σκέπῃ ἔστω καὶ σκέπεσθαι[221] ὅταν καθαίρηται.

Ἐπὴν δέ σοι δοκέῃ ἀρκούντως ἔχειν, παύεσθαι· ἢν
δὲ μή, αὖτις προστιθέναι ἕτερον· μετὰ δὲ τοῦτο ἀμφε-
λίξασα εἴριον μαλθακόν, στρογγύλον δὲ ποιεῖν, καὶ
ἐμβάπτειν ἐς νέτωπον, καὶ προστίθεσθαι μίαν ἡμέ-
ρην· τῇ δ᾽ ὑστεραίῃ ἔλαιον ῥόδινον ἐν εἰρίῳ· ἔπειτα
ἐλάφου στέαρ τήξας ἐν εἰρίῳ. λούσθω δὲ θερμῷ ὕδατι
αἰεὶ πρὸ τῶν προσθετῶν, καὶ πυριᾶσθαι εὐόδμοισι
βληχρῶς[222] πολὺν χρόνον· μετὰ δέ, μίαν διαλιπὼν
ἡμέρην, κλύσαι, ἢν μὲν φλεγματώδης ᾖ, τῷ σὺν τῷ
κόκκῳ, ἢν δὲ χολώδης, τῷ ἀπὸ τῆς σκαμμωνίης·
μετακλύζειν δὲ καὶ ἐπὶ τούτοισιν ἀμφοτέροισι· τῇ δ᾽
ὑστεραίῃ τῷ ναρκισσίνῳ ἐλαίῳ λευκῷ[223] καὶ οἴνῳ γλυ-
κεῖ· ἢν δὲ μὴ ᾖ τὸ ναρκίσσινον, ἀνθίνῳ, ἢ ἰρίνῳ ὡς
καλλίστῳ· ἔστω δὲ τοῦ ἐλαίου τρίτη μοῖρα τοῦ οἴνου·
ἔπειτα διαλιποῦσα δύο ἡμέρας, προσθέτω τὸ σὺν τῇ
γλήχωνι[224] ἡμέρην μίην. ἢν δὲ γένηται τὰ ἐπιμήνια,
342   πινέτω | νῆστις τὸν κάστορα τρεῖς ἡμέρας ἐν οἴνῳ
εὐώδει λευκῷ· ἐπὴν δὲ ἀπολήγῃ καὶ παραμόνιμα ᾖ,
λούσθω, καὶ διανιψαμένη ὕδατι ψυχρῷ κυκεῶνα ἄναλτον πινέτω, σίτου δὲ μὴ ἁψάσθω· συνευδέτω δὲ τῷ
ἀνδρὶ δύο ἡμέρας ἢ τρεῖς. ὁπόσον δ᾽ ἂν χρόνον καθ-

[219] -ρότερον Θ: -ρόν MV.   [220] -μένη Potter: -μένη codd.
[221] καὶ σκέπεσθαι om. MV.   [222] βληχρῶς om. Θ.

378

making the mixture dryish rather than moist. Smear this on to a clean, soft piece of wool, dip into white Egyptian oil, and apply to the patient as a suppository after she has taken a bath. She should wear a bandage, sit in the shade, and be sheltered while her cleaning is taking place.

When you think the patient is sufficiently restored, stop the treatment: otherwise, apply another suppository. After this she should twist together a piece of soft wool so that it is solid, dip it into oil of bitter almond, and insert it for one day. On the next day infuse rose oil into the wool, and then next melt deer's fat and apply it to the wool. Have the woman wash herself in hot water every time before she applies the suppositories, and foment herself gently with fragrant agents over a longer time. After that desist for one day, and then apply—if she is phlegmatic—an infusion of (sc. Cnidian) berries or—if she is bilious—an infusion with scammony. After each of these infusions she should make another infusion: on the following day infuse with white narcissus oil in sweet wine; if no narcissus is available, then use the finest lily or iris oil: there should be one portion of the oil to three portions of the wine. Then, after waiting for two days, have her apply a suppository made with pennyroyal, for one day. If the menses appear, have her drink in the fasting state castoreum in fragrant, white wine for three days; when they stop and the patient's condition is stable, have her bathe, and after she has been thoroughly washed with cold water drink an unsalted cyceon, but not touch food; she should sleep with her hus-

---

223 ἐλ. λ. om. MV.
224 γλήχωνι MV: μήκωνι Θ.

αἴρηται, τὴν λινόζωστιν ἑψοῦσα ἐν ὕδατι, καὶ ἐπειδὰν
ἑφθὴ ᾖ, ὑποτρῖψαι²²⁵ δὲ σκόροδα καὶ κύμινον καὶ
ἅλας καὶ ἔλαιον, καὶ ἀνακυκᾶν ἐν τούτοισι, καὶ ὀλίγον ἐπιχέαντα τοῦ χυλοῦ ἀναζέσαι· τοῦτο πρὸ τῶν
σιτίων ἐσθίειν· σκόροδα δὲ ἑφθὰ καὶ ὀπτὰ λαμβάνειν
ὡς πλεῖστα· ἢν δ' ἡ λινόζωστις μὴ ἁπαλὴ ᾖ λίην,
συνέψειν χρὴ κράμβην, καὶ ἡδύνειν, καὶ προσηνὲς
εἶναι· καὶ ἢν ἐν γαστρὶ λάβῃ, ὑγιὴς γίνεται, ἢν πάντα
οἱ κατὰ κόσμον εἴη.

54. (163 L.) Ἢν δὲ τὸ στόμα τῶν μητρέων ξυμμύσῃ,
γίνεται ἰσχυρὸν ὡς ἐρινεός· καὶ ἢν ἐπαφήσῃς τῷ δακτύλῳ, ὄψει καὶ σκληρὸν καὶ ξυνιλλόμενον, καὶ τὸν
δάκτυλον οὐκ ἐσίησι· καὶ τὰ ἐπιμήνια κεκρύφαται,
καὶ τὴν γονὴν οὐ δέχονται τούτου τοῦ χρόνου, καὶ
ὀδύνη ἴσχει τὴν νειαίρην γαστέρα καὶ τὴν ὀσφὺν καὶ
τοὺς κενεῶνας· ἔστι δ' ὅτε καὶ ἄνω προσίσταται καὶ
πνίγει.

Ὅταν οὕτως ἔχῃ, φάρμακον χρὴ πεῖσαι κάτω, καὶ
λούειν τῷ θερμῷ πολλῷ, καὶ προστιθέναι ἅσσα μαλ-
θάσσει τὸ στόμα, καὶ ὑπάλειπτρον καθιέναι καὶ ἀνα-
στομοῦν, τόν τε δάκτυλον ὡσαύτως, καὶ αἰονᾶν· ὅταν
δὲ μαλθακὸν ᾖ, προστιθέναι ἅσσα καθαίρει αἷμα, καὶ
τῶν ποτημάτων διδόναι, καὶ πειραθῆναι ὅ τι ἂν προσ-
δέχηται· καὶ τὴν κράμβην ἐσθιέτω, καὶ τὸν χυλὸν
ῥυφείτω.

55. (164 L.) Ἢν ξυμμύωσι αἱ μῆτραι καὶ τὰ ἐπι-
344 μήνια μὴ φαίνηται, | κολοκυνθίδα ἀγρίην καὶ κύμι-
νον²²⁶ Αἰθιοπικὸν καὶ λίτρον καὶ ἅλας Θηβαϊκὸν καὶ

band for two or three days. As long as the cleaning is taking place, have her boil mercury herb in water, and when it is boiled knead garlic, cumin, salt, and olive oil, stir these up together, pour a little of the mercury decoction over this, and boil again: this she should eat before taking food, and consume as much boiled and baked garlic as she can. If the mercury herb is not especially tender, have her boil cabbage with it, season it, and prepare it mild. If the woman becomes pregnant, she will recover if everything goes as it should with her.

54. If the mouth of a woman's uterus closes, it will become tough like a wild fig, and if you palpate her with your finger, you will see that it is hard and compressed, and your finger will not go in. The menses disappear, and the uterus will not take up the (sc. male) seed at this time; pain occupies the lower belly and back, and the flanks; sometimes it is also felt higher up and causes suffocation.

When the case is such, you must give a purgative medication, bathe the patient in copious hot water, and apply suppositories to soften the (sc. uterine) mouth; also introduce a probe and dilate the mouth, do the same with a finger, and foment. When the mouth becomes soft, apply suppositories that will clean blood and give drinks tested for their acceptability. Have the woman eat cabbage and drink its juice.

55. If a woman's uterus closes and her menses are suppressed, boil a wild gourd, Ethiopian cumin, soda, Theban

---

225 ὑποτρῦψαι Θ: ἐκθλῦψαι· ὑποτρίβειν δὲ MV.
226 κύμινον MV: φύλλον Θ.

νεφρίδιον καὶ ἄλευρα καὶ σμύρναν καὶ ῥητίνην, ζέσας ταῦτα ὁμοῦ, μίξας λεῖα, καὶ βάλανον ποιέων, προστίθει.

56. (165 L.) Ἢν θρομβωθῶσιν αἱ μῆτραι, τὸ στόμα αὐτῶν γίνεται οἷον ὀρόβων μεστόν, καὶ ἢν ἐσαφήσῃ,[227] ὄψει οὕτως ἔχον, καὶ τὰ ἐπιμήνια οὐ γίνονται, οὐδὲ ἡ γονή. τέως ἂν ὧδε ἔχῃ, τῆς κυκλαμίνου τὸν φλοιὸν περιλέψας, καὶ σκόροδον, καὶ ἅλας, καὶ σῦκον, καὶ μέλι ὀλίγον, ταῦτα τρῖψαι καὶ συμμίξαι, καὶ ποιεῖν βάλανον, καὶ προσθεῖναι πρὸς τὸ στόμα τῶν μητρέων· καὶ τῶν ἄλλων προσθετῶν, ὅσα τε δριμέα ἐστὶ καὶ κατεσθίει καὶ ὑφ᾽ ὧν καθαίρεται αἷμα, καὶ τῶν ποτημάτων ἅσσα ὑστέρας καθαίρει.

57. (166 L.) Ἢν παρὰ φύσιν αἱ μῆτραι χάνωσι, τὰ ἐπιμήνια χωρέει πλέονα καὶ γλίσχρα καὶ θαμινά,[228] καὶ ἡ γονὴ οὐκ ἐμμένει· καὶ τὸ στόμα κεχηνός ἐστι, καὶ οὐχ οἷόν τε εἰρύεσθαι τὴν γονήν, καὶ πῦρ καὶ ῥῖγος λαμβάνει τὴν κάτω κοιλίην καὶ τὰς ἰξύας. ἡ δὲ νοῦσος λάζεται ἐκ ῥόου αἱματώδεος, γίνεται δὲ καὶ ὅταν τὰ ἐπιμήνια ἐξαπίνης ἱστάμενα ῥαγῇ.

Δίαιτα δὲ πρόσθεν εἴρηται. χρὴ δὲ καὶ προσθέτοισιν ἐν ἀρχῇ μὲν καθαρτηρίοισιν ἅμα καὶ μαλθακοῖσιν, ἔπειτα ἠρέμα στύφουσι· καὶ σπόγγοισι τὰ κάτω τοῦ ὀμφαλοῦ πυριᾶν μυρσίνης ὕδατι, ἢ βάτον ἐναφέψειν, ἢ ἐλαιῶν φύλλα, ἢ ῥόδον, ἢ οἰνάνθην ἀμπέλου.

58. (167 L.) Ἢν τὸ στόμα τῶν μητρέων ἀναχάνῃ <μᾶλλον>[229] ἢ ὡς πέφυκεν, τὰ ἐπιμήνια γίνονται πλείω καὶ κακίω πάντα καὶ ὑγρότερα καὶ διὰ πλείονος

salt, kidney fat, meal, myrrh and resin; knead these smoothly together, form into a suppository, and apply.

56. If a woman's uterus produces clots, its mouth fills up with what looks like vetches, and if you palpate it, you will find that this is how it really is; the menses do not pass, nor does the (sc. male) seed. As long as this condition persists, take peeled cyclamen bark, garlic, salt, a fig, and a little honey, crush these and knead them together, form this into a suppository, and apply it against the mouth of the uterus. Apply other pungent suppositories, and give foods that will clean blood, and drinks that will clean the uterus.

57. If a woman's uterus gapes open unnaturally, the menses will pass in a greater amount, sticky, and without intermission, and the seed will not remain inside. The mouth (sc. of the uterus) gapes open and is unable to retain the seed, and fever and chills invade the lower cavity and around the loins. This disease follows after a bloody flux, and also occurs when stagnant menses suddenly break forth.

The appropriate regimen has been described above: you must use suppositories, in the beginning using cathartics which are at the same time emollients, and then mild astringents. Also foment the region below the navel with sponges using myrtle juice, or apply boiled brambles, or leaves of olive plants or roses, or grapevine blossoms.

58. If the mouth of a woman's uterus gapes open more than is natural, her menses all pass in greater quantity, in worse quality, more watery, and at longer intervals, and

---

227 ἐσαφ. Θ: ἐπαφ. MV.   228 Add. καὶ πυκινά Θ.
229 Add. Foes in note 205.

346 χρόνου, καὶ ἡ γονὴ οὐχ ἅπτεται, οὐδὲ μένει, | ἀλλὰ
πάλιν ἔξεισι· κὴν ἐπιμένῃ, καὶ τὸ στόμα εὑρήσεις
διαπεπλιχὸς· καὶ ἀδυναμίη ἴσχει ὑπὸ τῶν ἐμμηνίων,
καὶ κούφη ἐστὶ καὶ ἄτονος, καὶ πρόσω χωρέει, τὰ δὲ
κράτεα χαλᾶται, καὶ πῦρ βληχρόν, καὶ ῥῖγος, καὶ
ὀδύνη ἴσχει τὴν νειαίρην γαστέρα καὶ τοὺς κενεῶνας
καὶ τὰς ἰξύας. πάσχει δὲ ταῦτα μάλιστα, ἤν τι ἐν
αὐτῇ διαφθαρὲν σήπηται καὶ παγῇ, πάσχουσι δὲ καὶ
ἐκ τόκων, αἱ δὲ καὶ ἄλλως.

Ὅταν ὧδε ἔχῃ, φάρμακον χρὴ πιπίσκειν, καὶ χρη-
σιμωτέρη ἔσται· καὶ ἢν ὀδύνη ἔχῃ, τῶν χλιασμάτων
προστιθέναι, καὶ τῷ ψυχρῷ λούειν, καὶ διαλείποντα
κλύζειν, καὶ τῶν ποτημάτων διδόναι ὅ τι ἂν μάλιστα
προσδέχηται, καὶ ὑποθυμιᾶν ὅσα ξηραίνει, καὶ πουλύποδας ἐσθιέτω καὶ τὴν λινόζωστιν. ἢν δὲ μὴ με-
μύκωσιν αἱ μῆτραι ὡς χρή, καθίννυσθαι ἐν ὕδατι
μυρσίνης ἐναφεψημένης, ἢ σχίνου, ἢ ἀμπέλου, ἢ
ἐλαίης φύλλοισιν, ἢ ῥόδων. δίαιτα ἢ λέγεται ἐπὶ τοῦ
ἐρυθροῦ ῥόου· ἄριστον δέ δίψα, ἔμετοι θαμέες, λουτρῶν εἴρξεις. ὅταν ἀναστομωθῶσι μᾶλλον τοῦ δέοντος αἱ ὑστέραι καὶ μὴ μεμύκωσι, καθάρσιος δέονται
καὶ κλυσμῶν καὶ θυμιημάτων.

59. (168 L.) Ἢν δὲ λεανθέωσιν αἱ μῆτραι,[230] τὰ
ἐπιμήνια πλείω γίνονται καὶ κακίω καὶ ὑγρότερα καὶ
πυκνά, καὶ ἡ γονὴ οὐκ ἐμμένει, ἀλλὰ πάλιν οἴχεται,
348 καὶ τῷ δακτύλῳ ὄψει τὸ στόμα λεῖον, | καὶ ἀδυνασίη
αὐτὴν λαμβάνει ὑπὸ τῶν ἐμμηνίων, καὶ πυρετὸς καὶ
ῥῖγος, ὀδύνη τε ἐς τὴν νειαίρην γαστέρα, καὶ τὰς

the seed cannot take hold and remain in place but runs
back out: if this condition goes on, you will also discover
that the mouth is wide open. Weakness (sc. of the uterus)
accompanies the menses, along with lightness and slack-
ness; it moves forward, its strong ligaments give way, mild
fever sets in with chills, and pain occupies the lower belly,
flanks, and loins. A woman suffers these things, in most
cases, if something inside her withers, decomposes, and
congeals; some women also suffer them after child birth,
and others in other ways too.

When the case is such, give a purgative potion and the
patient will improve. If pain is present, apply warming
poultices, bathe with cold water, after leaving an interval
apply a douche, give whichever drink is most readily received, fumigate from below to dry it, and have her eat
octopus and mercury herb. If the uterus fails to close as it
should, have the patient take a sitz bath in water with
boiled myrtle berries, or mastic, vine, olive, or rose leaves.
Regimen: as described for the red flux: best are to thirst,
to vomit frequently, and to avoid baths. When the uterus
has dilated more than it should and fails to close, it requires cleaning, injections, and fumigations.

59. If a woman's uterus becomes slippery, her menses
will increase in amount and deteriorate in quality, and be
more fluid and more frequent; the seed will not remain
inside but will run back out, and with your finger you will
feel that the uterine mouth is slippery. Weakness befalls
this patient during her menses, along with fever, chills,
and pain in her lower abdomen, loins, and flanks, espe-

---

230 λ. αἱ μ. Θ: ἀνθῶσιν MV.

ἰξύας καὶ τοὺς κενεῶνας, μάλιστα δὲ ἤν τι ἐν αὐτῇ
διαφθαρὲν σαπῇ, καὶ ἐκ τόκου καὶ ἄλλως. ὅταν οὕτως
ἔχῃ, θεραπεύειν χρὴ, ὅπου ἂν ἡ ὀδύνη ἔχῃ, ὡς ἐπὶ τῶν
πρόσθεν γέγραπται.

60. (169 L.) Ἢν αἱ μῆτραι φλεγμήνωσι, τὰ ἐπιμήνια ἐπηλυγίζονται, καὶ ὁ τράχηλος συναρθμοῦται,
καὶ πυρετὸς ὀξὺς καὶ γνώμης ἁπτόμενος, καὶ πονηρὰ
καὶ ὀλίγα οἱ ἐπιφαίνεται· καὶ ὅταν νῆστις ᾖ, ἔμετος
αὐτῇ ἐπέρχεται· ὁκόταν δέ τι φάγῃ, ταῦτα ἐμέει· καὶ
ὀδύνη τὴν νειαίρην γαστέρα ἴσχει καὶ τὰς ἰξύας, καὶ
ἀποψύχει, ἢ[231] καὶ περίψυξις ὅλου τοῦ σώματος· ἡ δὲ
γαστὴρ τοτὲ μὲν σκληρή,[232] τοτὲ δὲ μαλθακή, καὶ
ἐμπίμπραται, καὶ ἀείρεται, καὶ δοκέει ἐν γαστρὶ ἔχειν·
καὶ ἔστι δ' ὅτε κενὸν φαίνεται τὸ πλήρωμα τῆς γα-
στρός· καὶ ἐμπίμπραται ἡ κοιλίη ὕδατος, καὶ ὁ ὀμφαλὸς ἐξίσχει, καὶ τὸ στόμα ἰσχνόν, καὶ ἐξαπίνης
ἐφάνη τὰ ἐπιμήνια τρύζοντα, ὀλίγα καὶ πονηρά· καὶ
λεπτύνει τε τὰς κληῖδας καὶ τὴν δειρήν, καὶ οἱ πόδες
οἰδίσκονται καὶ αἱ πέζαι μάλιστα.

Ὅταν ὧδε ἔχῃ, φάρμακον πῖσαι κάτω, καὶ[233] πυριᾶσαι τὰς ὑστέρας [ὡς][234] εὐώδεσι, τὰς ἡμέρας τοῖσι
μολυβδίοισι, καὶ λοῦσαι θερμῷ ὕδατι πρὸ τῆς προσθέσιος μὴ δάκνοντι· καὶ μετὰ τὴν πρόσθεσιν κλύζειν
τὰς ὑστέρας·[235] κνήστρου δύο πόσιας ἑψῆσαι ἐν κοτύλῃ ὕδατος, καὶ ἀποχέαντα τὸ ὕδωρ κλύσαι, κοτύλην
συμμίξαντα μέλιτος καὶ ἐλαίου ναρκισσίνου ἢ ἀνθι-
νοῦ· μετὰ δὲ διαλιπεῖν | ἡμέρας τρεῖς, εἶτα κλύσαι τῷ

350

cially if something inside her has withered and is decomposing; but also after child birth, or otherwise. When the case is such, you must apply your treatment at the site of the pain, in the way described in the cases above.

60. If a woman's uterus becomes inflamed, her menses will be suppressed, the uterine neck will be contracted, she will have acute fever which brings delirium; a painful, scanty flux will appear. When the woman fasts, she has retching, and when she eats something, she vomits it back up; pain occupies her lower abdomen and loins, she is faint, and her whole body feels cold. The patient's belly is at one time constipated, and at another time relaxed; at one time it fills up and swells, so that she seems to be pregnant, while at other times the fullness of the belly recedes again. The belly fills up with fluid, the navel protrudes, and the mouth of the uterus withers, and then the menses suddenly return and squirt out noisy, scanty, and painful. The woman becomes thin around her collarbones and neck, and her feet swell, especially in the insteps.

When the case is such, have the patient drink a purgative medication, and foment her uterus with fragrant vapor baths; during the day insert lead sounds, after first bathing her with a hot, nonirritating fluid; after the application, flush her uterus with a douche: boil two drafts of cnestron in a cotyle of water, decant the water, and infuse it after adding a cotyle of honey and oil of narcissus or lily. After that, leave three days, and then flush again

---

231 λιποψυχέει add. ΘM.
233 Add. προσθέτοισι MV.
235 τὰς ὑστέρας om. MV.

232 -ρή Θ: -ροτέρη MV.
234 Del. Fuchs (note 117).

387

σὺν τῷ ὄξει· καὶ πρὸ τοῦ σιτίου τὴν λινόζωστιν
ἑψοῦσα σὺν οἴνῳ ποσῷ, ἐσθιέτω τὴν κράμβην καὶ τὸν
χυλὸν ῥυφείτω. εἰ δ' ἐπιμένοι, φάρμακα πιπίσκειν,
ὑφ' ὧν ὕδωρ καθαίρεται τὰς ὑστέρας. σιτίοισι δὲ
χρῆσθαι ἄρτοισι καὶ λαχάνοισιν ἐφθοῖσι τακεροῖσι,[236] θαλασσίοισι δὲ μᾶλλον ἢ κρέασιν, ἢ ἀκρο-
κωλίοισιν ἐφθοῖσι τακεροῖσι. καὶ ταλαιπωρέειν ὡς
πλεῖστα καὶ πρὸ τοῦ σιτίου καὶ μετὰ τὰ σιτία, καὶ
λοῦσθαι ὡς ἐλάχιστα ψυχρῷ, καὶ γλυκέων καὶ λι-
παρῶν εἰργέσθω· τὰς δὲ διὰ μεσηγὺ τῶν καθαρσίων
ἀδίαντον πινέτω, ἐν ὄξει κεκρημένῳ νῆστις. ἡ δὲ νοῦσος θανατώδης, ἐκφυγεῖν δ' ὀλίγαι δύνανται, ἢν μὴ
ἐν γαστρὶ ἔχωσιν.

61. (170 L.) Φλεγμασίης μητρέων· τὰ ἐπιμήνια ἐπηλυγίζονται, καὶ ὅταν ἄσιτος ᾖ, ἐμέει, ἢν δὲ βεβρώκῃ,
ὀδύνη ἴσχει τὴν νειαίρην γαστέρα καὶ τὰς ἰξύας, καὶ
ὅλη ἡ κοιλίη τοτὲ μὲν σκληρή, τοτὲ δὲ μαλθακὴ γίνεται, καθίσταται δὲ οὐ πάνυ· κοιλίη μεγάλη γίνεται
καὶ οὐ καθαίρεται, καὶ[237] δοκέει κυεῖν, καὶ πάσχει ὅσα
περ καὶ αἱ ἐγκύμονες· καὶ ἢν θιγγάνῃς τῆς κοιλίης,
κοῦφον τὸ οἴδημα ὡς ἀσκοῦ, καὶ ὅταν δοκέῃ τόκου
ὥρη εἶναι, αἱ μῆτραι ξυμπίπτουσι, καὶ τὰ ἐπιμήνια
ὀλίγα καὶ κακίονα. ταύτην φάρμακον πιπίσκειν κάτω,
ἢ προστιθέναι τῶν καθαρτηρίων, καθαρθεῖσα δὲ
ὑγιὴς γίνεται.

62. (171 L.) Ἢν φλεγμαίνωσιν αἱ μῆτραι, οὐ

---

[236] Om. τακεροῖσι Θ.   [237] οὐ add. MV.

with a douche of vinegar; before the patient eats have her boil mercury herb in a suitable amount of wine, and eat cabbage, and drink its juice. If the condition goes on, give potions to clean water from the uterus. As foods employ breads, vegetables boiled until they are soft, sea foods rather than meats, or the extremities of animals boiled until they are tender. The patient should exert herself as much as she can both before and after eating, bathe in a very little cold water, and avoid all sweet and fat foods. On the days between her cleanings, have her drink maiden-hair in diluted vinegar while fasting. The disease is mortal, and few women are able to escape from it, unless they become pregnant.

61. Inflammation of a woman's uterus: the menses are suppressed, when she goes without food she vomits, but when she eats pain occupies her lower belly and loins: her whole cavity is at one time harder while at another time it becomes soft, but it never returns to its proper state. The belly becomes enlarged and is not emptied, and the patient seems to be pregnant and suffers the same things pregnant women do. If you palpate the abdomen, the swelling is light like in a bladder; when the time of delivery arrives, the uterus will collapse and scanty very unhealthy menses will be passed. Such a patient should be given a purgative medication to drink, and cleaning suppositories: if she is cleaned, she will recover.

62. If a woman's uterus becomes inflamed, it does not

ψαύει·[238] ἢν δέ τι σφακελίζῃ τε[239] καὶ παλιγκοταίνῃ,
352 καὶ πῦρ ἔχει ὀξὺ καὶ μέγα, καὶ φρίκη σκληρή· | τὰ
ἀμφὶ τὰ αἰδοῖα χωρία[240] ἐκπάγλως αἴθεται καὶ δάκνεται καὶ ὀργᾷ,[241] καὶ εἴ τις ἐπαφήσει τῷ δακτύλῳ, καὶ
αὖτις κάκιον ἴσχει καὶ ἀδάξεται· καὶ τὴν κεφαλὴν
ἀλγέει καὶ τὸ βρέγμα, καὶ ἀχλύς, ἱδρώς τε περὶ μετω-
πίδιος· ἄκρεα ψύχονται καὶ τετραμαίνουσι, καὶ κῶμα
ἔχει ἄλλοτε καὶ ἄλλοτε,[242] καὶ ἐσακούειν οὐκ ἐθέλει·
οὐδὲ ἡ ὑστέρη ἐνεργεῖ· ἀσιτίη πολλή, καὶ στόμαχος
οὐ πάμπαν εἴρυαται τὴν τροφὴν καὶ ἡ κοιλίη· καὶ
βοᾷ, καὶ ἀναΐσσει, καὶ ὀδυνᾶται πᾶσα καὶ ἦτρον καὶ
βουβῶνας καὶ ἰξύας καὶ παραφύσιας· καὶ ταχὺ θνήσκουσιν.

Ἢν δὲ αἱ ὀδύναι καταιγίζωσι, σπόγγοισι θερμοῖς
ἐξ ὕδατος ἢ ἐλαίου ἐκπεπιεσμένοισι πυριᾶν· καὶ
προσθέτοισι μαλθακοῖσι, μυελὸν ἐλάφου καὶ χηνὸς
ἄλειφα καὶ κηρὸν λευκὸν καὶ ᾠοῦ τὸ πυρρὸν ἢ κηρωτὴν πισσηρὴν προστιθέναι σὺν ῥητίνῃ· καθαίρειν δὲ
ὀνείῳ γάλακτι ἢ αἰγείῳ, ἢ ζωμῷ ὄρνιθος· οἶνον δὲ μὴ
πίνειν, καὶ πτισάνης χυλὸν ῥοφείτω.

63. (172 L.) Μητρέων ὀδύνης ἔγχυτον· ἢν ὀδύνη
μοῦνον ἢ σπερχνὴ καὶ βίαιος, ἀχλὺς ἐν τῇσι μήτρῃσιν ἔνι, καὶ οὐκ ἔξεισι τὸ πνεῦμα, ἀλλ᾽ αὐτόθι μένει,
κακόν· ἀνδραφάξιος ἀγρίης καρπὸν καὶ τεῦτλα τρίψας ὁμοῦ λεῖα, χλιήνας τε, ἔγχεον ἐς τὰς μήτρας.

---

[238] οὐ ψαύει Θ: ψαύεται MV.   [239] τε om. MV.
[240] χωρία om. MV.   [241] ὀργᾷ Θ: ὁρμᾷ MV.

touch. . . .[32] If the lesion becomes gangrenous and aggressive in some way, violent, acute fever sets in along with unremitting chills. The genital region burns fiercely, and is angry and swollen; if it is palpated with a finger, it becomes even worse in its irritation. This patient has headache, a pain in her bregma, mistiness of her vision, and sweating over her forehead; her extremities become cold and tremulous, coma is present from time to time, and she is unable to pay attention. The uterus is inert, there is a great distaste for food, the gullet refuses steadfastly to admit nutriment, and so too does the cavity. The patient shouts out loud, leaps up, and suffers pains everywhere in her loins, lower abdomen, groins, and at the attachments (sc. of the muscles). Such patients soon succumb.

If the pains are pressing forcefully, foment with sponges immersed in warm water or olive oil, and squeezed out. Also apply emollient suppositories containing deer's marrow, goose grease, white wax and egg yolk, or a wax salve with pitch and resin. Clean with ass's or goat's milk, or chicken soup, forbid wine, and give barley gruel to drink.

63. Infusion for a pain in the uterus: if the only sign is an intense, violent pain, there is a mass in the uterus and the wind in it will not come out, but rather remains inside: this is a bad state. Treat thus: grind seeds of wild orach together with beets until smooth, warm it, and infuse into the uterus.

---

[32] Fuchs conjectures a gap in the text at this point.

---

[242] καὶ ἄλλοτε om. MV.

Ὑστερέων ὀδύνης· οἶνον ὡς ἥδιστον ἴσον ἴσῳ κεράσας τρία ἡμίχοα Ἀττικά, καὶ μαράθου ῥίζας καὶ
τοῦ καρποῦ τριτημόριον, καὶ ῥοδίνου ἐλαίου ἡμικο-
354 τύλιον· ταῦτα ἐμβαλεῖν ἐς ἐχῖνον καινὸν | καὶ τὸν οἶ-
νον ἐπιχέαι, εἶτα πυριάσαι. προστιθέναι σκίλλαν,
ἔστ᾽ ἂν τὸ στόμα μαλθακὸν ᾖ καὶ φαρκιδῶδες.

64. (173 L.) Καὶ ὅταν ἐν τῇ καθάρσει φλύκταιναι
ἀνὰ τὸ στόμα τῆς ὑστέρης θύωσιν· σάρκα βοὸς ἢ
πικερίῳ <ἢ>[243] χηνείῳ στέατι καὶ ἀννήσῳ χρίσασα
λείοις, τὴν σάρκα ἐς τὸ αἰδοῖον ἐντιθέναι.

65. (174 L.) Ἢν ἐρυσίπελας ἔχῃ ἐν τῇσι μήτρῃσιν,
οἰδέει τοὺς πόδας καὶ τοὺς μαζοὺς καὶ τὸ σῶμα
ὅλον,[244] καὶ πόνος λαμβάνει μιν, καὶ ὀρθοπνοίη γίνεται, καὶ ἀλγέει[245] τοὺς κενεῶνας καὶ τὸ ὑπογάστριον
καὶ τὰ στέρνα καὶ τὴν κεφαλήν· τρόμος τ᾽ ἔχει, καὶ
τὰς χεῖρας ναρκᾷ καὶ τοὺς βουβῶνας, καὶ τὰς ἰγνύας
τρέμει· ἐνίοτε δὲ καὶ ἐν τῇσιν ἰγνύῃσι πελιδναὶ γίνονται, καὶ κουφίζει πολλὸν[246] χρόνον· καὶ ἡ χροιή,
μάλιστα δὲ καὶ οἱ μαζοὶ ἀείρονται κατὰ τὴν ὁμο-
εθνίην, ἀλλὰ γὰρ καὶ οὐ πάνυ τι ἀλγέει· καὶ πυρετὸς
καὶ ῥῖγος λαμβάνει, καὶ ἐρυθρὸν τὸ πρόσωπον, καὶ
δίψα ἰσχυρή, καὶ τὸ ἧπαρ ξηραίνεται. ταῦτα ἢν ἐγ-
κύμονι περιπέσῃ, θνήσκει, καὶ οὐκ ἂν ἐκφύγοι.

(174bis L.) Ἢν ἐρυσίπελας ἐν τῇσι μήτρῃσι γίνηται, οἰδήματα γίνεται πλεννωδέστατα ἀπὸ τῶν ποδῶν

---

[243] ἢ add. Linden after Foes, note 232.   [244] ὅλον om. MV.
[245] Om. ἀλγέει Θ.   [246] πολλὸν Θ: ὀλίγον MV.

To relieve a pain in the uterus: three Attic hemichoa of very pleasant wine diluted one to one (sc. with water), one third part of roots and seeds of fennel, and a half cotyle of rose oil: place the ingredients in a new, large, widemouthed jar, add the wine, and foment with this. Also apply a squill suppository until the uterine mouth becomes soft and wrinkled.

64. When during (sc. menstrual) cleaning blisters form around the mouth of a woman's uterus, have her anoint a piece of beef with butter or goose grease and anise, and insert the beef into her vagina.

65. If erysipelas develops in a woman's uterus, her feet, breasts, and body in general will swell up, she becomes distressed and suffers from orthopnea, she has pains in her flanks, hypogastrium, sternum, and head, she trembles, and she has numbness in her hands and groins, and spasms in her hams; sometimes livid spots also appear in the hams, which provide relief for a long time. The skin, and in particular the breasts, become raised as the result of sympathy, but the patient does not suffer any special pain. Fever and chills are present, the face becomes red, there is violent thirst, and the liver[33] dries out. If this happens in a woman who is pregnant, she will die and not escape.

If erysipelas develops in a woman's uterus, very turgid swellings form beginning from her feet and extending all

[33] The ancient glossators and Galen read the otherwise unknown word ἴκταρ for "liver" here, with the possible meanings of "female genitalia" or "moisture."

ἀρξάμενα ἐς τὰ σκέλεα πάντα καὶ ἐς τὴν ὀσφύν. ὅσῳ
δ᾽ ἂν ὁ χρόνος πλείων γίνηται, καὶ ὁ θώρηξ ἐπαΐει·
καὶ οἰδίσκεται, καὶ περιψύχεται πᾶσα, καὶ πῦρ ἔχει
μέγα, καὶ | ῥῖγος ἐπιλαμβάνει, καὶ πνεῦμα πύκινον,
καὶ λιποθυμίη, καὶ ἀσθενείη, καὶ ὀδύνη παντὸς τοῦ
σώματος· δυσθυμέει τε καὶ αἰολᾶται τῇ γνώμῃ, καὶ
τὸ πάθος ἀνέρχεται ἐκ τῆς κάτω κοιλίης ἐς τὰς ἰξύας
καὶ ἐς τὰ νῶτα καὶ τὰ ὑπὸ τὰ ὑποχόνδρια καὶ τὰ
στέρνα καὶ τὸν τράχηλον καὶ τὴν κεφαλὴν καὶ τὸν
στόμαχον,[247] καὶ δοκέει θανεῖσθαι· ὅταν δὲ λύηται ἡ
ὀδύνη, νάρκα ἴσχει τὰς ἰξύας καὶ τοὺς βουβῶνας καὶ
τὰ σκέλεα, καὶ τὰ ἐν τῇσιν ἰγνύῃσι πελιδνὰ γίνονται,
καὶ ὀλίγον χρόνον δοκέει ῥάων εἶναι· ἔπειτα αὖτις πονέεται, καὶ ὁ χρὼς φλυκταινέων καταπίμπλαται, καὶ
τὸ πρόσωπον ἐρυθήματα λάζεται—προφανέα καὶ δη-
λεόμενα—καὶ ὁ φάρυγξ αὖος· γλῶσσα τρηχείη.

Αὕτη ἡ νοῦσος ἐγκύμονα κτείνει. ἢν δὲ μὴ κύῃ,
ἰητρείην χρὴ προσάγειν· ὄνου γάλα πιπίσκειν καὶ
καθαίρειν· ἢν δὲ μὴ λύηται ὧδε, ψύχειν τὴν κοιλίην
μαλθακοῖσι ψύγμασι, καὶ προσθέτοισι μὴ περισκε-
λέσι· καὶ καθῆραι κούφοισιν ἐκ τοῦ κατ᾽ ὀλίγον, καὶ
ἐμέειν· ἀγαθὸν δὲ ἀκτῆς φύλλα λαμβάνειν ἑφθὰ σὺν
ὀριγάνῳ ἢ θύμῳ ἢ πηγάνῳ· ἢν δὲ τὸ πῦρ μεθίῃ, καὶ
οἶνον καὶ σιτία διδόναι γλυκέα. παῦραι δὲ ὑγιαίνον-
ται.

66. (175 L.) Ὕδερος ἐγγίνεται ἐν τῇσι μήτρῃσι, τὰ
ἐπιμήνια χωρέει ἰσατώδεα[248] καὶ οὐ πάνυ αἱματώδεα·
οἰδέει ἡ ὑστέρη καὶ φλέβια ὅσα[249] ἄγχιστα, καὶ οὐ[250]

along her legs up to her sacrum; with the passage of time, her thorax too becomes involved. This patient fills with edema, becomes cold all over, and suffers great fever along with chills; she breathes rapidly, loses consciousness, is weak, and has pain throughout her body; she becomes depressed and restless in her mind, the disease spreads up from her lower cavity into her loins, back below the hypochondria, sternum, throat, head, and gullet, and she seems to be on the point of death. When the pain relents, numbness occupies her groins, legs, and loins, livid lesions appear in her hams, and for a short time she seems to be better. But then her suffering returns, her skin is covered with blisters, red spots appear on her face—prominent and malignant—and her throat is dry and her tongue is rough.

This disease is fatal for a woman who is pregnant; if she is not pregnant, the following treatment must be applied: give ass's milk to drink, and clean the body; if this brings no relief, cool the cavity with gentle cold compresses, and use nonirritating suppositories; clean gradually with light agents, and give an emetic. Elder leaves are good to take boiled together with marjoram, thyme, or rue. If the fever relents, also give wine and sweet foods. Few women recover their health.

66. Dropsy arises in a woman's uterus, and her menses pass with the appearance of woad, and not very bloody; the uterus and its closest vessels swell, the woman fails to

---

247 καὶ τὴν κεφαλὴν . . . στόμαχον om. Θ.

248 ἰσατώδεα Θ: ἀσώδεα καὶ ὑδατώδεα MV.

249 φλέβια ὅσα Potter: φλειβει ὅσα Θ: φλέβες καὶ ὅσα M: φλέβες V.   250 οὐ om. MV.

κυΐσκεται, ἔπειτα πνίγεται· καὶ οἱ μαζοὶ ῥέουσι, καὶ ἡ
νειαίρα γαστὴρ σκληρή ἐστι καὶ οἰδέει πᾶς ὁ ἀμφὶ
πέριξ χῶρος, καὶ ἀλγέει, εἴ τις ψαύσειε· καὶ πυρετὸς
καὶ βρυγμὸς ἴσχει, καὶ ὀδύνη σπερχνὴ ἐς τοὺς κενεῶ-
358   νας καὶ ἐς | τὰς ἰξύας, καὶ ἐξονειροῖ, καὶ κάκιον ἴσχει.

Λούειν δὲ θερμῷ καὶ χλιαίνειν, φάρμακον δὲ πιπίσκειν, καὶ πυριᾶν βληχρῇσι πυρίῃσι· καὶ κυκλάμινον
τριώβολον ἐν ὀθονίῳ προστιθέσθω ἀποδήσασα· καὶ
κυπαρίσσου ξύλον[251] βρέχειν ἐν ὕδατι, καὶ κανθαρίδας τρίβουσα ἐντιθέσθω ὀλίγον χρόνον, καὶ διὰ πλείονος ἐνεργεῖ. προσθετόν, κύμινον ὅσον χήμη, σταφὶς
λευκή. ἄλλο· κνίδης καρπός, ἄρου ῥίζα, τούτων ὃ
βούλει ἐπιεικέως προστιθέναι. ἐπειδὰν δ᾿ ἐκκαθήρῃς,
κλύζειν καὶ ἀφαιρέειν καὶ ἐγκλύζειν τὰ αἰδοῖα, καὶ
συγκοιμᾶσθαι,[252] καὶ ἢν διενέγκῃ τὸ ἔμβρυον, ἐκ-
καθαίρεται πᾶσα καὶ ὑγιὴς γίνεται.

67. (176 L.) Ὑδέρου μητρέων· ὕφαιμον ῥέει ἰχωρο-
ειδές, καὶ καθαίρεται, καὶ δάκνει σφόδρα καὶ ἑλκοῖ ὡς
ἅλμη τὰ αἰδοῖα καὶ τὰ πέριξ, καὶ ὅπου ἂν ἐπιστάξῃ
ἑλκοῖ, καὶ ἡ χροιὴ ἱκτεροειδής· τὰ δ᾿ ἄλλα καθαίρεται
πλῆθος, ὡς ἐν τοῖσιν ἄλλοισι ῥόοις. ἡ δὲ νοῦσος βληχροτέρη τε καὶ ἄλλως θανατώδης γίνεται, ἢν ἐξελκω-
θῶσιν αἱ μῆτραι.

Ταύτην θεραπεύειν ὡς τὴν ὑπὸ λευκοῦ ῥόου ἐχομέ-
νην, καὶ γάλα ὄνειον πιπίσκειν, καὶ ἰσχναίνειν, καὶ
ἰᾶσθαι φαρμάκοισι τοῖς εἰρημένοισιν. Ὑδέρου ἐλλει-

251 ξύλον Potter: χυλὸν codd.   252 Add. ἅμα MV.

become pregnant, she suffocates, her breasts secrete, her lower belly becomes hard, and the whole area there becomes edematous and on being touched feels pain. Fever and chattering of the teeth follow, violent pain invades the flanks and loins, the woman has nocturnal emissions, and her condition deteriorates.

Bathe the patient in hot water, warm her, give her a purgative potion, and foment her with mild fomentations. She should apply a suppository of three obols of cyclamen she has bound up in a piece of linen; also have her soak cypress wood in water, grind blister beetles, (sc. and apply these as a suppository) for a short time: their action will continue for a while longer. Suppository: a cheme of cumin with white raisins. Another: take Cnidian berries and arum root, and apply as much of this as you think is reasonable. When you have completed the cleaning, infuse, remove the suppository, flush the vagina, and have the woman sleep with her husband; if she carries the fetus to term, she is completely clean and will recover.

67. Dropsy of the uterus: the menstrual flux is between bloody and serous, it causes great irritation as it passes, and it ulcerates the genitalia and the area around them as brine would: wherever it drips it causes erosion, and makes the skin look yellowish. Otherwise, the volume of cleaning is as with fluxes in general. This disease is very stealthy, and especially dangerous if the uterus becomes ulcerated.

Treat this patient as you would treat one suffering from a white flux: have her drink ass's milk, bring down the swelling, and apply the medications indicated above. An

κτόν, ἢν ὕδωρ ἐκ τῶν μητρέων ῥέῃ, θεῖον, χηνὸς
ἄλειφα, λείχειν.

68. (177 L.) Ἢν ἄνεμος ἐν τῇσι μήτρῃσιν ἐνῇ, φῦσα
ἔξεισι καὶ τρύζει, καὶ οἰδέει πᾶσα, καὶ πῦρ ἔχει καὶ
κάματος· πρὸς τοῖς δ' ἀΐσσει ὑπὸ τῆς ὀδύνης, καὶ τὸν
360 ἄνδρα οὐ προσίεται· ἄχθεται σφόδρα | τὴν εὐνήν, καὶ
ἀδυνατέει ὀρθοῦσθαι· καὶ ὡς βαρέα μιν ἔγκειται ἐν
τῇσι μήτρῃσι, καὶ κεφαλὴν ἀλγέει, καὶ ἀλύει, καὶ
ἄναυδός ἐστιν· ἢν δὲ ἡ ὀδύνη προστῇ, βοᾷ τε καὶ
ἀλγέει πάντα καὶ ἰξύας καὶ ἐπίσιον καὶ τὴν ἕδρην· καὶ
τὸ οὖρον ἴσχεται καὶ ἡ κοιλίη, καὶ πνίγεται, καὶ θανεῖν ἐρᾶται· καὶ ὑποχόνδριον τιταίνεται, καὶ στόμαχος δάκνεται μέγα, καὶ στόμα πικρόν, καὶ ἐμέει χολώδεα ὀξέα ἄκρητα, καὶ ἐρεύγεται θαμινά, καὶ ῥαΐζει·
ἢν δὲ μή, ἀνοιδίσκεται, καὶ ἢν ἐπαφήσῃ, ἀντιτυπέει
καὶ ἀλγέει.

Κλύζειν οὖν[253] χρὴ τὴν μήτρην μελικρήτῳ καὶ ὀξυμέλιτι καὶ ἐλαίῳ· κύμινον τριπτόν, ἢ ἄννησον, καὶ
λίνου πέταλα τρίβειν, καὶ ὄρνιθος πάτον σὺν ᾠοῖς,
καὶ ὕδατι ἐνιέναι· προσθέτοισι δὲ οἷς ἂν ἐγὼ γράψω,
καὶ ποτήμασι χρῶ· καθίννυσθαι δὲ ἐν ἐλαίῳ θερμῷ,
καὶ ἀρώματα ἐμβάλλειν, σχοίνου ἄνθος, ἢ ἐν ὕδατι
δάφνης ἢ θαλάσσης· ἄριστον δὲ καθαίρειν κλυσμοῖς
μαλθακοῖσι τὴν κοιλίην· ἢ βάλανον προστιθέναι, ὡς
νηπίῳ κοιλίη λύεται, εἴριον ἄπλυτον σὺν μέλιτι· ἢν δὲ
γεραιτέρη ᾖ, κρόμμυον ἐς ἔλαιον ἐμβάπτειν,[254] ἢ ἐς

---

253 οὖν om. Θ.   254 ἐμβάπτειν om. MV.

electuary for dropsy, if watery fluid runs out of the uterus: sulfur, and goose grease: give it to be licked.

68. If wind is present in a woman's uterus, the air will be expelled with a noise, her whole body will become edematous, and fever and lethargy will be added. Besides this, the patient leaps up because of the pain, she has no inclination for her husband, she suffers severe pain during intercourse, and she is unable to hold her body straight. She also feels a kind of weight in her uterus, has headache, is restless, and loses her ability to speak. When pain is present, the patient shouts out loud and feels pain in the whole area of her loins, pubes, and seat. Her urine is held back, as is her cavity, she feels suffocation, and she asks to die. Her hypochondrium is stretched tight, her gullet is greatly irritated, her mouth is bitter, she expels concentrated bilious, acidic vomitus, and she has frequent eructations followed by relief. If this does not happen, she swells up with edema, and if she is palpated, the swelling is firm and painful.

Flush the uterus with melicrat, oxymel, and olive oil; knead ground cumin or anise together with flax leaves, and chicken droppings together with eggs: inject with water. Employ the suppositories I will describe below,[34] as well as potions. Have the patient take a sitz bath in hot olive oil to which you have added aromatics such as rush flowers, or in laurel juice, or in brine. It is best to clean the cavity with mild injections; or apply a suppository made with unwashed wool and honey like the one given to infants to relax their cavity. If the woman is older, immerse onions in olive oil or honey, and apply this as a suppository; or use

---

[34] Cf. chs. 96–97 below.

μέλι, καὶ προστιθέναι· ἢ ταύρου χολήν, ἢ νίτρον ξὺν
μέλιτι, ἢ ῥοιῆς ὀξείης χοίνικα²⁵⁵ σὺν μέλιτι καὶ ἀλήτῳ
κριθίνῳ.

69. (178 L.) ῍Ην δὲ μύλη ἐμφύηται ὑπὸ πάχεος
γονῆς ἐνεχομένης, θύμβραν λειήνας ἐν ὄξει καὶ ὕδατι,
τοῦτο πίνειν ἔνυγρον· ἢ ὑοσκυάμου τὸν καρπὸν λεῖον,
καὶ κλύζειν ἄλμῃ καὶ ὀπῷ καὶ ὄξει· ἢν δὲ δέῃ, ξὺν
ὕδατι· κράτιστον μελίκρητον ἐνιέναι ξὺν ὕδατι φακῶν, ἢ ὀρόβων, ἢ ἴου ἄνθος. ἢν δὲ καθαρθῇ οἷα τὰ
362   πυριφλεγέθη, μυρσίνην | ἕψειν καὶ διανίζεσθαι, σμύρ-
ναν καὶ νέτωπον ἐν εἰρίῳ προστίθεσθαι.

70. (179 L.) ῍Ην ἄνεμος ᾖ ἐν τῇσι μήτρῃσι καὶ
δάκνῃ, καὶ τῇδε διεξιὼν πῦρ ποιέει· καὶ οἰδέει καύματι,
καὶ ὀδυνῇ, καὶ τὸν ἄνδρα ἀπαναίνεται, καὶ ἄχθεται
σφόδρα τῇ συνουσίῃ, καὶ τείνεται, καὶ οἰδέει τὸ
ἦτρον, καὶ οὐ δύναται ὀρθοῦσθαι, καί θύει.²⁵⁶

Γνωστὸν ὅτι ἄνεμος καὶ γονὴ ἔνι ἐν τῇσι μήτρῃσι,
καὶ ἡ γονὴ ἐμπέφυκε·²⁵⁷ διὰ τοῦτ᾽ οὖν κάμνει. λαβὼν
μέλι, κηρόν, λίνου πέταλα, τρίψας λεῖα καὶ ὄρνιθος
στέαρ, οἴνῳ εὐώδει χλιήνας, ἔγχεον ἐς τὰς μήτρας
κλυστῆρι· πινέτω δὲ λίνου πέταλα· ἢ τὸν καρπὸν τρίψαι καὶ ἐς εἴριον ἐνελίξαι, πρὸς τὸ στόμα τῆς μήτρης·
εἰ δὲ μή, ἐνεργοτέρῳ²⁵⁸ χρῆσθαι· λίριον καὶ κρόκον,
λίνου πέταλα, ὄρνιθος στέαρ τρίψας λεῖα, διεὶς γάλα-

---

²⁵⁵ ἢ νίτρον . . . χοίνικα om. Θ.
²⁵⁶ θύει Θ: ἰθύειν Μ: ἰθύνειν V.   ²⁵⁷ ἐμπέφυκε om. MV.
²⁵⁸ -γοτέρῳ Θ: -γῷ MV.

bull's gall, soda with honey, or a choinix of bitter pomegranate peel with honey and barley meal.

69. If a mole forms because of the retention of thick
seed, crush savory in vinegar and water, and give this
liquid to drink; or crush henbane seed fine and flush (sc.
the uterus) with this in brine, (sc. silphium) juice, or vinegar, and—if necessary—with water. The most effective
injection is melicrat together with fluid from lentils or
vetches, or violet flowers. If the cleaning produces material that looks like it has been burned in a fire, boil myrtle
berries and wash with this decoction, and apply a suppository of myrrh and oil of bitter almond on a piece of
wool.

70. If wind is present in a woman's uterus and causes
irritation, in passing out through there it will also produce
a fever; she swells up as a result of the burning, and suffers
pain. She will reject her husband, and she suffers severe
pain during intercourse; she is distended, she swells up
with edema in her lower abdomen, she is unable to hold
her body straight, and she rages.

Recognize that wind and seed are present in the patient's uterus, and that the seed has implanted: this is why
she is ill. Take honey, wax, and flax leaves, knead these
smooth with chicken's fat, warm this in fragrant wine, and
infuse it into the uterus with a syringe. Have her drink the
fluid from the flax leaves, or grind linseed, wrap it in a
piece of wool, and apply it against the mouth of the uterus.
If not this, then use a more powerful suppository: take
Madonna lily, saffron, flax leaves and chicken's fat, knead
these smooth, dissolve this in woman's milk, sponge it up

κτι γυναικείῳ, ἀποσπογγίσας ἄχνη ἀπὸ ὀθονίων λε-
πτῶν, ἐνδῆσαι· προστιθέσθω δὲ πρὸς τὸ στόμα τῆς
μήτρης.

71. (180 L.) Καὶ ἢν ὑποφύηται πιμελὴ σαρκοειδε-
στέρη, τηκεδόνα ἐντιθέναι, καὶ λεπτύνειν μέσως· αἱ
γὰρ λίην λεπτυνόμεναι ἀραιαί εἰσι καὶ ἐκτιτρώσκου-
σιν.

72. (181 L.) Εἰ δὲ[259] οὐ δέχεται ἡ ὑστέρη, ἀλλὰ
ἀφίησι καὶ θερμὸν οὐκ ἔχει ἐν ἑωυτῇ, ὄργανον χρὴ
μηχανοποιέεσθαι, ἐφ᾽ ὃ ἑζομένη εἴσεισιν ἀτμὸς ἐς τὰς
μήτρας, ἀμφὶ δὲ τοῦτο εἵματα κυκλόσε τιθέναι· ὑποθυμιᾶν δὲ κασίην, κιννάμωμον, σμύρναν, ἴσον ἑκά-
στου, ἐν οἴνῳ | φυρᾶν σιραίῳ καὶ ἐπιβάλλειν, ὀλίγον
καὶ λοῦσθαι· σῖτα ὀλίγα. ἀρήγει δὲ καὶ προσθετὸν·
σμύρναν ἀπαλὴν ξὺν μέλιτι· ἔστω δὲ προμήκης ὡς
βάλανος· καὶ ταῦτα ποιέειν ὡς πλειστάκις πρὸς τὴν
δύναμιν ὁρῶν. καὶ οἱ ἐν σίτῳ πελεκῖνοι τριπτοὶ ξὺν
σμύρνῃ ὠφελέουσιν. ἕψειν δὲ μέλι, καὶ σὺν τῇσι[260]
δαισὶν ἀνακυκᾶν, καὶ ὅσον Αἰγύπτιον κύαμον προστιθέναι. καὶ ταύρου χολὴ καὶ ῥοὸς ἐρυθρῆς ὡς ἡ κό-
νυζα ποίη, ὅμοιον δ᾽ ἐστὶ σελίνῳ οὐλῷ, φύεται δὲ
ἄγχιστα θαλάσσης ἐν χωρίοις ψαμμώδεσιν, ὀδμὴ
δύσοιστος· σὺν μέλιτι ἢ οἴνῳ πρόσθες. ἢ βόλβιον, ἐν
πυροῖσι δὲ θεωρεῖται Αἰγυπτίοισι δὲ μάλιστα, δριμὺ
ὅμοιον κυμίνῳ Αἰθιοπικῷ· τοῦτο καὶ σκόροδον, λίτρον
ἐν τῷ αὐτῷ[261] προστίθει, προλούσθω δέ.

73. (182 L.) Ὅταν γυνὴ τὴν κεφαλὴν ἀλγέῃ καὶ τὸ
βρέγμα καὶ τὸν τράχηλον καὶ ἰλιγγιᾷ πρὸ τῶν ὀμμά-

with lint from fine linen, and tie it together; then have the patient apply this against the mouth of her uterus.

71. If fat of a quite fleshy kind builds up, insert a pessary to dissolve it, and gently thin the patient; for women who are thinned too forcefully become soft, and risk having abortions.

72. If a woman's uterus does not take up (sc. the male seed), but expels it due to a lack of warmth, you must fashion a device for her to sit on so that vapor will enter her uterus, and place coverings all around her in a circle. Fumigate her below with cassia, cinnamon, and myrrh—an equal amount of each mixed into boiled-down new wine, and placed (sc. in a vessel); the patient should bathe only rarely, and eat little. The following application also helps: mix soft myrrh with honey, and elongate this into a suppository; apply very often, keeping an eye on the (sc. patient's) strength. Pounded axeweed in food with myrrh is also beneficial. Or boil honey, stir it with pine sticks, and insert to the amount of an Egyptian bean. Or bull's gall and red sumac like the fleabane plant—fleabane is like curly celery, it grows in sandy places close to the sea, and it has an unpleasant smell: and apply them with honey and wine. Or the small bulb found in wheat, especially from Egypt, which is pungent like Ethiopian cumin: apply this, garlic, and soda all together as a suppository, but first have the patient wash herself.

73. When a woman has a headache and pain in her bregma and throat, sees things spinning before her eyes,

---

259 Εἰ δὲ recc.: Καὶ ΘΜV.
260 τῇσι Linden: τοῖς codd.
261 αὐτῷ ΜV: ὑγρῷ Θ.

των καὶ φοβῆται καὶ στυγνὴ ᾖ, καὶ οὖρα μέλανα καὶ
δι᾿ ὑστέρης ὅμοια, καὶ ἄση ἔχῃ καὶ δυσθυμέῃ, μέλαινα χολὴ ἐν τῇσι μήτρῃσιν ἔνι· ἐντεριώνην ἐνιαυ-
σίην, ταύρου χολήν, ἄνθος χαλκοῦ τρίβειν ξὺν βακκάρει, καὶ προσθετὰ ποιέειν, καὶ φάρμακον πιπίσκειν,
καὶ λούειν.

74. (183 L.) Ὁκόταν δὲ δάκνηται τὰς μήτρας καὶ
ἀλγέῃ καὶ ἀδάξηται, καὶ χολὴν οὐρέῃ ξανθήν, καὶ ἡ
μήτρη χάνῃ, καὶ οἱ ὀφθαλμοὶ ἰκτερώδεες, ἐν τῇσιν
366 μήτρῃσιν χολὴ ἔνι. ἄριστον | ἐκκαθαίρειν τὸ σῶμά τε
καὶ αὐτὰς προσθέτοισιν, ἃ χολὴν ἄγει.

75. (184 L.) Ψύχεται ἡ ὑστέρη, καὶ βάρος δοκέει
ἐγκεῖσθαι, καὶ τὸ χρῶμα οὐ λαμπρόν, καὶ πέπηγεν ἡ
ὑστέρη· καθαίρειν ὅ τι φλέγμα ἄγει, καὶ λεπτύνειν
χρὴ καὶ ἐμείτω.

76. (185 L.) Ὅταν γυναικὶ ὄζῃ κακὸν ἐκ τοῦ στόματος, καὶ οὖλα μέλανα[262] ᾖ καὶ πονηρά, κεφαλὴν
λαγωοῦ καὶ μύας τρεῖς κατακαῦσαι χωρίς, καὶ τῶν
δύο μυῶν ἐξελεῖν κοιλίην, ἧπαρ δὲ καὶ νεφροὺς μή·
καὶ ἐν θυίῃ λιθίνῃ τρίβειν μάρμαρον ἢ λίθον λευκήν,
καὶ διασῆσαι· μίσγειν δὲ ἴσον ἑκάστου, καὶ τοὺς
ὀδόντας[263] τρίβειν· χρὴ δὲ καὶ τὰ ἐν στόματι χωρία·
κἄπειτα εἰρίῳ πινωδεστάτῳ τρίβειν, καὶ διακλύζεσθαι
ὕδατι· βάπτουσα δὲ τὸ πινῶδες εἴριον μέλιτι, ἀνατρι-
βέτω τοὺς ὀδόντας καὶ τὰ οὖλα[264] καὶ τὰ ἔνδον καὶ τὰ
ἔξω. τρίβειν δὲ καὶ ἄννησον, καὶ ἀνήθου[265] καρπόν,
καὶ σμύρνης ὁλκὴν ὀβολοὺς δύο, διέναι οἴνῳ λευκῷ

becomes sullen and afraid, passes dark urines and the same kind of fluid from her uterus, has nausea, and feels depressed, then there is dark bile in her uterus. Knead together year-old insides of a bottle gourd, bull's gall, and flower of copper ground with bacchar, and make this into suppositories; also have her drink a purgative medication, and bathe.

74. When a woman feels irritation in her uterus, suffers pain and a biting sensation, passes yellow bile in her urine, her uterus gapes open, and she becomes jaundiced in her eyes, there is bile in her uterus. Best is to clean out her body and her uterus with suppositories that draw bile.

75. A woman's uterus becomes cold and seems to be holding a weight, her color is dull, and her uterus becomes fixed: you must clean with a medication that draws phlegm, thin the patient, and have her vomit.

76. When a woman smells badly from her mouth, and her gums are dark and painful, incinerate the head of a hare and three mice each one by itself, remove the intestines from two of the mice but not their liver and kidneys, and grind these in a mortar of marble or white stone, sieve, mix together an equal amount of each, and rub this on to the patient's teeth. You should also rub it on to the insides of the mouth; then rub with a piece of very greasy wool, and rinse out the mouth with water. Have the patient soak the greasy wool with honey, and have her rub it on both the insides and the outsides of her teeth and gums. Also knead together anise, dill seed, and a two-obol draft of

---

262 οὖλα μέλανα MV: οὖρα πελιδνά Θ.
263 Add. καὶ τὰ οὖλα καὶ τὰ ἔνδον Θ.
264 Om. καὶ τὰ οὖλα Θ.   265 ἀνήθου MV: ἀννήσου Θ.

ἀκρήτῳ ἡμικοτυλίῳ, τούτῳ διακλύζεσθαι, καὶ ἐν τῷ
στόματι πολὺν ἐχέτω χρόνον· θαμινὰ δὲ δρᾶν, καὶ
ἀναγαργαρίζεσθαι νῆστιν καὶ μετὰ τὴν τροφήν· ἄριστον δὲ ὀλιγοσιτέειν, κράτιστα δὲ χρὴ προσφέρεσθαι. τοῦτο τὸ φάρμακον ὀδόντας λευκαίνει καὶ εὐώ-
δεας ποιέει· καλέεται δὲ Ἰνδικὸν φάρμακον.

77. (186 L.) Ὅταν γυναικὶ ὁ μαζὸς τριχιήσῃ,[266]
στοιβῆς καρπόν, ἢ βάτου ἕψειν ἐν ὕδατι σὺν ἀλφίτοισι λεπτοῖσι[267] σὺν ἐλαίῳ, τοὺς μαζοὺς καταπλάσ-
σαι, καὶ τεύτλου | φύλλα ἐπιρρίπτειν· ἔπειτα ράψαι ἐκ
ράκους ὡς κυρβασίην, τεκμαιρόμενος ὅσον τὸν μαζὸν
ἐκχωρήσει, καὶ οὕτως ἐντιθέναι τὸν τιτθόν· ἢν δὲ δια-
πύῃ, ἄμεινον[268] τάμνειν, καὶ εἰρίῳ ρερυπωμένῳ μοτῶ-
σαι, καὶ ἐπιβάλλειν τοῦτο· καὶ μετέπειτα λύσαντα[269]
φακῷ ἑφθῷ ξὺν ἀλφίτῳ μῖξαι καὶ καταπλάσσειν.

78. (187 L.) Ἐπὴν γυναικὶ ἐν τῷ αἰδοίῳ ἢ ἐν τῷ
ἀρχῷ ἀσκαρίδες ἐγγένωνται, λύγου καρπὸς μίσγεται[270] ἢ φύλλα, καὶ βοὸς χολὴ παραμίσγεται ὅσον
ὀβολός· κεδρίνῳ δὲ ἐλαίῳ φυρᾶν, καὶ εἰρίῳ πινόεντι
εὐειροτάτῳ ἀναλαβεῖν· ἐντιθέσθω δὲ διὰ τρίτης νύκτα
καὶ ἡμέρην, τῇ δ' ὑστεραίῃ ἀφελομένη λούσθω θερμῷ,
καὶ σκόροδα ἑφθὰ καὶ ὠμὰ τρωγέτω. καὶ αἱ ἀσκαρίδες ἐξίασι καὶ θνήσκουσιν· ἅλμῃ δὲ κλύζειν χρή.

266 τριχ. Μ: τρηχ. ΘV.
267 σὺν ἀλφίτοισι λεπτοῖσι om. MV.
268 ἄμεινον om. MV.
269 λύσαντα om. MV.
270 μίσγεται om. Θ.

myrrh, dissolve this in a half cotyle of undiluted white wine, and rinse with this, having the patient hold it in her mouth for a long time; have her repeat this frequently, and gargle both in the fasting state and after her meals. Best is to eat little food, but to take the most powerful ones. This preparation whitens the teeth and gives them a pleasant fragrance: its name is "the Indian medication."

77. When a woman's breast suffers with "hair,"[35] boil the seed of thorny burnet or bramble in water with fine meal and olive oil, and plaster the breasts with this; also apply a plaster of beet leaves. Then stitch together a piece of cloth in the form of a Persian bonnet adjusted to the size of the breast, and put the breast inside it. If suppuration takes place, it is best to make an incision, to pack this with a tent of greasy wool, and to place some of the same wool over it. Later remove this and apply a plaster of boiled lentils mixed with barley meal.

78. When worms are bred in a woman's genitalia or anus, mix agnus castus seeds or leaves together with an obol of bull's gall, dissolve these in cedar oil, and soak this up in a piece of greasy very fine wool: have the woman insert this suppository every other day—both by day and by night—removing it the following day and washing herself with hot water; she should also eat garlic, both boiled and raw. The worms will leave her body and die; you must flush with brine.

---

[35] The reading of ΘV means "becomes rough." Erotian's gloss on M's reading explains: "$\tau\rho\iota\chi\iota\alpha\sigma\iota\varsigma$ means an apostasis (see Loeb Hippocrates vol. 8, 10) involving the breast." Cf. Aristotle, Historia Animalium 587b 24–27.

79. (188 L.) Πρόσωπον ἀγλαΐζειν· ἧπαρ σαύρου,[271]
τρίβειν ξὺν ἐλαίῳ, ἀλείφειν δὲ ἐν ἀκρήτῳ οἴνῳ· χολὴ
δὲ χλωροῦ φθείρει· λαμπρύνει καὶ πτισάνης χυλός,
καὶ ᾠῶν τὸ λευκόν, καὶ ἄλητον θέρμων καὶ ὀρόβων,
καὶ σῦκον καταπλάσσειν, καὶ κράμβης ῥίζα καὶ
σπέρμα· ταῦτα καὶ φακοὺς αἴρει, καὶ ἀλκυόνιον· καὶ
ἢν κονιορτὸς λυπέῃ τὸ πρόσωπον,[272] κηρωτῇ ὑγρῇ ῥοδίνῳ προχρίειν, καὶ ὕδωρ προσχέειν ψυχρόν. καὶ ῥυ-
τίδας ἐκτείνει, ἐν θυίῃ λιθίνῃ μολίβδαιναν τρίβειν,
μηνιαῖον ὕδωρ παραχέοντα πλάσαι κυκλίσκους· καὶ
ἐπειδὰν ξηροὶ γένωνται, ἐλαίῳ διεὶς χρῶ. |

370   80. (189 L.) Καὶ ἢν ῥέωσι τρίχες, λήδανον μετὰ
ῥοδίνου ἢ ἀνθινοῦ μύρου τρῖβε, καὶ μετ᾽ οἴνου χρίετω·
ἢ τὴν σμηκτρίδα γῆν σὺν οἴνῳ, ἢ ῥοδίνῳ, ἢ ὀμφακίῳ,
ἢ ἀκακίῃ· καὶ ἢν μαδήσῃ, κύμινον ἔμπλασσαι, ἢ πε-
λιάδων κόπρον, ἢ ῥάφανον τριπτήν, ἢ κρομμύῳ, ἢ
τεύτλῳ, ἢ κνίδῃ.

81. (190 L.) Τὰς δ᾽ ἐφηλίδας λεγομένας αἴρει τὸ
ὀρόβιον, τεύτλου χυλός, ᾠῶν τὸ λευκόν, πτισάνη, ἡ
σικύου ἀγρίου ῥίζα ξηρή, μετ᾽ οἴνου τρυγὸς τρι-
φθεῖσα, καὶ ἐπαλειφομένη, καὶ συκῆς φύλλα προστι-
θέμενα· σησάμῳ τριπτῷ σμήχεσθαι, ἢ ἀμυγδάλαις
οὐ γλυκείαις· καὶ κνίδης σπέρμα, σκορόδων[273] κέλυφος ἐπιδεόμενον, λεπίδιον.

82. (191 L.) Λειχῆνας ἐξάγει πάντας, ὄξος, μάννα,
κίσηρις, θεῖον μετ᾽ ὄξους, κάρδαμον ἄγριον καὲν καὶ
σποδωθέν, ἐχίδνης λεβηρίς,[274] καὶ λαπάθου ἀγρίου

79. To beautify the face: take lizard's liver, knead it with olive oil, and apply this on the face together with undiluted wine—bile of the green (sc. lizard) is deleterious. Barley water also makes the face radiant, as do the white of eggs, meal of lupines and vetches (plastered on with a fig), roots and seeds of cabbage (these also relieve lentigo), and bastard sponge. If dust has damaged the face, anoint it with a moist rose oil cerate, and pour cold water over it. To smooth out wrinkles, grind sulphuret of lead in a stone mortar, add month-old water, and form this into disks; when these have dried, dissolve them in olive oil, and apply them.

80. If hair falls out, knead gum ladanum with rose or lily oil, and rub this on with wine. Or take fuller's earth in wine or rose oil, or oil from unripe olives, or gum. If all the hair falls out, apply a plaster with cumin, pigeon excrement, ground radish, onion, beet, or stinging nettle.

81. What are called freckles are removed by vetch or beet juice, egg white, barley gruel, or dried root of squirting cucumber kneaded in wine lees, and anointed; also fig leaves applied as a poultice. Wash with ground sesame or bitter almonds; also seeds of stinging nettle, husks of garlic applied (sc. on the face), pepperwort.

82. All types of lichen are removed by vinegar, frankincense powder, pumice stone, sulfur with vinegar, wild cress burned and reduced to ashes, viper skin, and root of

---

271 σαύρου Θ: ταύρου MV.

272 τὸ πρόσωπον om. Θ.

273 σκορόδων MV: καὶ ῥόδων Θ.

274 λεβηρίς H: λεβήριδος ΘMV.

ῥίζα· τρίβειν δὲ μετ᾽ ὄξους οἰνώδεος· φλυκταινοῦται,
καὶ λιθαργύρου χρῶ.

83. (192 L.) (1) Ῥόου ἐρυθροῦ ποτὸν ἀγαθόν· ἐλάφου κέρας κατακαύσας, ὠμήλυσιν κριθέων συμμῖξαι
διπλασίην, ἐπ᾽ οἶνον Πράμνιον ἐπιπάσσουσα πινέτω,
καὶ ἵσταται. ἕτερον·[275] ἀδιάντου ῥίζαν | τρῖψαι, καὶ
ἐρεβίνθους φῶξαι, καὶ λέκιθον ποιέειν, ἐν μέλιτι ὡς
ποτὸν διδόναι.

(2) Ἢ ἀλήτου σητανίου ὀξύβαφον, κόμμεως λευκοῦ
ἥμισυ, μάννης τρίτον μέρος· καὶ σχίνου δὲ ὀλίγον, ἢ
πίτυος, ἢ κυπαρίσσου διεὶς ὕδατι πίνειν δίδου δὶς τῆς
ἡμέρης.

(3) Ἢ ἐλάφου κέρας κατακαίειν, τρίβειν δὲ καὶ
ὠμήλυσιν σὺν κεδρίσι πέντε· οἶνος σὺν τοῖσδεσσιν
αὐστηρὸς μέλας μίγεται.

(4) Ἢ ῥοιὴν γλυκείην ὀπτήσας, τὸν χυλὸν σὺν
οἴνῳ μέλανι πίνειν.[276]

(5) Ἢ κυπαρίσσου καρπὸς ὅσον τρία ἢ τέσσερα,
καὶ μύρτα μέλανα καὶ ὁμοῦ καὶ αὐτὰ καθ᾽ ἑωυτά,
πρὸς ἰσχὺν τοῦ σώματος ὁρῶν τῆς γυναικός, σὺν
οἴνῳ δὲ ἡ πόσις.

(6) Ἢ καστορίου ὀβολὸν καὶ σμύρνης ὀβολὸν ἐν
οἴνῳ τρίβειν αὐστηρῷ μέλανι καὶ πιπίσκειν.

(7) Ῥόου καὶ πάσης νούσου ποτόν, ὅσαι ἀπὸ ὑστερέων γίνονται· γλυκυσίδης καρπόν, καὶ τοῦ ῥόου τὰς
ῥίζας, καὶ κύμινον Αἰθιοπικόν, καὶ μελάνθιον ἐν οἴνῳ
λευκῷ διδόναι.

(8) Ἢ νάρθηκα ξύσας, ὅσον ὀξύβαφον, καὶ πρά-

wild monk's rhubarb: knead with wine vinegar. If blisters form, use litharge as well.

83. (1) A good potion for the red flux: incinerate deer's horn, mix with this a double amount of bruised meal of raw barley, and have the patient sprinkle this over Pramnian wine and drink the wine: the flux will come to a stop. Another: crush root of maidenhair, mix this with chick peas, make into gruel and give in honey as a potion.

(2) Or an oxybaphon of meal of spring wheat, half that amount of white gum, and a third as much frankincense powder; also give a little mastic or pine or cypress wood dissolved in water to drink twice daily.

(3) Or incinerate deer's horn, and also triturate bruised meal of raw grain with five juniper berries: dry dark wine is mixed with these ingredients.

(4) Or bake sweet pomegranate peel, and have the patient drink the juice together with dark wine.

(5) Or take three or four cypress berries and some dark myrtle berries—both together and each for itself—and observing the patient's strength, make a potion from these with wine.

(6) Or knead an obol each of castoreum and myrrh in dry dark wine, and give this to drink.

(7) A potion for every flux and disease that arises from the uterus: give peony seed, sumac roots, Ethiopian cumin, and black cumin in white wine.

(8) Or crush an oxybaphon of giant fennel with leek

---

275 ἕτερον Θ: ποτὸν ἕτερον MV.
276 πίνειν MV: καὶ ἴσχεται Θ.

411

σου χυλόν, ἐν οἴνῳ λευκῷ κεκρημένῳ, τοῦτο καὶ ἐκ
ῥινῶν αἷμα ῥέον παύει.

(9) Ἢ σίδην ἕψησαι ἐν οἴνῳ μέλανι, καὶ περιλέψαι,
καὶ τὰ ἔνδον τρίβειν, ἐν οἴνῳ μέλανι σὺν πάλῃ ἀλφίτου πίνειν.

(10) Ἢ λίνου σπέρμα, ἢ ἐρυσίμου φῶξαι, καὶ
ἐλαίης φύλλα χλωρῆς, καὶ μέλαιναν ῥίζαν, μήκωνα
ἁδράν· ταῦτα τρίψας ἐν τῷ αὐτῷ, ἐν οἴνῳ κεκρημένῳ
δίδου πίνειν.

(11) Ἢ τάμισον ὄνειον καὶ σίδης γλυκείης ῥίζαν
καὶ κηκίδα, ἴσα ταῦτα, καὶ ῥοιῆς γλυκείης χυλὸν σὺν
οἴνῳ πίνειν.

(12) Ἢ λαπάθου καρπόν, σὺν τῷ τῆς κηκίδος ἔξω
περιεξυσμένῳ· ταῦτα τρίβειν ἄμφω, καὶ ἐν οἴνῳ πί-
374 νειν,[277] καὶ μετέπειτα | κυκεῶνα.

(13) Ἢν αἷμα ῥέῃ λάβρον ἐξ ὑστερέων, ἄγνου
φύλλα σὺν οἴνῳ διδόναι[278] μέλανι· τὰ στρυφνὰ ῥόον
ἵστησιν· οἴνῳ μέλανι μίσγειν.

(14) Ῥόου καὶ ὀδύνης· κάχρυος ῥίζαν ἐν οἴνῳ μέλανι[279] πίνειν.

(15) Ἢν δὲ πλέον εἴη, τερμίνθου καρπὸς τριβόμε-
νος, χρὴ δ᾽ ἐν οἴνῳ καὶ ὕδατι διεῖναι καὶ πίνειν.

(16) Ἢν ῥόος ἐγγένηται, καρκίνους ποταμίους ἀπο-
πνίξας ἐν οἴνῳ, πίνειν διδόναι ξὺν ὕδατι δὲ τὸν οἶνον.

(17) Ἢν δ᾽ ἔτι φέρηται ὁ ῥόος, πρόμαλον φώξας
τρίψας ἐν οἴνῳ δίδου, ἢ τῶν πράσων τὸν χυλόν.

(18) Ἢν ῥόος ἐγγένηται πολύς, ἡμιόνου ὀνίδα
κατακαίειν, καὶ λειῆναι καὶ σὺν οἴνῳ δοῦναι.

juice in diluted white wine: this also stops blood flowing from the nostrils.

(9) Or boil a pomegranate in dark wine, strip off its skin, and crush the insides: give this with the finest barley meal in dark wine to drink.

(10) Or roast linseed or hedge mustard seed along with green olive leaves, black root, and ripe poppy: grind these together and give in diluted wine to drink.

(11) Or take ass's rennet, root of sweet pomegranate, and oak gall—an equal amount of each—and juice of sweet pomegranate peel, and give with wine to drink.

(12) Or monk's rhubarb seed with the grated exterior of an oak gall: grind the two of them, give in wine to drink, and then give a cyceon.

(13) If blood flows violently from the uterus, give leaves of the chaste tree in dark wine: astringents will also staunch the flow: mix them in dark wine.

(14) For a flux with pain: a potion of frankincense root in dark wine.

(15) If the flow is more violent, grind fruit of the terebinth tree, dissolve this in wine and water, and give it to drink.

(16) If a flux occurs, boil river crabs in wine, and give this wine to drink with water.

(17) If the flux persists, grind parched willow in wine and give this or leek juice to drink.

(18) If a violent flux occurs, burn mule's excrement, grind it fine, and give with wine.

---

277 ἄμφω . . . πίνειν Θ: ἅμα MV.
278 διδόναι om. M.
279 μέλανι om. Θ.

413

(19) Ἦν δὲ πολυχρόνιος ὁ ῥόος ᾖ, σπόγγος κατακαεὶς ἀρήγει, τρίβειν δὲ λεῖον καὶ σὺν οἴνῳ εὐώδει δοῦναι.

84. (193 L.) Κατάπλασμα ῥόων·

(1) Σκόροδα καὶ ἀνδράχνην καὶ σέλινον καὶ λωτοῦ πρίσματα καὶ κέδρου πρίσματα λεῖα ὁμοῦ μῖξαι, διεὶς δ᾽ ἐν μελικρήτῳ, καὶ κατάπλασμα ποιεῖ.

(2) Ἢ βάτου φύλλα καὶ ῥάμνου καὶ ἐλαίης, ὁμοῦ λεῖα μῖξας διεὶς μελικρήτῳ, σὺν ἀλφίτοισι καταπλάσαι.

(3) Ἢ ἀκτῆς καὶ μυρσίνης φύλλα κατάπλασσε.[280]

(4) Ἢ λωτοῦ πρίσματα, συκαμίνου φύλλα καὶ ῥοῦν σὺν ἀσταφίδι.

Πυριήσιες ῥόων·

(5) Αἰρῶν ἄλευρα πεφωσμένα ἕψε ἐν ὀξυκρήτῳ ἀκρητεστέρῳ καὶ ἐς ὀθόνιον ἐγχρίων πυρία.

(6) Ἢ φακοὺς φώξας, καὶ περιπτίσας, ποιεῖν ἄλευρα χονδρότερα, ἐν ὕδατι ἕψειν, καὶ ὁμοίως κατάπλασσε· ἢ ὀρόβους ὡσαύτως. ἀγαθὸν δὲ καὶ ἐλελίσφακος.

(7) Ἢ ἄχυρα | κριθέων ἐν ἀφεψήματι ἐλελισφάκου καὶ ὑπερικοῦ ἕψειν καὶ καταπλάσσειν.

(8) Ἢ λωτοῦ πρίσματα καὶ κυπαρίσσου ἐναφέψων ἐν σταφίδος ἀποβρέγματι, ἐς ὀθόνιον ἐπιχρίων, πυρία.

(9) Ἢ ἐλαίης φύλλα, ἢ κισσοῦ, ἢ μυρσίνης, ἐν ἀφεψήματι τούτων κριθέων ἄχυρα ἕψειν.

(10) Ἢ ἀρωμάτων ὕδατι συνέψειν πίτυρα πύρινα.

376

(19) If the flux becomes chronic, burned sponge is helpful: grind the sponge smooth and give it with fragrant wine.

84. Poultices for fluxes:

(1) Make a smooth mixture of garlic, purslane, celery, sawdust of nettle tree and cedar, dissolve in melicrat, and prepare poultices.

(2) Or make a smooth mixture of bramble, buckthorn, and olive leaves, dissolve this in melicrat, and prepare poultices with barley meal.

(3) Or apply poultices of elder and myrtle leaves.

(4) Or sawdust of nettle tree, mulberry leaves, and sumac with raisins.

Fomentations for fluxes:

(5) Take toasted meal of darnel, boil it in quite concentrated oxycrat, smear this on to a piece of linen, and foment with this.

(6) Or roast lentils, remove their hulls and make a coarse meal: boil this in water and apply poultices in the same way. Or vetches in the same way; salvia is also good.

(7) Or boil barley bran in a decoction of salvia and hypericum, and apply in poultices.

(8) Boil sawdust of nettle tree and cypress in an infusion of raisins, smear this on to a linen cloth, and use to foment.

(9) Take leaves of olive, ivy and myrtle, and make a decoction of these with barley bran.

(10) Or boil wheat bran in aromatic water.

---

280 Ἢ ἀκτῆς . . . κατάπλασσε om. Θ.

(11) Ἢ ἀσταφίδος[281] ἀποβρέγματι πίτυρα πυρῶν ἕψειν, ἢ λευκοΐου καρπόν ἢ τὰς ῥίζας ἀφέψειν,[282] καὶ τῷ ὕδατι σὺν πιτύροισι πυρίνοισιν ἐπιρρίπτειν· ἢ τῷ ἀφεψήματι πίτυρα πυρῶν τοῦτο ποιήσασα, θερμῷ ἐνελιξαμένη εἰρίῳ, πυριάσθω· ἢ ἑρπύλλου ἀφεψήματι πίτυρα τὸν αὐτὸν τρόπον.

(12) Πυρία δὲ καὶ σπόγγοις θερμοῖς καὶ εἰρίοισι μαλθακοῖς, ἢν περιωδυνέῃ, καὶ τοῖς ὀστρακίνοισιν ἀγγείοις, ὕδατος ἐγχέων· ἢ ἐν κύστεσιν ἐλαίῳ θερμῷ.

85. (194 L.) Κλυσμὸς ῥόων·

(1) Μυρσίνης φύλλα καὶ δάφνης καὶ κισσοῦ, ἐν ὕδατι ἀφέψειν· τούτῳ κλύζε χλιηρῷ.

(2) Ἢ ἀκτῆς φύλλα καὶ σχίνου ἀφέψειν ἐν ὕδατι, ἀποχέας, ἀκροχλιέρῳ κλύζειν.

(3) Ἢ οἰνάνθην καὶ κύπαιρον καὶ ἀσταφίδα ἐνέψειν ἐν μελικρήτῳ καὶ κλύζειν. ἢ τήλεως ὕδατι, ἢ βάτου ἀφεψήματι, ἢ ἐλαίης χλωρῆς, ἢ κυπαρίσσου, ἢ ἑρπύλλου, ἢ ῥοιῆς, ἢ λευκοΐου ῥίζης, ἢ σχίνου, ἀκροχλίαρον.

(4) Ἢ τὸ διὰ τοῦ βουτύρου καὶ ῥητίνης καὶ χηνείου ἐλαίου, ἢ τὸ διὰ μυελοῦ καὶ στέατος ὑείου.

86. (195 L.) Ὑποθυμιήσιες ῥόων·

(1) Κριθὰς πεφωγμένας ἐπ᾽ ἄνθρακας ὑποθυμιᾶν, ἢ ἐλάφου κέρας σὺν ἐλαίῃσιν ὀμφακίτισιν, ἢ ῥόον τὴν ἐρυθρήν, καὶ ἄλφιτα πεφρυγμένα σὺν ἐλαίῳ ἢ οἴνῳ διπλασίῳ· ἢ | ἄχυρα κριθέων, καὶ βόλβιτον ὁμοίως, ἢ λωτοῦ πρίσματα, ἢ ῥόον, ἢ κυπάρισσον σὺν οἴνῳ μέλανι αὐστηρῷ ξηρὴν ὑποθυμιᾶν· ἢ χαλ-

378

(11) Or boil wheat bran in a decoction of raisins; or boil white violet seeds or their roots, and apply as a plaster with the water containing wheat bran. Or have the woman make this with a decoction with wheat bran, wrap it in hot wool, and foment herself with it. Or bran with a decoction of tufted thyme in the same way.

(12) Also foment with hot sponges and pieces of soft wool, if the pain is severe, prepared by using potsherds covered with water; or in bladders with hot olive oil.

85. Douches for fluxes:

(1) Leaves of myrtle, laurel, and ivy boiled in water: infuse this while warm.

(2) Or boil leaves of elder and mastic in water, pour the water off, and infuse it lukewarm.

(3) Or boil grapevine blossoms, galingale, and raisins in melicrat, and infuse this. Or (sc. infuse with) water of fenugreek or a decoction of brambles, or of green olive wood, cypress, tufted thyme, pomegranate peel, white violet root, or mastic, this lukewarm.

(4) Or an infusion with butter, resin, goose grease, or swine's marrow and lard.

86. Fumigations for fluxes:

(1) Apply a fumigation from below made with toasted barley over hot coals, or deer's horn with unripe olives, or red sumac, or toasted barley meal with a double amount of olive oil or wine; or wheat bran and cow's excrement prepared in the same way. Or foment with sawdust of the nettle tree, sumac, or dry cypress in dry dark wine, or all-

---

[281] ἀποβρέγματι (414, 24) . . . ἀσταφίδος om. V.
[282] ἢ λευκοΐου . . . ἀφέψειν om. Θ.

βάνην, ἢ μάνναν, ἢ ῥητίνην οἴνῳ δεῦσαι, ἢ αἰγὸς
κέρας καὶ κηκίδα, καὶ ὁ ῥόος ἵσταται.

(2) Θυμιητά· ὀρύξαι χρὴ βόθρον, καὶ φῶξαι ὅσον
δύο χοίνικας Ἀττικὰς γιγάρτων, τῆς σποδιῆς ἐπιβαλὼν ἐπὶ τὸν βόθρον, οἴνῳ ἐπιψεκάσαι εὐώδει, καὶ
ἀμφικαθεζομένη καὶ διαπλίξασα θυμιήσθω.

(3) Ἢ τὸ λεγόμενον οἰσύπη αἰγὸς ξηρὰ κόψαι καὶ
φῶξαι σὺν κριθέων ἐρίγματι, ἐλαίῳ φυρήσας, θυμιᾶν.

(4) Ἢ ἄνθρακας † πολιων[283] κριθέων ἢ ἄχυρα †
ὑποβάλλων, ἢ πρίσματα κυπαρίσσου, μύρῳ δεύων,
θυμία.

(5) Ἢ κώνειον, ἢ σμύρναν, ἢ λιβανωτόν, μύρον δὲ
περιχέων,[284] θυμιᾶν. ἢ ἄσφαλτον καὶ κριθέων ἄχυρα
ὁμοίως.

(6) Ἢ κυπαρίσσου ῥίζαν ἀλείφατι ῥοδίνῳ περιχέας
θυμία.

(7) Ἢ καλάμῳ, κυπαίρῳ, σχοίνῳ, σελίνου σπέρ-
ματι, ἀννήσου, ῥόδινον ἔλαιον περιχέας, θυμία.
ὁμοίως δὲ καὶ ῥητίνην ὑποβάλλειν καὶ κιννάμωμον
καὶ σμύρναν σὺν βάτου φύλλοις, ἢ ῥόδων φύλλοις
ἡδυόδμοις σὺν ποσῷ κρόκῳ καὶ στύρακι· ταῦτα ἐν τῷ
αὐτῷ τρίβειν, καὶ θυμιᾶν ὀβολῷ Ἀττικῷ σταθμῷ ἐπὶ
σποδιῇ οἰναρέῃ, ἢ ἐπὶ βολβίτου πλαστοῦ ὡς ἐμβά-
φιον·[285] τὸ δὲ πῦρ κληματινον ἔστω· ἐπιτιθέναι δέ τι[286]
πρότερον ὡς μὴ ὀδμὴ εἴη· ἐπεὶ ἄμεινον μὴ θυμιᾶν.

---

283 πολιων Θ: πλεῖον MV.   284 Add. ὑποχέων Θ.
285 ἐμβάφιον Froben: ἐμβαφίου codd.   286 τι Θ: ἤδη MV.

heal juice, frankincense powder, or resin dissolved in wine, or goat's horn or oak gall, and the flux will stop.

(2) Fumigants: dig a pit, incinerate two Attic choinixes of grape stones, throw the ashes into the pit, and sprinkle this with fragrant wine; have the woman sit down over this with her legs apart to receive the fumigation.

(3) Or knead what is called goat's grease until it is dry, roast this with pounded barley, moisten the mixture with olive oil, and use this to fumigate.

(4) Or spread out some coals and sprinkle barley bran over them,[36] or knead sawdust of cypress into an unguent, and fumigate with this.

(5) Or pour an unguent on to hemlock, myrrh or frankincense and use this to fumigate; or asphalt and barley bran in the same way.

(6) Or soak cypress root in rose oil, and fumigate with this.

(7) Or take some reeds, galingale, or rushes, or celery seed or anise, pour rose oil over them, and use to fumigate. In the same way mix resin, cinnamon, and myrrh with bramble leaves, or fragrant rose petals with some saffron and storax: then knead these together and fumigate with one Attic obol by weight of it burned on vine ashes or cow's dung shaped like a saucer: the fire should be of wine twigs, and you should set something over it before in order to prevent any odor, since (sc. otherwise) it is better not to fumigate.

[36] As it is transmitted, this clause makes no sense: the translation is supplied from a parallel passage in *Nature of Women* ch. 34(3).

87. (196 L.) Ῥόου ἐρυθροῦ προσθετά·

380   (1) Σμύρνα καὶ βολβίον σὺν μέλιτι | τριφθὲν προσθετὸν ἄριστον.

(2) Ἢ ῥόδα ἐψήσας ἐν ὕδατι, καὶ τρίψας λεῖα ἐν μύρῳ ῥοδίνῳ, ἐν εἰρίῳ ἑλίξας, προστίθει.

(3) Ἢ τοῦ λωτοῦ τὸ ἄνθος ἑψῆσαι ἐν ὕδατι, κἄπειτα τρίβειν ἐν ῥοδίνῳ μύρῳ· ἐν εἰρίῳ πρὸς τὸ στόμα τῆς ὑστέρης προσάγειν.

(4) Ἢ κύπαιρον καὶ ἴριν καὶ ἄννησον ἴσον ἑκάστου ἐν μύρῳ ῥοδίνῳ, λεῖον, ἐν εἰρίῳ, μάλιστα πρὸς τὸν στόμαχον εἴσω.

(5) Ἢ μυρσίνης φύλλα μελαίνης ἐν οἴνῳ λευκῷ· πίτυος φλοιὸν παραμίσγειν ὁμοίως χρή.

(6) Ἢ κυπαρίσσου καρπὸν καὶ λιβανωτὸν ἴσον ἑκατέρου τρίβειν ὁμοῦ ἐν μύρῳ ῥοδίνῳ, ἐν εἰρίῳ προστίθει.

(7) Ἢν ξηρῆναι δέῃ ῥόον, καλαμίνθην ἐν οἴνῳ μέλανι ἕψειν, καὶ ἐς ὀθόνιον βάπτοντα,[287] ἐπιτιθέναι. ἢ ὑοσκυάμου τῶν φύλλων καὶ κώνειον ἕψειν. ὡσαύτως πράσα καὶ μαλάχην καὶ κηρὸν καὶ χηνὸς ἄλειφα μῖξαι, χλιαρὸν προστίθεσθαι πρὸς τὰ αἰδοῖα.

(8) Ἢ οἶνον ἄκρητον σὺν ῥητίνῃ καὶ σιδίῳ ἑφθῷ τρίβειν, καὶ προστιθέναι.

(9) Ἢ κνῆκον σὺν οἴνῳ τρίβων προστίθει. ἢ λωτοῦ πρίσματα ὡσαύτως.

(10) Ἢ σχίνου φύλλα ἢ ῥόον, μέλιτι καθέφθῳ, μίξας πρόσθες.

420

87. Suppositories for the red flux:

(1) Myrrh and a small bulb kneaded with honey make the optimal suppository.

(2) Or boil roses in water, knead them smooth in rose unguent, wrap this in a piece of wool, and insert it.

(3) Or boil nettle tree flowers in water, knead them smooth in rose unguent, and apply in a piece of wool against the mouth of the uterus.

(4) Or galingale, iris, and anise—an equal amount of each—kneaded smooth in rose unguent, and applied in a piece of wool inside right against the orifice.

(5) You must mix leaves of the black myrtle in white wine together with pine tree bark, and apply in the same way.

(6) Or cypress cones and frankincense—an equal amount of each—knead together in rose unguent and apply in a piece of wool.

(7) If a flux needs to be dried up, boil mint in dark wine, soak this up on a piece of linen, and insert it. Or boil henbane leaves and hemlock together. Mix leeks, mallow, wax, and goose grease in the same way, and apply this warm against the genitalia.

(8) Or mix undiluted wine with resin and boiled pomegranate peel, and apply.

(9) Or scrape oak gall into wine and apply this, or sawdust of the nettle tree in the same way.

(10) Or mix leaves of mastic or sumac into boiled honey, and apply.

---

287 βάπτοντα Θ: βρέξαντα MV.

(11) Ἢν δὲ μὴ λήγῃ, σούσινον, ἢ βλίτον ὡς γλῶσσαν ἐν εἰρίῳ προσθετόν.

88. (197 L.) (1) Ῥόου ὕδατος προσθετὸν καὶ ἔγχυτον· ἢν γυναικὶ ὕδωρ ῥέῃ ἐκ τῶν αἰδοίων, λαβὼν ῥητίνην ξηρὴν καὶ μυρίκης πέταλα καὶ λίνου καρπὸν τρῖψαι ἐν οἴνῳ, καὶ μῖξαι ὄρνιθος στέατι, καὶ ἔγχεον δ᾽ ἐς τὰ αἰδοῖα κλυστῆρι.

(2) Ἄλευρα χηραμύδα ὡς καθαρώτατα, ἢ ἀμύλιον ἐν εἰρίῳ προστίθεσθαι πρὸς τὸν στόμαχον.

382   (3) Ῥόου ὕδατος ἔγχυτον· | ὅταν ὕδωρ ἐκ τοῦ αἰδοίου ῥέῃ, μυρίκης πέταλα καὶ λιβανωτοῦ καρπὸν ἐν χηνὸς στέατι τρῖψαι καὶ ῥητίνην, ἐς τὰ αἰδοῖα κλυστῆρι ἐνιέναι.

89. (198 L.) Ῥόου ὑδατοείδεος τὸ ἦτρον ἀλγέει· λίνου[288] καρπὸν τρῖψαι ἐν μέλιτι, καὶ καταπλάσαι τὸ ἦτρον.

90. (199 L.) (1) Ῥόος αἱματώδης ἢ λευκὸς ἢ ὁποῖος· θεῖον καὶ μανδραγόρου τὸ ἄκρητον[289] ἀναλαβόντα εἰρίῳ προσθεῖναι, καὶ ὑπτίη εὑδέτω, καὶ ἀκίνητος.

(2) Ἢ σίδια ξηρὰ τρίβειν ἐν οἴνῳ Πραμνίῳ καὶ διδόναι πίνειν.

(3) Ῥόου λευκοῦ ποτόν· κισσοῦ λευκοῦ τὸ σπέρμα καὶ πίτυος φλοιὸς ἐν οἴνῳ αὐστηρῷ ποτόν.

(4) Ἢ ἐλάφου κέρας κατακαῦσαι μοῖραν, ὠμηλύσιος δὲ δύο μοίρας καὶ κεδρίδας πέντε, ἐν ὕδατι τρίβειν, καὶ πίνειν.

91. (200 L.) (1) Ὅταν πνίγηται ἀπὸ ὑστερέων· κάστορα καὶ κόνυζαν χωρὶς ἐν οἴνῳ καὶ ἐν τῷ αὐτῷ

(11) Or, if the flux does not stop, take lily oil or tongue-shaped blite and make a suppository in a piece of wool.

88. (1) Suppository and infusion against an aqueous flux: if water flows out of a woman's genitalia, take dry resin, leaves of tamarisk, and linseed, crush them in wine, mix this with chicken's fat, and inject it into the genitalia with a syringe.

(2) A cheramys of the purest wheat flower, or starch: apply on a piece of wool against the orifice.

(3) Infusion for an aqueous flux: when water flows out of a woman's vagina, knead tamarisk leaves and fruit of the frankincense tree in goose grease together with resin, and inject into the genitalia with a syringe.

89. The lower abdomen becomes painful together with an aqueous flux: knead linseed in honey, and apply as a poultice to the lower abdomen.

90. (1) A bloody flux, or a white one, or any other kind: take up some sulfur and undiluted mandrake in a piece of wool, and apply it as a suppository: the woman should sleep on her back and remain still.

(2) Or crush dry pomegranate peel in Pramnian wine and give it to drink.

(3) A potion for a white flux: dissolve white ivy seeds and pine tree bark in dry wine as a potion.

(4) Or take one portion of incinerated deer's horn, two portions of bruised meal of raw grain, and five Syrian cedar berries, beat these in water, and give to drink.

91. (1) When a woman suffocates from her uterus: have her drink castoreum and fleabane, both separately and

---

288 λίνου Θ: ἄγνου MV.
289 τὸ ἄκρητον recentiores: τῷ ἀκρήτῳ ΘV: τὸ ἀκρήτῳ M.

πινέτω. ἢ ἀσφάλτου ὅσον τριώβολον, ἢ φώκης στέαρ ὅσον δὶς τῷ δακτύλῳ λαβεῖν <ἢ>[290] ῥίζαν γλυκυσίδης ὅσον ἥμισυ πόσιος ἐν οἴνῳ εὐώδει.

(2) Ὅταν πνίγωσι καὶ ὁμοῦ βήσσῃ, σανδαράχης ὅσον ὀβολὸν καὶ θείου ἀπύρου ἴσον καὶ ἀμύγδαλα πικρὰ καθήραντα συμμίσγειν τρία ἢ τέσσερα, καὶ διδόναι ἐν οἴνῳ εὐώδει.

(3) Ἢν δὲ προσίσχωσιν[291] αἱ ὑστέραι, κριθὰς τρίψας λεῖα σὺν τοῖσιν ἀχύροισι[292] καὶ ἐλάφου κέρας· ἐλαίῳ δεύσασα ὑποθυμιήσθω.

(4) Ὅταν ἄνω ᾖ, καὶ φλίβηται καρδίη καὶ στόμαχος καὶ ὑπομένωσι, σμύρναν, ἢ ῥητίνην, ἢ νέτωπον, ἢ κάστορα, ἢ ὀπὸν σιλφίου πῖσαι. |

384   92. (201 L.) Ὅταν ὑστέρα πνίγῃ, πνεῦμα δὲ σεύηται ἁλὲς ἄνω, καὶ βάρος ἔχῃ, καὶ γνώμη καταπλήξ, ἀναυδίη, περίψυξις, πνεῦμα προσπαῖον, ὄμματα ἀμαλδύνηται, κεφαλὴν ξυρᾶν ὅτι τάχος, καὶ ταινίῃ ἀποδιωθέειν, ὑπὲρ ὀμφαλοῦ δὲ εἰλέειν· διδόναι δὲ καστόριον καὶ κόνυζαν, πηγάνου ὕδωρ, κύμινον Αἰθιοπικόν, ῥαφάνου σπέρμα, θεῖον, σμύρναν· πρὸς δὲ τὰς ῥῖνας τὰ κάκοδμα, εὔοδμα δὲ ἐς τὰς ὑστέρας.

Ἢν δὲ ἀνίσχηται, κρόκου[293] τὸ ἔνδον τὸ λευκὸν μέλιτι μίξας, ἀλείφειν τὴν ῥῖνα· ἢ σχῖνον τριπτήν,[294] ἵνα δάκνηται. ὅταν ὀδύνη ἔχῃ καὶ πνίγηται, μαλάχης ῥίζαν, ἢ ὀξύμελι ἢ φλοιὸν μαράθου καὶ κρῆθμον ἐν ὕδατι δοῦναι πιεῖν, ἄριστον ἐρυγγάνειν καὶ δ' ἀνακα-

[290] ἢ add. Linden.   [291] προσίσχ. MV: προΐσχ. Θ.

together, in wine. Or take three obols of asphalt or a pinch of seal's fat, or half a draft of peony root in fragrant wine.

(2) When patients suffocate and simultaneously cough, mix together an obol of realgar, the same amount of unburned sulfur, and three or four hulled bitter almonds, and give this in fragrant wine.

(3) If a woman's uterus becomes fixed somewhere, grind barley together with its bran fine and deer's horn: have the patient dissolve this in olive oil and use it to fumigate herself from below.

(4) When the uterus has moved upward and the cardia and the orifice of the stomach are compressed, and it stays there, give myrrh, resin, oil of bitter almond, castoreum, or silphium juice to drink.

92. When a woman's uterus suffocates her, while her breath rushes up in a wave, she feels a heaviness, her mind is overwhelmed, she loses her speech, she shivers, her breathing is broken, and her eyes become weak, you must shave her head at once, and dislodge the uterus by winding a bandage around her over her navel. Give castoreum, fleabane, rue water, Ethiopian cumin, radish seed, sulfur, and myrrh, and apply ill-smelling substances to the nostrils and fragrant ones to the uterus.

If the suffocation persists, mix the white insides of saffron with honey, and anoint this to the nose; or apply ground mastic to irritate it. When pain is present and the woman is suffocating, take mallow root, oxymel, or fennel bark together with samphire, and give this to drink in water; best for the patient would be to belch and to sit

---

[292] ἀχύροισι Θ: ἀλεύροισι MV.   [293] κρόκου Θ: καὶ κόχλου MV.   [294] τριπτήν Θ: λεπτήν MV.

θίζειν· ἢ ἐλλεβόρῳ[295] πταρμὸν ποιέειν, καὶ προσίσχειν πρὸς τὰς ῥῖνας. ἢν δ' ὑπὸ φρένας δοκέωσιν
ἵζεσθαι, ἐξαπίνης ἄφωνος γίνεται, ὑποχόνδρια
σκληρά, καὶ πνίγεται, καὶ τοὺς ὀδόντας συνερείδει,
καὶ οὐχ ὑπακούει καλεομένη· ὑποθυμιᾶν ὑπὸ τὰς ῥῖνας, εἴριον κατακαίων, ἄσφαλτον ἐς πῦρ ἐμβάλλων,
καστόριον, θεῖον, πίσσαν· βουβῶνας δὲ καὶ μηροὺς
μύρῳ ἔνδοθεν εὐωδεστάτῳ χρίειν· ἢ ἀστέρας τοὺς θαλασσίους τοὺς μέλανας καὶ κράμβην μίξας ἐν οἴνῳ
εὐώδει πινέτω. ἢ σμύρνης ὡς τριώβολον, κορίαννον
ὀλίγον, ῥητίνην, γλυκυσίδης ῥίζαν, κύμινον Αἰθιοπικόν, ταῦτα τρίψας ἐν οἴνῳ λευκῷ, ὕδατι ἢ μελικρήτῳ
διέντα, πιεῖν θερμαίνοντα ἀκροχλίαρον. βοηθέει δὲ
καὶ πευκέδανον, ἀριστολοχία, κρομμύου δάκρυον, |
386 πάνακες, ἐν οἴνῳ ἢ ὕδατι, ἀκροχλίαρον διδόναι.

Ὑστέρας ἐς χώρην ἄγει, τοῦ κρότωνος ῥίζα πινο-
μένη, ἢ κύμινον Αἰθιοπικόν, ἢ σέλινον, ἢ μαράθου
σπέρμα καὶ ἀννήσου, ἢ πέπερι ἢ σμύρνα, ἢ ὀπὸς
μήκωνος πινόμενος.

Ἢν ἡ καρδίη πνίγηται ὑπὸ ὑστέρης, ἀναθλίβεται
πνεῦμα ἢ ὑγρόν, καὶ ἢν πλεύμονα[296] ἔχῃ, καὶ ἆσθμα·
τὸν καρπὸν τῆς ἄγνου καὶ γλυκυσίδης ἐν οἴνῳ πίνειν,
ἢ ἀβρότονον, καὶ πάνακες, ἢ ἀμμωνιακόν, ἢ πήγανον,
ἢ ὑπνωτικὸν μηκώνιον. ἢν πνιγμὸς ἔχηται ἀπὸ ὑστέ-
ρων, μελάνθιον τρίβειν λεῖον, μέλιτι δεύειν, καὶ οἷον
βάλανον ποιέειν, καὶ[297] τῷ πτερῷ προστιθέναι· ἢ φιλίστιον ὁμοίως προστίθει· ἢ τηλέφιον, ἢ ἀνεμώνης

upright. Or make her sneeze by holding hellebore up to her nose. If the woman's uterus seems to be lodged under her diaphragm, she will suddenly lose her speech, her hypochondria will become hard, she will suffocate and clench her teeth, and she will not respond when she is called. Fumigate her under her nostrils with a piece of burning wool, and throw asphalt on to a fire, with castoreum, sulfur, and pitch. Anoint the insides of her groins and thighs with a very fragrant unguent; or mix some black starfish and cabbage in fragrant wine, and have her drink this. Or take three obols of myrrh with a little coriander, resin, peony root, and Ethiopian cumin, pound them in white wine, dissolve this in water or melicrat, and give it lukewarm to be drunk. Also beneficial are sulphurwort, aristolochia, onion juice, and all-heal in wine or water: give this lukewarm.

To drive the uterus to its proper place: a portion of castor tree root, Ethiopian cumin, celery, fennel and anise seeds, or pepper, myrrh, or poppy juice potion.

If a woman's heart is suffocated by her uterus, breath or fluid is forced up, and if it involves her lung, she has shortness of breath, too. Have her drink chaste tree and peony blossoms in wine; or southernwood and all-heal, or gum ammoniacum, rue, or opium poppy. If there is suffocation originating from the uterus, grind black cumin fine, dissolve it in honey, shape this into a suppository, and apply it with a feather. Or apply cleavers in the same way;

---

295 -βόρῳ Θ: -βορον καὶ MV.
296 ἢ ὑγρόν . . . πλεύμονα om. MV.
297 Add. πρὸς MV.

φύλλα τρίψας, ἔνθες ἐς τρυχίον, καὶ σμύρναν σμικρὴν ξυμμίσγειν.

Ἢν ἐς τὴν ὀσφῦν αἱ ὑστέραι καταστηρίξωσι, μὴ ψαύῃ δὲ τῆς κεφαλῆς ἡ πνίξ, ἐσθιέτω πουλύποδας ἐφθούς, ὀπτούς,[298] καὶ οἶνον πινέτω μέλανα εὐώδεα ἄκρητον πλεῖστον.

Ὅταν δ' ὡς πρὸς τὰ σπλάγχνα τραπεῖσαι πνίγωσιν, οἶνον κέδρινον καὶ κύμινον[299] Αἰθιοπικὸν πινέτω, καὶ θερμῷ λούσθω, πυρία καὶ[300] τὰ εὐώδεα.

93. (202 L.) Ἢν ἀνεμωθῶσιν αἱ ὑστέραι, ἡδύσματα πάντα ἐς τὸ μύρον ἐμβάλλεται, καὶ δάφνη, καὶ μυρσίνη, καὶ ἐλελίσφακος, καὶ κέδρου πρίσματα καὶ φύλλα[301] κυπαρίσσου· ταῦτα κόψαι καὶ κατασῆσαι λεῖα, καὶ ἐπ' οἶνον εὐώδεα ἐπιπάσσειν, καὶ ἐπιχέαι ῥόδινον ἔλαιον.

94. (203 L.) Ὅταν μετακινηθεῖσαι φλίβωσιν αἱ ὑστέραι, κριθὰς σὺν | τοῖσι κυρηβίοισι καὶ πρόμαλον καὶ ἐλάφου κέρας οἴνῳ δεύσας, ὑποθυμία.

Ὅταν προσιστάμεναι πνίγωσιν, ἐλλύχνιον ἅψαι καὶ ἀποσβέσαι ὑπὸ τὰς ῥῖνας, ⟨ὡς⟩[302] λιγνὺς καὶ αἰθαλὸς εἴσεισι· καὶ πίσσαν καὶ καστόριον καὶ πευκέδανον καὶ σμύρναν διεὶς μύρῳ, εἰρίῳ ἀναδήσασα,[303] προστιθέσθω· πίνειν δὲ ῥητίνην ἐλαίῳ διέντα.

298 ὀπτούς om. MV.
299 κύμινον Foes in note 311: οἶνον codd.
300 πυρία καὶ Potter: πυριαμα Θ: καὶ πυρία MV.

or grind telephion or anemone leaves, wrap them into a rag, and smear a little myrrh on to it.

If a woman's uterus becomes fixed against her sacrum, but no suffocation reaches to her head, have her eat boiled and baked octopus and drink a great amount of undiluted, fragrant, dark wine.

When a woman's uterus turns against her viscera and causes suffocation, have her drink wine mixed with cedar oil and a potion of Ethiopian cumin, and bathe in hot water, and foment her with fragrant agents.

93. If a woman's uterus becomes inflated, all kinds of fragrant substances are to be put into an unguent, laurel, myrrh, salvia and sawdust of cedar and cypress leaves: chop all these and sieve them fine, sprinkle them over wine, and add rose oil.

94. When a woman's uterus becomes displaced and exerts pressure, soak barley together with its bran, willow, and deer's horn in wine, and use this as a fumigation from below.

When the uterus causes suffocation by pressing against some part, light a lamp wick and extinguish it under the patient's nostrils in such a way that the smoke and soot will go in. Also dissolve pitch, castoreum, suphurwort, and myrrh in a unguent, and have the patient soak this up on a piece of wool and apply it as a suppository; as drink, resin dissolved in olive oil.

---

301 φύλλα om. MV.

302 Add. Littré following Calvus' *uti* and Cornarius' and Foes' *quo*.

303 -δήσασα Littré: -δεύσασα codd.

Ἢν πνίγωσι λίην, ποτὸν χελώνης παραθαλασσίης ὅσον τριώβολον τρίψας ἐν οἴνῳ λευκῷ, κοτύλην κυάθοις πίνειν· ἢ[304] λύχνον, ἐπιχέας ὅσον ἔλαιον,[305] ἅψαι τὸν λύχνον, καὶ ἐὰν ἀποσβεσθῇ, πρὸς τὰς ῥῖνας πρόσαγε· ἢ βόρβορον ὡς δυσωδέστατον ὁμοίως· ἢ εἴριον κατακαύσας·[306] ἢ ἀσφάλτου ὀλίγον τρίψας ἐν οἴνῳ λευκῷ πινέτω· ἢ ἐρυσίμου κόγχην καὶ καστορίου ἐν οἴνῳ λευκῷ δὸς πιεῖν, καὶ λοῦσον. ἢν δὲ[307] βήσσῃ, σανδαράχης ὀβολόν, θείου ὀβολοὺς δύο ἀπύρου, ἀμύγδαλα πικρὰ ἀποκαθήρας, καὶ καστορίου ὀβολὸν μίσγειν σὺν οἴνῳ εὐώδει, καὶ πιεῖν δίδου.

Ἢν δὲ πνίγωσιν αἱ ὑστέραι, καὶ τῆς καρδίης ψαύσῃ τὸ πνῖγμα, καὶ μεμύκῃ τὸ στόμα, ὄξος θερμὸν δίδου ῥοφεῖν—πασσάλῳ ἢ κερκίδι διάνοιγε—ἢ οἶνον ὁμοίως ἄνοδμον καὶ σὺν ὀξυμέλιτι.

Ἢν δὲ λίην[308] πνίγωσι, καὶ ἄφωνος ᾖ, κρόμμυον ἐν οἴνῳ χλιαρῷ τρίψας, ἐνστάζειν ἐς τὰς ῥῖνας, καὶ ἀνεγείρειν.

Ἢν δὲ ἄχρι ἥπατος ἀνίωσι, καὶ πνίγηται, ἄφωνος γίνεται, καὶ οὐδὲν ὁρᾷ, καὶ τοὺς ὀδόντας | συνερείδει, καὶ σκληρὴ γίνεται, καὶ οὐδὲν φρονέει, καὶ ἀναπνεῖ πυκινά, καὶ οὐδὲν ἀκούει·[309] ταύτην ὑπὸ τὰ ὑποχόνδρια λαβὼν τῇσι χερσὶ σείειν θαμινάκις, καὶ τοὺς ὀδόντας διαγαγὼν πασσάλῳ, οἶνον ἄκρητον χλιαρὸν ἐγχέειν, ἢν μή τι κωλύῃ, καὶ αὐτίκα ῥαΐζει ὡς τὰ πολλά.

390

---

304 ἢ om. Θ: κέρδον ἐς add. MV.　　305 Om. ὅσον ἔλαιον MV.　　306 Add. ὑπὸ τὰς ῥῖνας ὑποθυμιῆν τῆς γυναικὸς MV.

If the suffocation is extreme, employ a potion containing three obols of sea tortoise ground into white wine—a cotyle in all, a cyathos at a time. Or take a lamp, pour a little oil into it, light it, and when it goes out hold it up to the patient's nostrils. Or employ the most malodorous dirt in the same way, or a piece of wool you have incinerated. Or have the patient drink a little asphalt ground into white wine. Or give her a cheramys of hedge mustard and castoreum in white wine to drink, and then bathe her. If she has a fit of coughing, mix an obol of relgar, two obols of unburned sulfur, cleaned bitter almonds, and an obol of castoreum with fragrant wine, and give this to drink.

If the uterus causes suffocation involving the heart, and its mouth is closed, give warm vinegar to drink and open the mouth with a small stick or a peg, or give a bland wine prepared in the same way with oxymel.

If the suffocation is extreme and the patient loses her speech, mash an onion in warm wine, dribble this into her nostrils, and try to bring her to herself.

If a woman's (sc. uterus) moves up as far as her liver and causes suffocation, she loses her speech and vision, she clenches her teeth, her body becomes rigid, she loses her ability to understand anything, she breathes rapidly, and she loses her hearing. Take hold of this woman's body below her hypochondrium with your hands, shake her repeatedly, pry her teeth open with a stick, and pour warm undiluted wine into her mouth (unless there is some reason not to), and in most cases she will immediately improve.

---

[307] Add. μὴ M.

[308] δ. λ. Θ: λίην δὲ MV.

[309] οὐδὲν ἀκούει Θ: οὐχ ὑπακούει MV.

Ἢν ἐγκέωνται ἐς τοὺς βουβῶνας καὶ ἐρείδωσιν, αἰγὸς σπυράθους καὶ λαγῴας[310] τρίχας, ἐλαίῳ φώκης δεύσας, ὑποθυμία· ἢ τοῦ κυτίσου[311] τὸν καρπὸν ἢ τὰ φύλλα αὐαίνειν, ἢ τὸν φλοιὸν καὶ δρυὸς φύλλα καὶ ῥητίνην μίξας, ἐλαίῳ δεύειν, καὶ ὑποθυμιᾶν. ἢ φώκης τῆς πιτύης τὸ δέρμα κόψας λεῖον, καὶ σπόγγον καὶ βρύα λεῖα μίσγειν τῷ ἐλαίῳ τῆς φώκης, καὶ ὑποθυμιᾶν. αἰγὸς σπυράθους, καὶ φώκης πλεύμονα,[312] καὶ κέδρου πρίσματα ὑποθυμιᾶν. ἢ βόλβιτον, ἢ κεράτων ξύσματα βοὸς καὶ ἄσφαλτον, ἢ ἀκάνθης Αἰγυπτίης καρπὸν καὶ κέδρου πρίσματα, καὶ μυρσίνης φύλλα ξηρά, μύρῳ μαλθακῷ ταῦτα δεύσας, ὑποθυμιᾶν· ἀρώματα συχνὰ δ' ἐς τὸ μύρον ἐμβάλλειν. ἢ γίγαρτα κόψας λεῖα, καὶ κεδρίας καὶ ῥητίνην πιτυΐνην ὁμοῦ μίξας, γλυκεῖ ἑφθῷ δεύσας, ὑποθυμιᾶν.

Πυριήσιες ὡς ἀπίωσι·[313] βολβίτου κεκομμένου καὶ ὄξους ἥμισυ, καὶ ὀροβίου θαλάσσης ἢ ὕδατος ὁμοίως πυριᾶν τὰς ῥῖνας· πυριᾶν δὲ βληχρῶς, καὶ φάκιον πιεῖν, ἀπεμέειν δέ, καὶ ῥυφεῖν διδόναι ἄλητον καὶ ἔτι τὸν οἶνον· τῇ δ' ὑστεραίῃ κόκκος ἔστω κατάποτος· καὶ οὐρητικὸν δὲ ἀσταφίδα καὶ ἐρεβίνθους, δύο τρίψας ἀσταφίδος τῆς ἀρίστης, ἐπιχέας χόεα, ἕψε, ἔπειτα ἀποχέας πρὸς τὴν αἰθρίην θεῖναι, καὶ τῇ ὑστεραίῃ πίνειν, καὶ τὸ λοιπὸν ἐλελίσφακον, καὶ λίνου σπέρμα· 392 ἄλφιτον διδόναι | δὶς τῆς ἡμέρης ἐπ' οἴνῳ κεκρημένῳ, κοτύλας τέσσερας. ἐλαίου ἡμικοτύλιον, ἀκτῆς φύλλα

310 λαγῴας Θ: λαγωιοῦ Μ: λαγωοῦ V.

If a woman's uterus presses against her groins and fixes itself there, soak goat's excrement and hare's hair in seal's oil, and apply as a fumigation from below. Or dry fruit or leaves of the medick tree, or add bark and leaves of oak together with resin, moisten with olive oil, and fumigate with this from below. Or pound the skin of seal's rennet smooth, mix sponge and tree moss smooth with seal oil, and fumigate with this from below. Do the same with goat's excrement, seal's lung, and cedar sawdust. Or take cow's excrement, scrapings of cow's horn, and asphalt, or fruit of acacia and cedar sawdust, and dried myrtle leaves, soak these in a soft unguent, and fumigate with this from below: add many aromatics into the unguent. Or crush grape stones fine, mix this into cedar oil and pine resin, dilute with boiled, sweet (sc. wine), and fumigate with this from below.

Fomentations to make the uterus retract: Take a half amount each of powdered cow's excrement and vinegar, mix this with vetch flour in brine or water, and foment the nostrils in the same way. After fomenting gently, give lentil soup as an emetic, and have the patient drink barley gruel and more wine. On the following day have her take a (sc. Cnidian) berry as a pill. As diuretic employ raisins and chick peas: crush two portions of the best raisins, add a chous of water, boil, decant, and set in the open air; on the next day have the patient drink this, and furthermore salvia and linseed. Give barley meal twice a day in four cotyles of diluted wine. Or boil a handful of elder leaves

---

311 κυτίσου MV: κίσσου Θ.
312 -μονα Froben: -μονος ΘMV.
313 ὡς ἀπίωσι om. M.

χεῖρα πλέην, ταῦτα ἕψειν, καὶ πυριᾶν θερμῷ, ἢ ὀστρά-
κοισι θερμοῖσιν, ἐπὶ δίφρου καθίζεσθαι, ἀμφικαλύπτειν δ᾽ εἵμασιν. ἢ τῆς ἀκτῆς φύλλα σὺν μυρσίνῃ
ἕψειν, καὶ κριθέων ἄχυρα ἕψειν· καὶ πυριᾶν,[314] εἰ οἷά
τε εἴη ὑποφέρειν, ὄξος, ἔλαιον, μέλι, ὕδωρ, ταῦτα κεράσας καὶ ἀναμίξας, ἀναζέσαι σφόδρα, ἐς κύστιν
ἐγχέαι· ἢ τῆς πίτυος τὸν φλοιὸν καὶ τῆς ῥοιῆς τὰ
φύλλα ἐμβάλλειν ἐς ὕδωρ, ἀφέψειν δὲ ἰσχυρῶς· καὶ
ἐμβάλλειν ἐς τὸ ὕδωρ καὶ κριθῶν ἄχυρα, ἕψειν, ἔλαιον
ἐπιχέαντα· ἢ λωτοῦ πρίσματα καὶ κυπαρίσσου, ὕδωρ
ἐπιχέας καὶ ἔλαιον, ἕψε εὖ μάλα, καὶ πυρία σὺν ἀρώ-
μασι· μύρον δὲ ἐγχέειν, καὶ κηκίδα ἐμβάλλειν καὶ
ῥάμνου φλοιὸν καὶ πύρινα ἄλφιτα σὺν ὕδατι.

95. (204 L.) Ἢν δὲ προΐσχωσιν ἔξω, τὰ δὲ νεῦρα
τὰ καλεόμενα ὄχοι χαλᾶται· μύρτα, λωτοῦ πρίσματα,
βάτου, ἐλαίης φύλλα ἅμα ἕψειν, καὶ πυριᾶν ἀκροχλί-
αρος· ἢ οἴνῳ μετὰ τῶν κακωδέων, ὁμοίως· περιχρίειν
δὲ ᾠῶν λευκῷ τὰ ἔξω· ὅταν δὲ ψύχωνται καὶ πελιδναὶ
ἔωσιν, ὕδατι θερμῷ.

96. (205 L.) Μαλθακτήρια ὑστέρης·

(1) Τὸς[315] στέαρ, ᾠῶν λέκιθος, μέλι, ἔλαιον ῥόδινον, τούτοισιν ἀναφυρήσας ἄλητα, παραχλίαινε πυρὶ
μαλθακῷ, τὸ ἀποσταζόμενον ἐς εἴριον ἀναμαλάσσειν,
καὶ προστιθέναι.

(2) Ἢ στέαρ ὄϊος ἡδύ, νίτρον τὸ ἐρυθρόν, ἢ[316]
χηνὸς ἄλειφα, ῥόδινον ἔλαιον, συντήξας καὶ ἐς εἴριον
ἀναφυρήσας, προστιθέναι.

in a half cotyle of olive oil, and foment with this still hot
or heated with hot potsherds: seat the patient on a stool,
and cover her with blankets. Or boil elder leaves with
myrtle berries, or also barley bran. Also foment, if the
patient can tolerate it, with vinegar, olive oil, honey, and
water: mix and stir these together, bring to a rapid boil,
and pour into an (sc. animal's) bladder. Or put pine bark
and pomegranate leaves into water, and boil vigorously;
also add barley bran to water along with olive oil, and boil.
Or take nettle tree or cypress sawdust, add water and oil,
and boil vigorously; use this to foment together with aromatics. Add unguent, oak gall, buckthorn bark, and wheat
meal to water.

95. If the uterus moves to the exterior, the cords called
"ligaments" have been relaxed: boil together myrtle berries, nettle tree sawdust, and leaves of bramble and olive,
and foment with this lukewarm; or in wine with the addition of ill-smelling agents, and applied in the same way.
Smear the outside (sc. of the prolapsed uterus) with egg
white, and when the uterus cools off and becomes livid,
foment it with hot water.

96. Emollients for the uterus:

(1) Mix lard, egg yolk, honey, and rose oil into meal,
and warm this over a gentle fire: drip this on to a piece of
wool, knead it smooth, and apply as a suppository.

(2) Or melt together seasoned sheep's fat and red soda,
or goose grease and rose oil, soak it up on a piece of wool,
and apply.

---

314 πυριᾶν om. MV.
315 ὑὸς Θ: ὅιος MV.
316 ὅιος ἡδύ . . . ἢ MV: ἡδυντὸν τὸ ἐρυθρόν Θ.

394   (3) Ἡ χηνὸς ἄλειφα, | μήλειον στέαρ, κηρὸν λευ-
κόν, νέτωπον, ῥόδινον ἔλαιον, ὡς ἄριστα ταῦτα ἀναμὶξ
ποιέειν, καὶ ῥάκεα ἐγκατατίλλειν λεπτά, καὶ αὐτίκα
λουσαμένη, προστιθέσθω χλιαρὰ πρὸς τὸ στόμα.

(4) Ἡ ἐλάφου μυελὸν καὶ στέαρ τῆξαι ἐν ῥοδίνῳ
ἐλαίῳ, ἀναφορύξασα εἴριον μαλθακόν, προστιθέσθω.
Μαλθακὰ προσθετά·

(5) Ὕδωρ ἄγει καὶ μύξας καὶ δέρματα, καὶ οὐχ
ἑλκοῖ· σμύρναν ὡς ἀρίστην, καὶ ἁλὸς χόνδρον καὶ
πίσσαν ὡς ἡδυντήν, τρίβειν λεῖα.

(6) Ἕτερον προσθετόν· ἐκλέψας κόκκους τριήκοντα,
τὸ Ἰνδικὸν ὃ καλέουσιν οἱ Πέρσαι πέπερι, ἐν τούτῳ
δ’ ἔνι στρογγύλον ὃ καλέουσι μυρτίδανον, σὺν γάλακτι ὁμοῦ τρίβειν γυναικὸς καὶ μέλιτι διέναι· ἔπειτα
εἴριον μαλθακὸν καὶ καθαρὸν ἀναφυρήσας, περὶ
πτερὸν περιελίξας προσθεῖναι, καὶ τὴν ἡμέρην ἐᾶν·
ἢν δὲ ἰσχυρότερον βούλῃ, σμύρναν ὀλίγην παραμίσγειν ὅσον τριτημόριον, καὶ εἴριον μαλθακὸν καθαρὸν ἢ ἡμίρρυπον.

(7) Ἄγει ἐξ ὑστερέων, στόμα μαλθάσσει· νάρκισσος, κύμινον, σμύρνα, λιβανωτός, ἀψίνθιον, κύπαιρος, καθ’ ἑωυτὰ ξὺν ῥοδίνῳ ἢ ἐλαίῳ λευκῷ, προσθέσθω δὲ λουσαμένη.

(8) Καὶ ἰχῶρα καὶ ὕφαιμον ἄγει· σὺν τοῖσδεσι
μίσγειν σμύρναν, ἅλας, κύμινον, χολὴν ταυρείην,
μέλι, ἐν εἰρίῳ προστιθέναι· καὶ κυμίνου φύλλα ἠδελφισμένως σὺν οἴνῳ· ἢ ὀπὸν σιλφίου σύκῳ μίσγειν,
καὶ βάλανον ποιήσαντα προσθεῖναι· τὸ δ’ αὐτὸ δρᾷ

(3) Take the finest goose grease, sheep's fat, white wax, oil of bitter almond, and rose oil, mix together, soak this up with soft rags, and have the patient insert them warm against the mouth (sc. of her uterus) immediately after having bathed.

(4) Or melt deer's marrow and fat in rose oil, and have the patient soak it up on a piece of soft wool, and insert it.

Emollient suppositories:

(5) Water carries off mucus and pieces of skin without causing ulceration; take the finest myrrh, a lump of salt, and some seasoned pitch, and knead them smooth.

(6) Another suppository: shell thirty corns of the Indian plant the Persians call "pepper" which has a spherical part they call "myrtidanum," grind these up in woman's milk, and dissolve this in honey; then soak it up on a piece of clean, soft wool, wind this around a feather, insert, and leave in place for the day. If you want something stronger, add a little myrrh—about a third in amount—with soft wool, either clean or half-greasy.

(7) A suppository to empty the uterus and soften its mouth: take narcissus, cumin, myrrh, frankincense, wormwood, and galingale by themselves with rose oil or white unguent , and have the woman insert this suppository after she has bathed.

(8) A suppository to attract both serum and what is more bloody: with the same ingredients just listed mix myrrh, salt, cumin, bull's gall, and honey, and apply on a piece of wool. Also cumin leaves prepared the same way with wine. Or mix silphium juice with a fig, form this into a suppository, and insert it. White root with honey has the

καὶ ἡ λευκὴ ῥίζα[317] ξὺν μέλιτι, καὶ μετέπειτα λούειν
396 σὺν[318] ῥοδίνῳ· ἐνεργεῖ δὲ καὶ σκορόδου | μώλυζα, λί-
τρον ἐρυθρόν, σῦκον πῖον ἴσον μίσγειν· μίσγειν δὲ
καὶ κηκίδα μικρήν, καὶ βάλανον ποιέειν, καὶ ἐμβάπτειν ἔς τι τῶν ὑγρῶν, καὶ προστιθέναι,[319] κἄπειτα
λουσαμένη ἐλάφου στέαρ ἐν εἰρίῳ ἐχέτω.

(9) Μᾶλλον δὲ αἱματώδεα τῶν πρόσθεν ἄγει καὶ
μαλθάσσει· πέπερι, ἐλατήριον, συμμίσγειν δὲ καὶ γυναικὸς γάλα, τρίβειν σὺν τοῖσδε καὶ μέλι καὶ ἄλειφα
λευκὸν ἢ ἐλάφου στέαρ.

(10) Τρηχὺ μέν,[320] σφοδρὰ δὲ ἄγει παντοῖα· σύκου
τὸ πῖαρ, ἐλατηρίου δύο πόσιας, λίτρον ἐρυθρόν, ὅσον-
περ ἐλατήριον, μέλι ὀλίγιστον, ἐν ῥάκει ἢ εἰρίῳ, βάλανον ποιέειν.

(11) Ἢ[321] νέτωπον, χολὴν ταύρου, λίτρον, κυκλάμινον, κηκίδα, τρίβειν καὶ σὺν μέλιτι, μετέπειτα λουσαμένη στέαρ ἐχέτω ἢ γλήχωνα· καὶ χολὴν ταύρου,
σμύρναν, μέλι προστιθέναι, καὶ λουσαμένη ἔλαιον
ῥόδινον.

(12) Ἢ χολὴν ταύρου[322] τριπτὴν περιπλάσσειν
πτερῷ, καὶ ἐς ἄλειφα ἐμβάψας Αἰγύπτιον, προστιθέ-
ναι· ἢ κυκλάμινον ὅσον ἀστράγαλον σὺν χαλκοῦ ἄνθει, ἢ ἀνεμώνης κεφαλὰς τρῖψαι σὺν ἀλήτῳ, πτερῷ
περιπλάσσειν, ἢ ἐς λευκὸν εἴριον ἐμβάπτεσθαι.

317 ἡ λευκὴ ῥίζα Θ: ησυκηριζα Μ: ηλσυκὴ ῥίζα V.
318 λούειν σὺν Θ: χρίειν MV.
319 καὶ προστιθέναι om. Θ.

same effect, and after that anoint with rose oil. A head of garlic mixed together with equal amounts of red soda and greasy fig is also effective: also mix in a little oak gall, make this into a suppository, soak it in a fluid, and apply; then have the patient bathe and receive some deer's fat in a piece of wool.

(9) The following medication will draw bloody fluids better than the ones above, and soften as well: mix pepper and squirting cucumber juice together with woman's milk, and knead this into honey and white unguent or deer's fat.

(10) The following is irritating, but it moves all sorts of fluids well: take the fat part of a fig, two drafts of squirting cucumber juice, red soda equal in amount to the juice, and a very little honey, apply this to a rag or a piece of wool, and form it into a suppository.

(11) Or take oil of bitter almond, bull's gall, soda, cyclamen, and oak gall, knead these together into honey, and, after the woman bathes, have her retain (sc. a suppository of) fat or pennyroyal. Also apply bull's gall, myrrh, honey, and, after the woman bathes, rose oil.

(12) Or plaster kneaded bull's gall around a feather, immerse this in Egyptian unguent, and apply; or cyclamen to the amount of a vertebra with flower of copper; or crush a head of anemone, mix it with meal, and plaster this around a feather or soak it on to a piece of white wool.

---

320 τρηχὺ μέν Littré: τρηχυνομένη codd.
321 ἢ Θ: αἱμαγωγόν· ἢ MV.
322 σμύρναν . . . ταύρου om. M.

(13) Ἄγει πάντα· σικύης[323] ἐντεριώνης τῆς μακρῆς
ἐξελὼν τὸ σπέρμα σὺν γάλακτι, καὶ σμύρναν, σὺν
αὐτοῖσιν ἄκρητον μέλι, ποσὸν ἔλαιον Αἰγύπτιον, ἐν
εἰρίῳ μαλθακῷ ἀναφορύξαι· ἢ τὴν ἐντεριώνην τῆς σικύης ξηραίνειν, καὶ μέλι παραχέαι, καὶ τρίβειν,
καὶ βάλανον ποιέειν· ἐπὴν δὲ λούσηται, στέαρ προσ-
τιθέσθω· ἢ ἐλατηρίου τρεῖς πόσιας ἐν στέατι[324] μαλ-
θακτηρίῳ, ᾠοῦ τὸ πυρρόν, ἄλητον, μέλι, κηρὸν λευ-
κόν· ταῦτα ὁμοῦ χλιαίνειν, τὸ ἀποστάζων εἰρίῳ
398 ἀνασπογγίσας, | πρόσθες· ἢ χηνὸς ἄλειφα, κηρὸν
λευκόν, ῥητίνην, μύρον ῥόδινον· ἢ μυελὸν ἐλάφου
τήκειν ἢ στέαρ ὄϊος καὶ αἰγός, ᾠοῦ τὸ λευκόν, ῥόδινον
μύρον· ἢ βάλανον ἢ εἰρίῳ ἀναλαβεῖν.

97. (206 L.) Ὑστέρας καθαίρειν, ἢν σκληραὶ ἔωσι
πυρίαι·

(1) Οἶνον χρὴ ὡς ἥδιστον ἴσον[325] κεράσαι, ὡς τρία
ἡμίχοα Ἀττικά· καὶ μαράθου ῥίζαν καὶ καρποῦ μαράθου τεταρτημόριον, καὶ ῥοδίνου ἀλείφατος ἡμικοτύλιον, ἐς ἐχῖνον ἐράσαι, οὗ τὸ ἐπίθεμα ὀπὴν ἔχει, καὶ
ἐπιχέαι τὸν οἶνον, καὶ ἐνθέντα κάλαμον πυριᾶσαι, καὶ
μετέπειτα τὴν σκίλλαν προστίθεσθαι.

(2) Ἢν ὀδύναι καταιγίζωσιν ἐξαπιναῖοι, καὶ ἀψυχίαι ἔωσι· ῥόδων φύλλα, κιννάμωμον, σμύρναν καθ-
αρήν, νέτωπον, ὀπὸν μήκωνος, τούτων φθοῖς ποιῆσαι
ὅσον δραχμιαίους, ἐπίθες ἐπὶ τρύφος ἀμφορέως, καὶ
ἢν διαφανὲς ᾖ, χρῶ ὑποθυμιήματι.

(3) Ἢ στύρακος, ὅσον ἐμβάλλουσιν ἐς τὸ ἔλαιον,
ὡς ἐπὶ τῆς προτέρης χρῆσθαι· ἢ πάντα ὅσα ἐς μύρα

(13) The following medications draw everything: take seeds removed from the insides of a long bottle gourd and mix them with milk, add myrrh, pure honey, and an amount of Egyptian oil, and soak this up on a piece of soft wool. Or dry the insides of a bottle gourd, add honey, knead, and form into a suppository; after the woman bathes she should apply this unguent. Or mix three drafts of squirting cucumber juice into emollient fat, egg yolk, meal, honey, and white wax, warm this mixture, dribble it on to a piece of wool, sponge it off clean, and apply. Or goose grease, white wax, resin, and rose unguent; or melt deer's marrow or sheep and goat's fat, egg white, and rose unguent: apply as a suppository, or soak it up on a piece of wool.

97. To clean the uterus, if it is hard: foment.

(1) Dilute three Attic hemichoes of very fragrant wine with an equal amount (sc. of water); place fennel root, a quarter as much fennel seed, and a half cotyle of rose unguent in a large widemouthed jar whose lid has an opening, pour in the wine, insert a reed into the lid, and with this foment; after that apply a suppository of squill.

(2) If sudden pains overwhelm patients and they lose consciousness: take rose leaves, cinnamon, pure myrrh, oil of bitter almond, and poppy juice, make one drachma of this into pastilles, and set it on the fragment of an amphora (sc. in a fire): when this becomes red-hot, introduce as a fumigation from below.

(3) Or employ as much storax as they add to olive oil as in the previous case; everything that is added in unguents

---

323 σικύης Θ: σίκνου MV.   324 στέατι Θ: ὕδατι MV.
325 Add. ἴσῳ MV.

ἐμβάλλεται, κόψαι καὶ κατασῆσαι· ἐπιβάλλειν δὲ ἐς
τὸ βόλβιτον καὶ στύρακα, καὶ περιχρίειν νετώπῳ, ῥόδινον ἔλαιον ὡς ἄριστον, καὶ Αἰγύπτιον λευκόν, θυμία
τοῦτο μετὰ τὰς καθάρσιας.

(4) Ἢ δάφνην καὶ μυρσίνην κόψας[326] καὶ κυπαίρου
καρπόν, ὀργάσασθαι Αἰγυπτίῳ λευκῷ μύρῳ καὶ νε-
τώπῳ· ἐπὶ βολβίτῳ θυμία.

(5) Ἀρήγει δὲ καὶ μάννα, κυπαρίσσου πρίσματα,
καὶ κυπαίρου ῥίζας κόψας, σῆσαι δὲ σχοῖνον τὸν
ἡδύοσμον, καὶ κάρδαμον, καὶ ἴριν, ταῦτα μίσγειν,
400   περιχέαι δὲ ῥόδινον ἔλαιον | καὶ νέτωπον, καὶ χρῶ ἐν
κρίμνοισι πυρίνοισιν.

(6) Ἢ ῥητίνην ἐπὶ νεοπτήτου κεραμίδος θυμία,
ἐπιβαλὼν καστόριον ἢ τῶν ἀρωμάτων· τὴν δὲ κεφαλὴν τέγγειν ἐλαίῳ ῥοδίνῳ, ἐς δὲ τὸ οὖς μύρσινον
ἢ μήλινον.

(7) Ἢ λευκοὺς ἐρεβίνθους καὶ ἀσταφίδας ἑψήσας
δοῦναι πιεῖν, καὶ ἐν ὕδατι θερμῷ καθίζεσθαι.

(8) Ἢ ἐλαίης λευκῆς, πρὶν ἂν ἔλαιον ἀνεῖναι, καπνίσαι[327] καὶ αὐῆναι, καὶ ἐπ' οἶνον τρίβειν εὐώδεα·
ἐμβάφιον Ἀττικὸν διδόναι.

(9) Ἢ κανθαρίδων τὰς γαστέρας, καὶ ἀδίαντον, καὶ
λίτρον ἐρυθρὸν Αἰγύπτιον, καὶ ῥίζαν νάρθηκος, καὶ
σελίνου λείου σπέρματα, ταῦτα διδόναι.

(10) Ἢν δὲ ἐπιλάβῃ στραγγουρίη, ἐν ὕδατι καθ-
ιζέσθω, καὶ γλυκὺν πινέτω.

(11) Ὀδύνης[328] ὑστερέων· οἶνον ὡς ἥδιστον ἴσον
ἴσῳ κεράσας, ὡς τρία ἡμίχοα Ἀττικά, καὶ μαράθου

should be crushed and sieved: put this into cow's excrement and storax, and coat it with oil of bitter almond, the finest rose oil, and white Egyptian unguent, and employ this as a fumigant after the menstrual cleanings.

(4) Or crush laurel and myrtle with galingale blossoms, and knead with white Egyptian unguent and oil of bitter almond: burn as a fumigant over a fire of cow's excrement.

(5) Also beneficial are frankincense powder, cypress sawdust, and pounded galingale root; sift very fragrant rushes, cress, and iris, mix these together, add rose oil and oil of bitter almond, and apply in wheat groats.

(6) Or a fumigation of resin to which castoreum and one of the aromatics have been added, made over a newly fired clay tile: moisten the patient's head with rose oil, and inject myrtle or quince oil into her ear.

(7) Or boil white chickpeas with raisins, give this to drink, and have the patient sit in hot water.

(8) Or take white olives before they have oil, smoke and dry them, and grind this over a fragrant wine: give one Attic oxybaphon.

(9) Or the bellies of blister beetles together with maidenhair, red Egyptian soda, giant fennel root, and fine celery seeds: give these.

(10) If strangury develops, have the patient take a sitz bath in hot water, and drink sweet (sc. wine).

(11) For pain of the uterus: three Attic hemichoes of very fragrant wine diluted with an equal amount of water;

---

326 δάφνην . . . κόψας Θ: δάφνης καὶ μυρσίνης φύλλα κόψαι MV.

327 καπνίσαι Θ: κατακνίσαι MV.

328 Ὀδύνης Θ: Ἢν πνίγηται ὀδύνη MV.

ῥίζας καὶ τοῦ καρποῦ τριτημόριον, καὶ ῥοδίνου ἐλαίου
ἡμικοτύλιον· ταῦτα ἐμβαλεῖν ἐς ἐχῖνον καινόν, καὶ
τὸν οἶνον ἐπιχέαντα πυριᾶν· καὶ τὴν σκίλλαν προστίθεσθαι, ἔστ᾽ ἂν φῇ τὸ στόμα μαλθακὸν εἶναι καὶ
φαρκιδῶδες καὶ εὐρύ.

(12) Καὶ ἢν ἑλκωθῇ, ἢ ὀλοφλυκτίδες ἔωσιν, ἄμεινον
στέατος χηνείου ἀλείφατι σὺν λιβανωτῷ.

(13) Ὑστερέων ὀδύνης· πευκέδανον, ἀριστολοχίαν,
πάνακες, ταῦτα πάντα μῖξαι γλυκεῖ οἴνῳ, καὶ χλιά-
ναντα πῖσαι· καὶ μήκωνος λευκῆς ῥυφεῖν, καὶ κνίδης
σπέρμα.

(14) Ἄλλο· πυρίη θυμιητή, ἢν ὀδύνη ἔχῃ· σίδια
ῥοιῆς γλυκείης, λωτοῦ πρίσματα, ἐλαίης φύλλα ξηρὰ
κεκομμένα, ταῦτα ἐλαίῳ ὀργάσασθαι χρὴ καὶ ἐπὶ
βόλβιτον πεπυρωμένον ἐπιρρίπτειν· ἢ χαλβάνην,
σμύρναν, λιβανωτόν, ἔλαιον [περιχέας]³²⁹ λευκὸν Αἰ-
γύπτιον ἐπ᾽ οἰναρίδων. |

402

(15) Ἢ διὰ τοῦ ἀσφάλτου στερροτέρη· ἀσφάλτου
Ζακυνθίης, λαγωοῦ τρίχας, πήγανον, κόριον ξηρόν·
ταῦτα τρίψας πάντα, φθόεις πλάσσε, θυμία.

(16) Ἢ πίσση, σανδαράκη, κόψας λείην, μῖξαι
κνίσματα κυπαρίσσου, καὶ τούτοις κηρόν³³⁰ καὶ μύρον
ἐπιχέας, φθόεις ποιέειν· ἐπὶ πυρὶ θυμιᾶν.

(17) Αἰγὸς κέρας καταπρίσας, ἐλαίῳ ἀνακυκᾶν, ἐπὶ
πυρὶ θυμιᾶν.³³¹

---

329 περιχέας (-χέαι MV) ΘMV: del. recentiores.
330 καὶ τούτοις κηρόν om. MV.

fennel roots, a third as much fennel seed, and half a cotyle of rose oil: put these into a new widemouthed jar, pour in the wine, and use to foment; then apply a suppository of squill until the patient says that the mouth (sc. of her uterus) is soft, wrinkled, and wide open.

(12) If a woman's uterus ulcerates or is all covered with blisters, it is preferable to use an ointment of goose grease and frankincense.

(13) For pain of the uterus: mix suphurwort, aristolochia, and all-heal together in sweet wine, warm, and give to drink; also have the patient drink white poppy gruel with stinging nettle seeds.

(14) Another: a fumigant fomentation, if pain is present: soak peel of sweet pomegranate, sawdust of nettle tree, and triturated dry olive leaves; knead this in olive oil, and throw it on to a fire of burning cow's excrement. Or all-heal juice, myrrh, frankincense, and white Egyptian oil burned over olive twigs.

(15) Or a fomentation made more solid with asphalt: take some of the asphalt from Zakynthos, hair of a hare, rue, and dry coriander: knead all these together, make pastilles, and burn as a fumigation.

(16) Or grind pitch and realgar smooth, mix in cypress sawdust, over these pour wax and unguent, and make into pastilles: fumigate over a fire.

(17) Saw up a goat's horn, stir this into olive oil, and use to fumigate over a fire.

---

331 αἰγὸς κέρας . . . θυμιᾶν om. M.

(18) Ἢ ὀρύξας βόθρον, γίγαρτα φρύγειν, καὶ τὴν σποδιὴν ἐμβάλλειν ἐς τὸν βόθρον, καὶ οἴνῳ νοτίσας εὐώδει τὰ γίγαρτα, περικαθεζομένην θυμιήσθω· ἔστωσαν δὲ τῶν γιγάρτων δύο μοῖραι, αὗται δὲ αὖαι ὡς μάλιστα.

98. (207 L.) Ἢν ὑστέρη ἀλγέῃ ἄχρι κύστιος, πράσου καρπὸν σὺν ὕδατι τριπτὸν πίνειν· ἢ κυκλαμίνου ῥίζαν ἐν οἴνῳ λευκῷ πιπίσκειν νῆστι· καὶ θερμῷ λούσθω καὶ ἀπὸ θερμῶν πινέτω νῆστις, καὶ χλιάσματα προστιθέναι· ἢ σκορόδου μώλυζαν καὶ λίτρον ὀπτὸν καὶ κύμινον, λεῖα ποιήσας, μέλιτι δεύων προστίθει, καὶ τῷ θερμῷ λούσθω· καὶ ἀπὸ θερμῶν πινέτω.

99. (208 L.) Ὑστερέων νούσου πάσης· λίνον τὸ σχιστὸν αὐτῇ τῇ καλάμῃ ὅσον δραχμὴν κόψας λεπτά, καταβρέξαι ἐν οἴνῳ λευκῷ ὡς ἡδίστῳ τὴν νύκτα, ἔπειτα ἀπηθήσας, χλιαίνειν, εἴριον ὡς μαλθακώτατον ἐμβάπτων· τὸ μὲν προστιθέναι, τὸ δὲ ἀφαιρέειν. βοηθεῖ δὲ καὶ κρόκος, σμύρνα, κάρυα ποντικά, ἄλευρον καθαρόν· ἐν χηνὸς στέατι καὶ μύρῳ ἰρίνῳ προστιθέναι. |

404   100. (209 L.) (1) Ἢν περιωδυνίη ἔχῃ ἐκ προσθέτων καθαιρομένην, μυρσίνης ἐμβάφιον, λιβανωτοῦ ἴσον, μελάνθιον, κύπειρον, σέσελι, ἄννησον, λίνον, νέτωπον, μέλι, ῥητίνην, ἔλαιον χηνός, ὄξος λευκόν,[332] καὶ μύρον Αἰγύπτιον, ἴσον ἑκάστου, τρίβειν ἐν οἴνῳ λευκῷ γλυκεῖ, δύο κοτύλῃσι, καὶ κλύζειν χλιαροῖσι κλυσμοῖσι.

(2) Κλυσμὸς ἢν ὀδύνη ἔχῃ μετὰ κάθαρσιν·[333] κύ-

(18) Or dig a pit, burn grape stones, pour the ashes into the pit, soak the stones with fragrant wine, and have the patient sit down over the pit, and foment herself; there should be two portions of the stones, and they should be very dry.

98. If a woman feels pain in her uterus that reaches to her bladder, give her leek seeds ground up in water to drink. Or give her cyclamen root in white wine to drink in the fasting state; and have her bathe in hot water, afterward take a drink in the fasting state, and apply warm poultices. Or knead a head of garlic, burned soda, and cumin smoothly together, soak in honey, and then insert; also have the woman bathe and after that take a potion.

99. For every disease of the uterus: crush a drachma of fine linen together with its stalk, and soak this in very fragrant white wine for a night; then sieve it, warm it, soak it on to a piece of very soft wool: alternately insert and remove this suppository. Also beneficial are saffron, myrrh, hazel nuts, clean wheat meal: apply these as a suppository with goose grease and iris oil.

100. (1) If violent pains arise in a woman who has been cleaned with suppositories: take an oxybaphon of myrrh, the same amount of frankincense, and equal amounts of black cumin, galingale, celery, anise, linen, oil of bitter almond, honey, resin, goose oil, white vinegar and Egyptian unguent, knead these into two cotyles of sweet white wine, and inject as a warm douche.

(2) Douche if pain is present after a cleaning: take

---

332 Add. καὶ μέλι MV.
333 μετὰ κάθαρσιν Θ: σφόδρα κανθαρίδας MV.

παιρος, κάλαμος, σχοῖνος καὶ ἴρις, ἐν οἴνῳ μέλανι
ἕψων χρῶ.

(3) Ἢν περιωδυνίη καὶ στραγγουρίη ἔχῃ· πράσου
χυλόν, ἀκτῆς καρπόν, σέσελι, ἄννησον τρίβειν, λιβανωτόν, σμύρναν, οἶνον χύλωσον,[334] καὶ μῖξον καὶ κλύσαι.

(4) Ἢ σμύρνης ὀξύβαφον, λιβανωτοῦ ἴσον, μελαν-
θίου τοῦ κυπρίου ἴσον, σέσελι, ἄννησον, σελίνου
σπέρμα, νέτωπον, μέλι, ῥητίνην, χηνὸς στέαρ, ὄξος
λευκόν, μύρον Αἰγύπτιον, τούτων ἴσον ἑκάστου διεῖναι οἴνῳ λευκῷ γλυκεῖ καὶ ἐγκλύζειν.

(5) Ἢ λινοζώστιος ὕδωρ ἀφεψήσας σὺν σμύρνῃ,
λιβανωτῷ, νετώπῳ, ἢ ἐλελίσφακον, ὑπερικόν, ἕψειν ἐν
ὕδατι καὶ κλύζειν.

(6) Ἢ λίνου καρπόν, ἄννησον, μελάνθιον, σέσελι,
σμύρναν,[335] κασσίης καρπὸν ἐν οἴνῳ ἕψειν καὶ κλύσαι.

(7) Κλυσμὸς ἢν ὀδύνη σφοδρὴ ἔχῃ μετὰ κάθαρσιν·
ἀκτῆς καρπὸν καὶ δαφνίδας ἕψειν ἐν οἴνῳ μέλανι καὶ
κλύσαι, ἢ ἀκτὴν ἑψήσας ἐν ὕδατι ἀποχέειν τὸ ὕδωρ,
οἶνον δὲ γλυκὺν παραχέας κλύσαι. ἢν μετὰ κλυσμὸν
ὀδύνη ἐγγένηται, ἕψειν τὰ θυώματα ἃ ἐς τὸ μύρον
ἐμβάλλεται, καὶ ἀποχέαι τοῦ ὕδατος δύο κοτύλας, μῖξαι δ᾽ ἔλαιον χηνὸς καὶ ῥόδινον, κλύζειν χλιαρῷ.
406 πλέον δὲ | κλύσμα δύο κοτύλαι μηδενὶ ὡς ἔπος εἰπεῖν.

(8) Ἢ λινόζωστιν ἕψειν ἐν ὕδατι, καὶ μῖξαι χυλὸν
μυρσίνης, λιβανωτόν, νέτωπον ἴσον, καὶ κλύσαι χλιαρῷ.

galingale, reeds, rushes, and iris, boil these in dark wine, and apply.

(3) If violent pain and strangury are present, knead leek juice, elder berries, celery, and anise; make an infusion of frankincense, myrrh, and wine: mix and inject.

(4) Or an oxybaphon of myrrh with an equal amount of frankincense, equal amounts of black cumin and galingale, and celery, anise, celery seed, oil of bitter almond, honey, resin, goose grease, white vinegar, and Egyptian unguent: disperse equal amounts of these in sweet, white wine, and inject as a douche.

(5) Or boil juice of mercury herb with myrrh, frankincense, and oil of bitter almond, or take salvia and hypericum, boil them in water, and inject.

(6) Or boil linseed, anise, black cumin, celery, myrrh, and cassia seed in wine, and inject.

(7) Douche if violent pain is felt after the menstrual cleaning: boil elderberries and bay berries in dark wine, and inject; or boil elder in water, decant the liquid, add sweet wine, and inject. If after the injection there is new pain, boil the fumigants which are put into an unguent, pour off two cotyles of the fluid, add goose and rose oil, and flush with this warm. No injection should ever be more than two cotyles, so to speak.

(8) Or boil mercury herb in water, add myrtle juice, frankincense with an equal amount of oil of bitter almond, and flush with this warm.

---

334 χύλωσον M: χυλῷ ἴσον ΘV.
335 -ναν recentiores: -νης ΘMV.

(9) Ἢν ἀλγέῃ τὰς ὑστέρας, κυκλαμίνου ῥίζαν ἐν οἴνῳ λευκῷ πιπίσκειν νῆστιν, καὶ τῷ θερμῷ λοῦσθαι, καὶ ἀπὸ θερμῶν πινέτω.

101. (210 L.) Ἢν ἀφθῇ τὰ αἰδοῖα· μύρον θερμὸν σὺν οἴνῳ διακλυζέσθω, καὶ μελίλωτον προστιθέναι· καὶ γλυκυσίδης ἐν οἴνῳ ἡψημένης ἢ τριπτῆς προστίθεσθαι· καὶ σμύρναν καὶ ῥητίνην ὁμοῦ μίξαι καὶ διεὶς οἴνῳ, ὀθόνιον ἐμβάπτων, προστιθέναι.

Καὶ ἢν στραγγουρίη λάβῃ, πυριᾶν καὶ χρίειν στέατι τὸ ἦτρον· ἐγκαθίννυσθαι δὲ χρὴ[336] ῥόδων ἀφεψήματι ἢ βάτου ἢ μυρσίνης ἢ ἐλαίης[337] ἢ ἑλίκων ἀμπέλων ἢ ἀρκευθίδων ἢ ἐλελισφάκων.

102. (211 L.) Ἢν δ᾽ ἄνεμος ἐγγένηται ἐν τῇ κοιλίῃ,[338] πόνος ἔνι σπερχνός, καὶ φῦσα οὐκ ἔξεισιν· κύμινον προστιθέσθω· ἢ ἐλελίσφακον καὶ κύπαιρον κόψας, καὶ τέγξας τὴν νύκτα, ἕωθεν ἀπηθῆσαι, τὸ διαγὲς δὲ ἐς ἄγγος ἐγχέαι, καὶ κρίμνα πύρινα ἐν ὄξει[339] λευκῷ φορύξαι, καὶ ὀπὸν σιλφίου ὡς κύαθον, κρᾶμα ἕψειν, ἐνωμότερον διδόναι ῥυφεῖν.

103. (212 L.) Ἢν κιὼν ἐν τοῖσιν αἰδοίοισιν γένηται, ὀδύνη ἴσχει· τῆς μὲν ὀδύνης σελίνου καρπὸς ἀλέξημα, καὶ κισσὸς δὲ καὶ ῥοιῆς γλυκείης τριπτῆς· σὺν οἴνῳ παλαιῷ σὺν ποταινίῃ σαρκὶ προστίθεσθαι, καὶ ἐπιπλάσαι τὰ φύλλα· τὴν δὲ νύκτα ὅλην ἐχέτω, κἄπειτα ἀφελομένη οἴνῳ διανιζέσθω.

---

[336] χρὴ Θ: ἐν MV.
[337] Add. ἀφεψήματι Θ.

(9) If a woman is suffering pain in her uterus, give her cyclamen root in white wine to drink in the fasting state, and have her take a hot bath and after that drink the potion.

101. If aphthae develop on a woman's genitalia: have the patient flush herself well with warm unguent mixed in wine, and apply a suppository of melilot; also have her apply boiled or ground peony in wine; also mix myrrh and resin together, dissolve it in wine, smear this on to a piece of linen, and insert it.

If strangury comes on, apply a fomentation and anoint the lower belly with fat; also give the patient a sitz bath in a decoction of roses, brambles, myrtle berries, olives, vine tendrils, Phoenician juniper berries, or salvia.

102. If wind arises in a woman's cavity, she will have a violent pain and the air does not come out. Have her apply a suppository of cumin. Or crush salvia and galingale, soak them for a night, sieve this at dawn, pour the transparent part into a vessel, soak wheat groats in white vinegar, add a cyathos of silphium juice, boil the mixture, and give this somewhat underdone for the patient to take as gruel.

103. If a wart forms on a woman's genitalia, it is painful: celery seed will allay the pain, as will ivy and grated sweet pomegranate peel: apply these with fresh meat in aged wine as a pessary; also apply the (sc. pomegranate) leaves as a poultice. The patient should retain this through the whole night, and then remove it and wash herself well with wine.

---

338 κοιλίη Θ: μήτρη MV.
339 ὄξει Θ: οἴνῳ MV.

LEXICON OF
THERAPEUTIC AGENTS[1]

Acacia, ἄκανθα Αἰγυπτία, *Acacia Arabica*. *See* Gum
Alexanders, ἱπποσέλινον, *Smyrnium Olusatrum*
Alkanet, σχεδιάς, *Anchusa tinctoria*, the lesser and the greater
All-heal, πάνακες, *Ferulago galbanifera*
All-heal juice, χαλβάνη
Almond, ἀμυγδάλη, fruit of the *Prunus Amygdalus*
Almond, bitter oil of, νέτωπον
Alum, στυπτηρίη, a white crystalline compound of aluminum and
    ammonium sulfate with marked astringent properties; Egyptian
    and Melian varieties are mentioned
Anemone, ἀνεμώνη, *Anemone coronaria*
Anise, ἄννησον, *Pimpinella Anisum*
Aphronitrum, ἀφρὸς νίτρου (= ἀφρόνιτρον), "a fungus-like growth
    that appears on recently built walls, and that consists of soda sul-
    phate or carbonate, sometimes of magnesium sulphate, and occa-
    sionally . . . of potash nitrate." *Sydenham Soc. Lex.*
Aristolochia, ἀριστολοχία, *Aristolochia rotunda*
Aromatic herb or spice, ἄρωμα (= ἡδύσματα)
Arsenic, red, σανδαράκη (= σανδαράχη), As₂S₂, also called realgar
    or orpiment
Artemisia herb, ἀρτεμισίη ποίη, *Artemisia arborescens*
Ashes, κονία
Asphalt, ἄσφαλτος

[1] In this lexicon the identifications range widely in their cer-
tainty. I have often referred with profit to the notes and "Lexique
des substances pharmaceutiques" (pp. 235–46) of F. Bourbon's
Budé edition of *Nature of Women*.

453

Ass's excrement, ὄνις

Ass's hair, ὄνου τρίχες

Ass's milk, ὄνου/ὄνειον γάλα

Ass's rennet, τάμισος ὄνειος

Axeweed, πελεκῖνος, *Securigera Coronilla*

Bacchar, βάκκαρις, an unguent perhaps made from a Lydian plant of the same name. *Cf.* Erotian B14 (βάκχαρις) and Galen (βάχαρις) vol. 19, 87

Baneberry, ἀκταίη, *Actaea spicata*

Barley, κριθή, *Hordeum sativum*

Barley, parched, κάχρυς

Barley, peeled, πτισάνη. *See also* Gruel (barley)

Barley bran, ἄχυρον

Barley cake, μᾶζα

Barley meal, κρίμνον (= ἄλφιτον); untoasted, ἄλφιτα προκώνια

Basil, field, πολύκνημον, *Zizyphora capitata*

Bastard sponge, ἀλκυόνιον (= ἀλκυόνειον), a zoophyte

Bay/laurel tree, δάφνη, *Laurus nobilis*; leaves of

Bayberry, δαφνίς, fruit of the bay tree

Beef, βοὸς σάρξ

Beet, σεῦτλον, *Beta maritima*

Bindweed, ἑλξίνη, *Convolvulus arvensis*

Black medication, φάρμακον μέλαν. See *Diseases of Women I*, ch. 96

Black root, ῥίζα μέλαινα, "*Genista acanthoclada*, said to be the μέλαινα ῥίζα of Hippocrates." *Sydenham Soc. Lex.*, s.v. Genista

Blister beetle, κανθαρίς, possibly *Cantharis vesicatoria* or *Meloë Cichorei*

Bottle gourd, σικύα, *Lagenaria vulgaris*

Bramble, βάτος, *Rubus ulmifolius*

Bread, leavened, ἄρτος ζυμίτης

Brine, ἅλς (= θάλασσα)

Bryony, ἐχέτρωσις, *Bryonia cretica*

Buckthorn, ῥάμνος, various species of *Rhamnus*

Bulb (small), βόλβιον; β. τὸ ἐκ τῶν πυρῶν. Cf. *Diseases of Women II*, 73, "the small bulb that is seen in wheat, especially in Egypt; it is sharp, like Ethiopian cumin."

Bull's gall, ταύρου (= βοὸς) χολή

Bull's liver, ταύρου ἧπαρ

Buprestis, βούπρηστις, perhaps a species of *Meloë*

Burnet, thorny, στοιβή, *Poterium spinosum*

Butter, βούτυρον (= πικέριον), literally, "cow's cheese"

Cabbage, κράμβη, *Brassica cretica*

Caper plant, ὄφις, *Capparis spinosa*

Cardamom, καρδάμωμον, *Elattaria Cardamomum*

Carding comb of fullers, cleanings (tow) from, καθάρσεις ἀπὸ τοῦ κνάφου τῶν κναφέων. Cf. Linen/flax, tow of

Cassia, κασίη, *Cinnamomum iners*

Castor oil tree, κροτών, *Ricinus communis*

Castoreum, κάστωρ (= καστόριος ὄρχις): "a reddish-brown unctuous substance, having a strong smell and nauseous bitter taste, obtained from two sacs in the inguinal region of the beaver." *Shorter Oxford English Dictionary*

Cedar oil, κεδρίη (= ἔλαιον κέδρινον), obtained from the *Juniperus excelsa*

Cedar tree, κέδρος, *Juniperus Oxycedrus* and other species

Celery, σέλινον, *Apium graveolens*; seed of

Cerate/wax salve, κηρωτή

Cereals, σιτία. See Barley; Rice-wheat; Spelt; Wheat

Chamomile, εὐάνθεμον, *Matricaria Chamomilla*

Chaste tree, ἄγνος (= λύγος), *Vitex Agnus-castus*

Chickpea, ἐρέβινθος, *Cicer arietinum*

Cinnabar, κιννάβαρι, mercuric sulfide, HgS

Cinnamon, κιννάμωμον, *Cinnamomum Cassia*

Cleavers, φιλίστιον, *Galium Aparine*

Clover, τρίφυλλον, *Psoralea bituminosa* or a member of the genus *Trifolium*

Cnestron, κνῆστρον, *Daphne oleoides*

Cnidian berry, Κνίδιος κόκκος, fruit of the θυμελαία, *Daphne Cnidium*

Cold compresses, ψύγματα

Copper, black, μέλαν τὸ κύπριον = copper scoria, σποδὸς κυπρίη, a byproduct of copper smelting, which consists mainly of zinc oxide. *See* Dioscorides 5.75, and Berendes, p. 507, n. 1

Copper, flower of, ἄνθος χαλκοῦ, small grains of cuprous oxide made by quenching heated copper; roasted, ὀπτόν

Copper filings, ῥινήματα χαλκοῦ
Coriander, κόριον (= κορίαννον), *Coriandrum sativum*
Cow's excrement, βόλβιτον
Cow's marrow, βοὸς μυελός
Cow's milk from a black cow, βοὸς μελαίνης γάλα
Crab, fluvial, καρκίνος ποτάμιος, possibly crayfish, κάραβος
Cress, κάρδαμον, *Lepidium sativum*
Cucumber, σίκυος, *Cucumis sativus*
Cumin, κύμινον, *Cuminum Cyminum*
Cumin, black, μελάνθιον, *Nigella sativa*
Cumin, Ethiopian, κ. Αἰθιοπικόν, *Carum copticum*
Cuttlefish bone, ὄστρακον, bone of the *Sepia*
Cuttlefish egg, σηπίης ᾠόν, egg of the *Sepia*
Cyceon, κυκεών, a mixed drink containing meal, cheese, herbs, and
    wine
Cyclamen, κυκλάμινος, *Cyclamen graecum*
Cypress, κυπάρισσος, *Cupressus sempervirens*
Cypress wood sawdust, κυπαρίσσου πρίσματα

Darnel, αἶρα, *Lolium temulentum*
Dauke, δαῦκος, perhaps *Athamanta Cretensis*
Deer's fat, ἐλάφου στέαρ
Deer's horn, ἐλάφου κέρας
Deer's marrow, ἐλάφου μυελός
Dill, ἄνηθον, *Anethum graveolens*
Dittany, δίκταμνον, *Origanum Dictamnus*
Dodder of thyme, (λευκὸν) ἐπίθυμον, *Cuscuta Epiththymum*, a par-
    asitic plant growing on thyme

Earth, γῆ; fuller's e., σμηκτρὶς γῆ
Earth-almond, ὁλοκωνῖτις, *Cyperus esculentus*
Eel, ἔγχελυς, *Anguilla vulgaris*; conger eel, γόγγρος, *Conger vul-
    garis*
Egg, ᾠόν (ἀλεκτορίδος), of the *Gallus gallinaceus*. See also Cuttle-
    fish egg
Egyptian oil (white), ἔλαιον Αἰγύπτιον (λευκόν)
Egyptian unguent (white), μύρον Αἰγύπτιον (λευκόν)
Elder, ἀκτῆ, *Sambucus nigra*; its berry, καρπός

Electuary, ἐκλεικτόν

Eledone, βόλβιον, a small polyp or octopus with a fetid smell, *Eledone muschata*

Emollient, μαλθατήριον

Ergot (of wheat), μελάνθιον τὸ ἐκ τῶν πυρῶν. Cf. v. Grot, pp. 123–27

Evil-smelling agents, δυσώδεα (= κακώδεα)s

Falcon/hawk, excrement of, ἴρηκος ἄφοδος

Fat, στέαρ (= λιπαρόν = πιμελή); seasoned, ἡδυντόν; seasoned red, ἡδυντὸν ἐρυθρόν. *See also* Deer's fat; Goat's fat; Goose grease; Lard; Sheep's fat

Feather (very fine), πτερόν (ἐλάχιστον)

Fennel, μάραθον, *Foeniculum vulgare*; seeds, roots of

Fennel, giant, νάρθηξ, *Ferula communis*

Fenugreek, βούκερας (= τῆλις), *Trigonella Foenumgraecum*

Ferula, σαγάπηνον, *Ferula persica*

Field basil, πολύκνημος, *Zizyphora capitata*

Fig, σῦκον; Phibalian, Φιβάλιον. Cf. *Suda* s. v.; white, λευκόν

Fig, dried, ἰσχάς

Fig, wild, ὄλονθος

Fig tree, συκῆ, *Ficus carica*

Fish thistle, ἄκανθα λευκή, *Acacia albida* or *Cnicus Acarna*

Flax. *See* Linen/flax

Fleabane, κόννζα, some species of *Erigeron* or *Inula*

Flour, ἄλητον

Fragrant agents, εὐώδεα

Frankincense, λιβανωτός (= λίβανος), gum of the *Boswellia Carteri*

Frankincense plant, κάχρυ, *Boswellia Carteri*

Frankincense powder, μάννα

Galena, μολίβδαινα, lead sulfide

Galingale, κύπαιρος (= κύπειρος), *Cyperus longus*

Garlic, σκόροδον, *Allium sativum*

Garlic head, μώλυζα (σκορόδου)

Goat's cheese, αἴγειον τυρόν

Goat's excrement, αἰγὸς σπύραθος

Goat's fat, αἰγὸς σταῖς (= στέαρ)
Goat's liver, αἰγὸς ἧπαρ
Goat's milk, αἴγειον γάλα
Goby/gudgeon, κωβιός, species of the genus *Gobius*
Gold scoria, χρυσῖτις σποδός (= λιθάργος), lead monoxide
Goose, Egyptian, χηναλώπηξ, *Chenalopex aegyptiaca*
Goose grease, χήνειον στέαρ (= χηνὸς ἄλειφα)
Goose marrow, χήνειος μυελός
Goose oil, χήνειον ἔλαιον
Gourd, κολοκύντη, *Cucurbita maxima*; wild g., κολοκυνθὶς ἀγρίη, *Citrullus Colocynthis*
Grape, unripe, ὄμφαξ
Grape stone, γίγαρτον
Grapevine, ἄμπελος, *Vitis vinifera*
Grapevine blossom, οἰνάνθη
Groats (wheat), πυρῶν κρίμνα
Groats (barley), untoasted, ἄλφιτα προκώνια
Gruel (barley), χυλὸς πτισάνης
Gum, κόμμι, obtained from *Acacia arabica*. Cf. Dioscorides, vol. 1, 93.
Gum ammoniacum, ἀμμωνιακόν
Gum ladanum, λήδανον, a gum exuded from plants of the genus *Cistus*, esp. *C. ladaniferus* and *C. creticus*
Gypsum, τίτανος, a white earth

Hare's hair, λαγῴη θρίξ
Hartwort, σέσελι, *Tordylium officinale*
Hazelnut, κάρυον Ποντικόν, fruit of the *Corylus Avellana*
Hedge mustard, ἐρύσιμον, *Sisymbrium polyceratium*
Hellebore, ἐλλέβορος, *Veratrum album*; black h., ἐλ. μέλας, *Helleborus cyclophyllus*; treat with h., ἐλλεβορίζειν
Hellebore, black, powder of, ἔκτομον
Hemlock, κώνειον, *Conium maculatum*; leaves of
Henbane, ὑοσκύαμος, *Hyoscyamus albus*; leaves, roots of
Henna, κυπρός (= κύπρινος), *Lawsonia inermis*; ἔλαιον κύπρινον, oil from the flower
Hog's gall, ὑὸς χολή
Hog's grease, ὑὸς ἔλαιον

LEXICON OF THERAPEUTIC AGENTS

Honey, μέλι; Attic, Ἀττικόν; boiled, ἐφθόν
Honeycomb, κηρίον
Horse fennel, ἱππομάραθον, *Prangos ferulacea*
Hulwort, πόλιον, *Teucrium Polium*
Hypericum, ὑπερικόν, *Hypericum Crispum* or *empetrifolium*

Iris, ἶρις, *Iris pallida*, etc.; fragrant, εὐώδης
Iris unguent, μύρον ἴρινον
Iron slag, σκωρία σιδήρου
Ivy, κισσός, *Hedera Helix*

Juniper, κεδρίς, *Juniperus communis*; Phoenician juniper, ἄρκευθος, *Juniperus phoenicea*
Juniper, dwarf, ἄρκευθος μικρά
Juniper berry, ἀρκευθίς (= κεδρίς), fruit of the *Juniperus phoenicea* or *Juniperus excelsa*
Juniper oil, κεδρίη

Kidney fat, νεφρίδιον (= στέαρ νεφρῶν)
Kid's meat, ἐρίφου κρέας

Lampwick, ἐλλύχνιον
Lard, ὑὸς στέαρ/πιμελή
Laurel. *See* Bay/laurel tree
Lead, μόλιβδος; hammered, ἐληλασμένος
Lead, white, ψιμύθιον, Pb(OH)₂.2PbCO₂
Lead sound/probe, μήλη μολιβδίνη
Leek, πράσον, *Allium Porrum*; boiled
Lentil, φακός, the *Ervum lens* and its fruit
Lentil decoction, φάκιον (= φακῶν ὕδωρ)
Lentil soup, φακῆ
Lettuce, wild, θρίδαξ (= θριδακίνη), *Lactuca Scariola*
Licorice, γλυκυρρίζη, *Glycyrrhiza glabra*
Lily oil, σούσινον (= ἄνθινον ἔλαιον), extracted from various species of *Lilium*
Linen/flax, λίνον, *Linum usitatissimum*; tow of raw flax left on the carding comb, ἐπικτένιον ὠμολίνου. *Cf.* Carding comb
Linseed, λίνου καρπός/σπέρμα, seed of the flax

Linseed and honey dish, χρυσόκολλα
Lizard's liver, σαύρου ἧπαρ
Lupin, θέρμος, *Lupinus albus*

Madder, ἐρευθέδανον, *Rubia tinctorum*
Madonna lily, λείριον, *Lilium candidum*
Maggot, εὐλή
Maiden hair, ἀδίαντον, *Adiantum Capillus-Veneris*
Mallow, μαλάχη, *Malva silvestris*
Marjoram, ὀρίγανον, various species of *Origanum*; crushed and
    dried leaves of
Mastic, σχῖνος, the *Pistacia Lentiscus* and its resin
Meal (barley), ἄλφιτον
Meal (wheat), ἄλευρον
Meal, bruised, of raw grain, ὠμήλυσις
Meat, κρέας. See also Beef; Mutton
Medlar, ἀμαμηλίς, *Mespilus germanica*
Melicrat, μελίκρητον, a mixed drink of honey and water
Melilot, μελίλωτον, *Trigonalla graeca*, a kind of clover so called
    from the quantity of honey it contains
Mercury (herb), λινόζωστις, *Mercurialis annua*
Mild agents, μαλθακά
Milk, γάλα. See also Ass's milk; Cow's milk; Goat's milk; Woman's
    milk
Millet, ἔλυμος (= μελίνη), *Panicum miliaceum*
Mint, ἡδύοδμον (= [καλα]μίνθη), *Mentha viridis* or a species of the
    genus *Calamintha*
Misy, μίσυ, a shining yellow copper or iron ore mined at the time of
    Galen (vol. 12, 226ff.) in Cyprus
Mouse's excrement, μυῶν ἀπόπατος
Mulberry, μόρον (= συκάμινος), *Morus nigra*; leaf of
Mule's excrement, ἡμιόνου ὀνίς
Mullein, φλόμος, *Verbascum sinuatum*; leaves of; plugs made of
Mutton, ὄϊος/μήλειον κρέας
Myrrh, σμύρνα, gum of an Arabian tree, *Balsamodendron Myrrha*
Myrtidanon, μυρτίδανον, seed of the Persian pepper tree
Myrtle, μυρσίνη, *Myrtus communis*
Myrtle berry, μύρτον

LEXICON OF THERAPEUTIC AGENTS

Narcissus, νάρκισσος, various species of *Narcissus*

Narcissus oil, ἔλαιον ναρκίσσινον

Narcissus unguent, μύρον ναρκίσσινον

Navelwort, κοτυληδών, *Cotyledon Umbilicus*; broad (πλατεῖα) and small closed (σμικρὰ συμμύουσα) varieties are mentioned

Nettle(-seed), κνίδη (= κνίδιον), *Urtica*. See Stinging nettle

Nettle tree and its wood, λωτός, *Celtis australis*; sawdust of, πρίσματα; shavings of, τορνεύματα

Nightshade, στρύχνος, various neurotropic plants including *Solanum nigrum*, *Datura Stramonium*, and *Withania somnifera*

Oak-fern, black, δρυοπτερίς, *Asplenium onopteris*

Oak gall, κηκίς, an excrescence produced on some species of oak by punctures of the gallfly (genus *Cynips*)

Octopus, πουλύπους

Oil, white. See Egyptian oil

Olive, wild, ἀγριελαία

Olive oil, ἔλαιον; of unripe olives, ὀμφάκιον

Olive scab, ψώρη ἐλαίης

Olive tree and its fruit, ἐλαίη, *Olea Europaea*; leaves of

Onion, κρόμμυον, *Allium Cepa*

Opium poppy, μηκώνιον ὑπνωτικόν, *Papaver somniferum*; ὄπιον, juice of

Orach, ἀνδράφαξις (= ἀτράφαξυς), *Atriplex rosea*

Oxeye, βοάνθεμον, *Anacyclus radiatus*

Oxycrat, ὀξύκρατον, sour wine mixed with water

Papyrus, βύβλιον

Pear, ἄπιος; fruit of, *Pyrus communis*

Pennyroyal, γλήχων, *Mentha Pulegium*

Peony, γλυκυσίδη, *Paeonia officinalis*; berry of, κόκκος

Pepper, πέπερι, *Piper nigrum*; peppercorn, στρογγύλον. Cf. Myrtidanon

Pepperwort, λεπίδιον, *Lepidium latifolium*

Persea, περσέα, an Egyptian tree, *Mimusops Schimperi*

Pigeon, πελ(ε)ιάς, *Columba livia domestica*. Cf. *Index Hipp.* s.v.

Pine tree, πίτυς, various species of *Pinus*

Pinewood, δαΐς; very oily, πιοτάτη; shavings of

LEXICON OF THERAPEUTIC AGENTS

Pipe, αὐλός

Pitch, πίσσα; boiled, ἐφθή; liquid/raw, ὑγρά/ὠμή

Polypody, πουλυπόδιον, *Polypodium vulgare*

Pomegranate, ῥοιή (= σίδη), *Punica Granatum* and its fruit; vinous, οἰνώδης; peel of, σίδιον; seed of, κόκκος

Poplar, black, αἴγειρος, *Populus nigra*; Cretan, αἴγ. Κρητική

Poppy, μήκων, various species of *Papaver* and other plants; white, λευκή; red, πυρρός

Poppy juice (= opium), ὄπιον

Pork, συὸς κρέας

Probe/sound/spatula, μήλη. *See also* Lead sound/probe; Tin sound/probe/spatula

Purslane, ἀνδράχνη (= ἀνδράχλη), *Portulaca oleracea* or *Arbutus Andrachne.*

Purslane, wild, πέπλιον (= πεπλίς), *Euphorbia Peplis*

Quince, μῆλον τὸ Κυδώνιον, *Cydonia vulgaris*

Radish, ῥάφανος (= συρμαία), *Raphanus sativus*, also its juice

Raisin, (ἀ)σταφίς; white, λευκή; wild, ἀγρίη

Ranunculus, βατράχιον, *Ranunculus asiaticus*

Reed, κάλαμος, *Arundo Donax*

Rennet, πυτίη (= τάμισος). *See also* Ass's rennet; Kid's rennet; Seal's rennet

Resin, ῥητίνη, a gum collected from the pine and other trees

Rhubarb, monk's, λάπαθον (= λαπάθιον), *Rumex Patientia*

Rice-wheat, ζειά, one-seed wheat, *Triticum monococcum*

Ringdove, φάσσα, *Columba palumbus*

Rockrose, κισθός, *Cistus villosus*

Rose, ῥόδον, *Rosa gallica*

Rose oil, ἔλαιον ῥόδινον

Rose unguent, μύρον ῥόδινον

Rue, πήγανον, *Ruta graveolens*

Rush, σχοῖνος, various species of *Juncus*, etc.

Safflower, κνῆκος, *Carthamus tinctorius*

Saffron, κρόκος, *Crocus sativus*

Sage. *See* Salvia/sage

462

LEXICON OF THERAPEUTIC AGENTS

Salt, ἅλς ; Cyprian, Κύπριον; Egyptian, Αἰγύπτιον; Theban, Θηβαϊκόν

Salvia/sage, ἐλελίσφακος, a member of the genus *Salvia*, possibly *S. officinalis*

Samphire, κρῆθμον, *Crithmum maritimum*

Savory, θύμβρα, *Satureia Thymbra*

Scammony, σκαμ(μ)ωνία, *Convolvulus Scammonia*; juice of

Sculpin, σκορπίου θαλασσίου, species of *Scorpaena*; bile of

Sea turtle, χελώνη θαλασσία

Sea urchin, ἐχῖνος θαλάσσιος

Seal's fat, φώκης στέαρ

Seal's lung, πλεύμων φώκης

Seal's oil, ἔλαιον φώκης

Seal's rennet, πυτίη φώκης

Seaweed, βρύον θαλάσσιον, *Ulva Lactuca*

Sesame, σήσαμον, *Sesamum indicum*

Sharp agents, δριμέα

Sheep's fat, ὄϊος σταῖς/στέαρ (= μήλειον στέαρ)

Sheep's liver, ὄϊος ἧπαρ

Shepherd's purse, θλάσπις, *Capsella bursa-pastoris*

Silphium, σίλφιον (= φύλλον τὸ Λιβυκόν), *Ferula tingitana*; juice of, ὀπός

Silver, flower of, ἄνθος ἀργύρου, probably λιθάργυρος, lead monoxide

Snake skin, (ἐχίδνης) λεβηρίς

Soapwort, στρούθιον, *Saponaria officinalis*

Soda, νίτρον (= λίτρον), native sodium carbonate; red, ἐρυθρόν; Egyptian, Αἰγύπτιον

Sole, ψῆσσα, *Pleuronectes solea*

Soot from the oven, αἴθαλος ἀπὸ τοῦ ἵπνου

Southernwood, ἀβρότονον, *Artemisia arborescens*

Sow's lard, ὑὸς θηλείης στέαρ

Spelt flour dough, σταῖς, from *Triticum spelta*

Spelt groats, χόνδρος

Spikenard, νάρδος, *Nardostachys Jatamansi*

Spring wheat, σητάνιον

Spurge, τιθύμαλλος (= μηκώνιον), various species of *Euphorbia*

Spurge flax, κνέωρον, leaves of the θυμελαίη, *Daphne Cnidium*

463

Squill, σκίλλα, *Urginea maritima*
Squirting cucumber, σίκνος ἄγριος (= σικνώνη), *Ecballium Elaterium*
Squirting cucumber juice, ἐλατήριον
Starch, ἀμύλιον
Starfish, ἀστὴρ θαλάσσιος
Stavesacre, ἀσταφὶς ἀγρία, *Delphinium Staphisagria*
Stinging nettle, κνίδη, *Urtica urens*
Stone sperage, ἀσπάραγος, *Asparagus acutifolius*
Storax, στύραξ, *Storax officinalis*
Sulfur, θεῖον
Sulphurwort, πευκέδανον, *Peucedanum officinale*
Sumac (red), ῥόος (ἡ ἐρυθρή), *Rhus Coriaria*; used in tanning, βυρσοδεψική

Tamarisk, μυρίκη, *Tamarix tetranda*; leaves of, πέταλα
Tassel hyacinth, βολβός, *Muscari comosum*
Telephion, τηλέφιον, *Andrachna telephioides*
Terebinth tree, τέρμινθος, *Pistacia Terebinthus*; also terpentine, τερμινθίνη ῥητίνη
Thapsia, θαψίη, *Thapsia garganica*
Thyme, θύμον, *Thymbra capitata*
Thyme, tufted, ἕρπυλλος, *Thymus Sibthorpii*
Tin sound/probe/spatula, μήλη κασσιτερίνη
Torpedo, νάρκη, *Torpedo marmorata* and allied species
Tree medick, κύτισος, *Medicago arborea*
Tree moss, βρύον, "any moss or moss-like plant that grows on trees; applied esp. to certain lichens." *Shorter Oxford English Dictionary*

Unguent, μύρον (= ἄλειφα); fragrant, εὐῶδες; very pleasant, ἥδιστον
Urine, οὖρον

Valonia oak, φηγός, *Quercus Aegilops*, also its acorn
Verdigris, ἰός, a green or greenish-blue copper compound
Vetch, ὄροβος, *Vicia Ervilia*

LEXICON OF THERAPEUTIC AGENTS

Vine twigs, κληματίνη; ashes of, σποδός; vine, tears of, δάκρυα
    ἀμπέλου
Vinegar, ὄξος

Water, ὕδωρ; cold, λούεσθαι ψυχρῷ; hot, θερμολουτεῖν
Water chestnut, τρίβολος παραθαλάσσιος, *Trapa natans*
Watercress, σίσυμβρον, *Nasturtium officinale*
Wax, κηρός
Wheat, *Triticum vulgare*, πυρός. *See also* Ergot; Groats; Meal;
    Spring wheat
Wheat bran, πίτυρον πύρινον
Whey, ὀρός
White root, ῥίζη λευκή, identified by Galen as dragon arum; the
    Sydenham Soc. Lex. names the *Angelica lucida*, *Asclepias tu-*
    berosa, and *Ligusticum actaeifolium*
White violet, λευκόϊον, *Matthiola incana* (gilliflower) or *Galanthus*
    nivalis (snowdrop)
Willow, πρόμαλος (= ἰτέα), various species of *Salix*
Willow weed, πολύκαρπον, *Polygonum Persicaria*
Wine, οἶνος; mixed with cedar oil, κέδρινος
Wine lees/must, τρύξ; wine sediment, ἰλύς (= σμῆγμα τῆς ἰλύος)
Wolf's excrement, λύκου κόπρος
Woman's milk, γάλα γυναικός; of a nursing mother, κουροτρόφου
Wool, εἴριον; soft Milesian, Μιλήσιον μαλθακόν
Wormwood, ἀψίνθιον, *Artemisia Absinthium*

GREEK NAMES OF
THERAPEUTIC AGENTS
INCLUDED IN LEXICON

ἀβρότονον	Southernwood
ἄγνος	Chaste tree
ἀγριελαία	Olive, wild
ἀδίαντον	Maiden hair
αἴγειον γάλα	Goat's milk
αἴγειρος	Poplar, black
αἰγὸς ἧπαρ	Goat's liver
αἰγὸς σπύραθος	Goat's excrement
αἰγὸς σταῖς	Goat's fat
αἴθαλος	Soot
αἶρα	Darnel
ἄκανθα Αἰγυπτία	Acacia
ἄκανθα λευκή	Fish thistle
ἀκταίη	Baneberry
ἀκτῆ	Elder
ἄλειφα	Unguent
ἄλευρον	Meal (wheat)
ἄλητον	Flour
ἀλκυόν(ε)ιον	Bastard sponge
ἅλς	Brine, Salt
ἄλφιτον	Meal (barley)
ἄλφιτα προκώνια	Meal (barley), untoasted

ἀμαμηλίς	Medlar
ἀμμωνιακόν	Gum ammoniacum
ἀμπέλιον	Grapevine (small branch)
ἀμυγδάλη	Almond
ἀμύλιον	Starch
ἀνδράφαξις/ἀτράφαξυς	Orach
ἀνδράχνη/ἀνδράχλη	Purslane
ἀνεμώνη	Anemone
ἄνηθον	Dill
ἄνθινον ἔλαιον	Lily oil
ἄνθος ἀργύρου	Silver, flower of
ἄνθος χαλκοῦ	Copper, flower of
ἄννησον	Anise
ἄπιος	Pear
ἀριστολοχία	Aristolochia
ἀρκευθίς	Juniper berry
ἄρκευθος	Phoenician juniper
ἀρτεμισίη ποίη	Artemisia herb
ἄρτος ζυμίτης	Bread, leavened
ἄρωμα	Aromatic herb or spice
ἀσπάραγος	Stone sperage
ἀσταφίς	Raisin
ἀσταφὶς ἀγρία	Stavesacre
ἀστὴρ θαλάσσιος	Starfish
ἄσφαλτος	Asphalt
αὐλός	Pipe
ἀφρὸς νίτρου/ἀφρόνιτρον	Aphronitrum
ἄχυρον	Barley bran
ἀψίνθιον	Wormwood
βάκκαρις	Bacchar
βάτος	Bramble

βατράχιον	Ranunculus
βοάνθεμον	Oxeye
βόλβιον	Bulb (small)
βόλβιτον	Cow's excrement
βολβός	Tassel hyacinth
βοὸς γάλα	Cow's milk
βοὸς μυελός	Cow's marrow
βοὸς σάρξ	Beef
βοὸς χολή	Bull's gall
βούκερας	Fenugreek
βούπρηστις	Buprestis
βούτυρον	Butter
βρύον	Tree moss
βρύον θαλάσσιον	Seaweed
βύβλιον	Papyrus
βυρσοδεψικὴ ῥόος	Sumac of tanners
γάλα	Milk
γῆ	Earth
γίγαρτον	Grape stone
γλήχων	Pennyroyal
γλυκυρρίζη	Licorice
γλυκυσίδη	Peony
γόγγρος	Eel, conger
γογγυλίς	Turnip
γυναικὸς γάλα	Woman's milk
δαΐς	Pine wood
δάκρυα ἀμπέλου	Vine, tears of
δαῦκος	Dauke
δάφνη	Bay/laurel tree
δαφνίς	Bayberry

469

δίκταμνον	Dittany
δριμέα	Sharp agents
δρυοπτερίς	Oak-fern
δυσώδεα	Evil-smelling agents
ἔγχελυς	Eel
εἴριον	Wool
ἐκλεικτόν	Electuary
ἔκτομον	Hellebore, black, powder of
ἐλαίη	Olive tree and its fruit
ἔλαιον	Olive oil
ἔλαιον Αἰγύπτιον (λευκόν)	Egyptian oil (white)
ἔλαιον κέδρινον	Cedar oil
ἔλαιον ναρκίσσινον	Narcissus oil
ἔλαιον ῥόδινον	Rose oil
ἐλατήριον	Squirting cucumber juice
ἐλάφου κέρας	Deer's horn
ἐλάφου μυελός	Deer's marrow
ἐλάφου στέαρ	Deer's fat
ἐλελίσφακος	Salvia
ἐλλέβορος	Hellebore
ἐλλύχνιον	Lampwick
ἐλξίνη	Bindweed
ἔλυμος	Millet
ἐπίθυμον	Dodder of thyme
ἐπικτένιον ὠμολίνου	Flax, tow of raw
ἐρέβινθος	Chickpea
ἐρευθέδανον	Madder
ἐρίφου κρέας	Kid's meat
ἔρπυλλος	Thyme, tufted
ἐρύσιμον	Hedge mustard
εὐάνθεμον	Chamomile

470

εὐλή	Maggot
εὐώδεα	Fragrant agents
ἐχέτρωσις	Bryony
ἐχῖνος θαλάσσιος	Sea urchin
ζειά	Rice-wheat
ἡδύοδμον	Mint
ἡμιόνου ὀνίς	Mule's excrement
θάλασσα	Brine
θαψίη	Thapsia
θεῖον	Sulfur
θέρμος	Lupin
θλάσπις	Shepherd's purse
θρίδαξ/θριδακίνη	Lettuce, wild
θρύον	Rush
θύμβρα	Savory
θύμον	Thyme
ἰλύς	Wine sediment
ἴον	Violet
ἰός	Verdigris
ἱππομάραθον	Horse fennel
ἱπποσέλινον	Alexanders
ἴρηκος ἄφοδος	Falcon/hawk, excrement of
ἶρις	Iris
ἰσχάς	Fig, dried
ἰτέα	Willow
κακώδεα	Evil-smelling agents
καλαμίνθη	Mint

κάλαμος	Reed
κανθαρίς	Blister beetle
κάραβος	Crab
κάρδαμον	Cress
καρδάμωμον	Cardamom
καρκίνος ποτάμιος	Crab, fluvial
κάρυον Ποντικόν	Hazel nut
κασίη	Cassia
κάστωρ (= καστόριος ὄρχις)	Castoreum
κάχρυ	Frankincense plant
κάχρυς	Barley, parched
κεδρίη	Cedar oil
κεδρίς	Juniper
κέδρος	Cedar tree
κηκίς	Oak gall
κηρίον	Honeycomb
κηρός	Wax
κηρωτή	Cerate/wax salve
κιννάβαρι	Cinnabar
κιννάμωμον	Cinnamon
κισθός	Rockrose
κισσός	Ivy
κληματίνη	Vine twigs
κνάφον τῶν κναφέων	Carding comb of fullers
κνέωρον	Spurge flax
κνῆκος	Safflower
κνῆστρον	Cnestron
κνίδη	Nettle, stinging
Κνίδιος κόκκος	Cnidian berry
κολοκύντη	Gourd
κολοκυνθὶς ἀγρίη	Wild gourd

κόμμι	Gum
κονία	Ashes
κόνυζα	Fleabane
κόριον (= κορίαννον)	Coriander
κοτυληδών	Navelwort
κράμβη	Cabbage
κρέας	Meat
κρῆθμον	Samphire
κριθή	Barley
κρίμνον	Barley meal
κρόκος	Saffron
κρόμμυον	Onion
κυκεών	Cyceon
κυκλάμινος	Cyclamen
κύμινον	Cumin
κύπειρος (= κύπαιρος)	Galingale
κυπάρισσος	Cypress
κύπρινον ἔλαιον	Henna oil
κύτισος	Tree medick
κωβιός	Goby
κώνειον	Hemlock
λαγώη θρίξ	Hare's hair
λάπαθον	Rhubarb, monk's
λεβηρίς	Snake skin
λείριον	Madonna lily
λεπίδιον	Pepperwort
λευκόϊον	White violet
λήδανον	Gum ladanum
λιβανωτός/λίβανος	Frankincense
λινόζωστις	Mercury (herb)
λίνον	Linen/flax

λίνου καρπός/σπέρμα	Linseed
λίτρον	Soda
λύκου κόπρος	Wolf's excrement
λωτός	Nettle tree
μᾶζα	Barley cake
μαλάχη	Mallow
μαλθακά	Mild agents
μαλθακτήριον	Emollient
μάννα	Frankincense powder
μάραθον	Fennel
μέλαν τὸ κύπριον	Copper, black
μελάνθιον	Cumin, black; ergot
μέλι	Honey
μελίκρητον	Melicrat
μελίλωτον	Melilot
μελίνη	Millet
μήκων	Poppy
μηκώνιον ὑπνωτικόν	Opium poppy
μήλειον κρέας	Mutton
μήλειον στέαρ	Sheep's fat
μήλη	Probe/sound
μῆλον τὸ Κυδώνιον	Quince
μίνθη	Mint
μίσυ	Misy
μόλιβδος	Lead
μολίβδαινα	Galena
μόρον	Mulberry
μυρίκη	Tamarisk
μύρον	Unguent
μύρον Αἰγύπτιον (λευκόν)	Egyptian unguent (white)
μύρον ἴρινον	Iris unguent

μύρον ναρκίσσινον	Narcissus unguent
μύρον ῥόδινον	Rose unguent
μυρσίνη	Myrtle
μυρτίδανον	Myrtidanon
μύρτον	Myrtle berry
μυῶν ἀπόπατος	Mouse's excrement
μώλυζα	Garlic head
νάρδος	Spikenard
ναρκή	Torpedo
νάρκισσος	Narcissus
νέτωπον	Almond, bitter oil of
νεφρίδιον	Kidney fat
νίτρον	Soda
ὀθόνιον	Linen cloth
οἰνάνθη	Grapevine blossom
οἶνος	Wine
ὄϊος ἧπαρ	Sheep's liver
ὄϊος κρέας	Mutton
ὄϊος σταῖς/στέαρ	Sheep's fat
ὁλοκωνῖτις	Earth-almond
ὄμφαξ	Grape, unripe
ὀμφάκιον	Oil of unripe olives
ὀνίς	Mule's excrement
ὄνου/ὄνειον γάλα	Ass's milk
ὄνου τρίχες	Ass's hair
ὄξος	Vinegar
ὀξύκρατον	Oxycrat
ὄπιον	Poppy juice
ὀπός (σιλφίου)	Silphium juice
ὀρίγανον	Marjoram

ὄροβος	Vetch
ὀρός	Whey
ὄστρακον	Cuttlefish bone
οὖρον	Urine
ὄφις	Caper plant
πάνακες	All-heal
πελ(ε)ιάς	Pigeon
πελεκῖνος	Axeweed
πέπερι	Pepper
πέπλιον	Purslane, wild
περσέα	Persea
πευκέδανον	Sulphurwort
πήγανον	Rue
πικέριον	Butter
πίτυρον πύρινον	Wheat bran
πίτυς	Pine tree
πόλιον	Hulwort
πολύκαρπον	Willow weed
πολύκνημος	Basil, field
πουλύπους	Octopus
πράσον	Leek
πρόμαλος	Willow
πτερόν	Feather
πτισάνη	Barley, peeled
πυρός	Wheat
πυτίη	Rennet
ῥάμνος	Buckthorn
ῥάφανος	Radish
ῥητίνη	Resin
ῥίζα λευκή	White root

ῥίζα μέλαινα	Black root
ῥινήματα χαλκοῦ	Copper filings
ῥόδον	Rose
ῥοιή	Pomegranate
ῥόος	Sumac
σαγάπηνον	Ferula
σαύρου ἧπαρ	Lizard's liver
σέλινον	Celery
σέσελι	Hartwort
σεῦτλον	Beet
σηπίης ᾠόν	Cuttlefish egg
σήσαμον	Sesame
σητάνιον	Spring wheat
σίδη	Pomegranate
σίδιον	Pomegranate peel
σικύα	Bottle gourd
σίκυος	Cucumber
σίκυος ἄγριος (= σικυώνη)	Squirting cucumber
σίλφιον	Silphium
σίσυμβρον	Watercress
σιτία (τὰ)	Cereals
σκαμ(μ)ωνία	Scammony
σκίλλα	Squill
σκόροδον	Garlic
σκορπίος θαλασσίον	Sculpin
σκωρία σιδήρου	Iron slag
σμηκτρὶς γῆ	Earth, fuller's
σμύρνα	Myrrh
σούσινον	Lily oil
σποδὸς κυπρίη	Copper, black
σπύραθος	Goat's excrement

σταῖς	Fat; spelt flour dough
σταφίς	Raisin
στέαρ (= σταῖς)	Fat
στοιβή	Burnet, thorny
στρογγύλον	Peppercorn
στρούθιον	Soapwort
στρύχνος	Nightshade
στυπτηρίη	Alum
στύραξ	Storax
συκάμινος	Mulberry
συκῆ	Fig tree
σῦκον	Fig
συὸς κρέας	Pork
συρμαία	Radish
σχεδιάς	Alkanet
σχῖνος	Mastic
σχοῖνος	Rush
τάμισος ὄνειος	Rennet, ass's
ταύρου ἧπαρ	Bull's liver
ταύρου χολή	Bull's gall
τερμινθίνη ῥητίνη	Terpentine
τέρμινθος	Terebinth tree
τηλέφιον	Telephion
τῆλις	Fenugreek
τιθύμαλλος	Spurge
τίτανος	Gypsum
τρίβολος παραθαλάσσιος	Water chestnut
τρίφυλλον	Clover
τρύξ	Wine lees/must

ὕδωρ	Water
ὑὸς ἔλαιον	Hog's grease
ὑὸς πιμελή/στέαρ	Lard
ὑὸς χολή	Hog's gall
ὑοσκύαμος	Henbane
ὑπερικόν	Hypericum
φακῆ	Lentil soup
φάκιον	Lentil decoction
φακός	Lentil
φάρμακον μέλαν	Black medication
φάσσα	Ringdove
φηγός	Valonia oak
φιλίστιον	Cleavers
φλόμος	Mullein
φώκης ἔλαιον	Seal's oil
φώκης πλεύμων	Seal's lung
φώκης πυτίη	Seal's rennet
φώκης στέαρ	Seal's fat
χαλβάνη	All-heal juice
χηναλώπηξ	Goose, Egyptian
χήνειον ἔλαιον	Goose oil
χήνειον στέαρ	Goose grease
χήνειον μυελός	Goose marrow
χηνὸς ἄλειφα	Goose grease
χόνδρος	Spelt groats
χρυσῖτις σποδός	Gold scoria
χρυσόκολλα	Linseed and honey dish
χυλὸς πτισάνης	Gruel (barley)

ψῆσσα	Sole
ψιμύθιον	Lead, white
ψύγματα	Cold compresses
ψώρη ἐλαίης	Olive scab
ὠμήλυσις	Meal, bruised, of raw grain
ᾠόν	Egg

INDEX

481